Statistical Analysis of Random Fields

T0325835

Mathematics and Its Applications (*Soviet Series*)

Managing Editor:

M. HAZEWINKEL
Centre for Mathematics and Computer Science, Amsterdam, The Netherlands

Editorial Board:

A. A. KIRILLOV, *MGU, Moscow, U.S.S.R.*
Yu. I. MANIN, *Steklov Institute of Mathematics, Moscow, U.S.S.R.*
N. N. MOISEEV, *Computing Centre, Academy of Sciences, Moscow, U.S.S.R.*
S. P. NOVIKOV, *Landau Institute of Theoretical Physics, Moscow, U.S.S.R.*
M. C. POLYVANOV, *Steklov Institute of Mathematics, Moscow, U.S.S.R.*
Yu. A. ROZANOV, *Steklov Institute of Mathematics, Moscow, U.S.S.R.*

VOLUME 28

Statistical Analysis of Random Fields

by

A. V. Ivanov
Cybernetic Center, Kiev, U.S.S.R.

and

N. N. Leonenko
University of Kiev, U.S.S.R.

KLUWER ACADEMIC PUBLISHERS
DORDRECHT / BOSTON / LONDON

Library of Congress Cataloging in Publication Data

Ivanov, A. V.
    Statistical analysis of random fields / A.V. Ivanov and N.N.
Leonenko ; translation edited by S. Kotz.
        p. . cm. -- (Mathematics and its application. Soviet series)
    Translated from Russian.
    Bibliography: p.
    Includes index.
    ISBN 9027728003
    1. Random fields. 2. Mathematical statistics.   I. Leonenko, N.
N. II. Title. III. Series: Mathematics and its applications
(Kluwer Academic Publishers). Soviet series.
QA274.45.I93 1989
519.2--dc19   88-8196

ISBN 90-277-2800-3

Published by Kluwer Academic Publishers,
P.O. Box 17, 3300 AA Dordrecht, The Netherlands.

Kluwer Academic Publishers incorporates
the publishing programmes of
D. Reidel, Martinus Nijhoff, Dr W. Junk and MTP Press.

Sold and distributed in the U.S.A. and Canada
by Kluwer Academic Publishers,
101 Philip Drive, Norwell, MA 02061, U.S.A.

In all other countries, sold and distributed
by Kluwer Academic Publishers Group,
P.O. Box 322, 3300 AH Dordrecht, The Netherlands.

This is an revised and updated edition of the original work
СТАТИСТИЧЕСКИЯ АНАЛИЗ СЛУЧАЯНЫХ ПОЛЕЯ
Published by Vysca Skola, Kiev, © 1986.

Translated from the Russian by A. I. Kochubinsky
Translation edited by S. Kotz.

printed on acid free paper

All Rights Reserved
This English edition © 1989 by Kluwer Academic Publishers.

No part of the material protected by this copyright notice may be reproduced or
utilized in any form or by any means, electronic or mechanical,
including photocopying, recording, or by any information storage and
retrieval system, without written permission from the copyright owner.

Contents

SERIES EDITOR'S PREFACE

'Et moi, ..., si j'avais su comment en revenir,
je n'y serais point allé.'

        Jules Verne

The series is divergent; therefore we may be
able to do something with it.

        O. Heaviside

One service mathematics has rendered the
human race. It has put common sense back
where it belongs, on the topmost shelf next
to the dusty canister labelled 'discarded non-
sense'.

        Eric T. Bell

Mathematics is a tool for thought. A highly necessary tool in a world where both feedback and non-linearities abound. Similarly, all kinds of parts of mathematics serve as tools for other parts and for other sciences.

Applying a simple rewriting rule to the quote on the right above one finds such statements as: 'One service topology has rendered mathematical physics ...'; 'One service logic has rendered computer science ...'; 'One service category theory has rendered mathematics ...'. All arguably true. And all statements obtainable this way form part of the raison d'être of this series.

This series, *Mathematics and Its Applications*, started in 1977. Now that over one hundred volumes have appeared it seems opportune to reexamine its scope. At the time I wrote

> "Growing specialization and diversification have brought a host of monographs and
> textbooks on increasingly specialized topics. However, the 'tree' of knowledge of
> mathematics and related fields does not grow only by putting forth new branches. It
> also happens, quite often in fact, that branches which were thought to be completely
> disparate are suddenly seen to be related. Further, the kind and level of sophistication
> of mathematics applied in various sciences has changed drastically in recent years:
> measure theory is used (non-trivially) in regional and theoretical economics; algebraic
> geometry interacts with physics; the Minkowsky lemma, coding theory and the structure
> of water meet one another in packing and covering theory; quantum fields, crystal
> defects and mathematical programming profit from homotopy theory; Lie algebras are
> relevant to filtering; and prediction and electrical engineering can use Stein spaces. And
> in addition to this there are such new emerging subdisciplines as 'experimental
> mathematics', 'CFD', 'completely integrable systems', 'chaos, synergetics and large-scale
> order', which are almost impossible to fit into the existing classification schemes. They
> draw upon widely different sections of mathematics."

By and large, all this still applies today. It is still true that at first sight mathematics seems rather fragmented and that to find, see, and exploit the deeper underlying interrelations more effort is needed and so are books that can help mathematicians and scientists do so. Accordingly MIA will continue to try to make such books available.

If anything, the description I gave in 1977 is now an understatement. To the examples of interaction areas one should add string theory where Riemann surfaces, algebraic geometry, modular functions, knots, quantum field theory, Kac-Moody algebras, monstrous moonshine (and more) all come together. And to the examples of things which can be usefully applied let me add the topic 'finite geometry'; a combination of words which sounds like it might not even exist, let alone be applicable. And yet it is being applied: to statistics via designs, to radar/sonar detection arrays (via finite projective planes), and to bus connections of VLSI chips (via difference sets). There seems to be no part of (so-called pure) mathematics that is not in immediate danger of being applied. And, accordingly, the applied mathematician needs to be aware of much more. Besides analysis and numerics, the traditional workhorses, he may need all kinds of combinatorics, algebra, probability, and so on.

In addition, the applied scientist needs to cope increasingly with the nonlinear world and the extra mathematical sophistication that this requires. For that is where the rewards are. Linear models are honest and a bit sad and depressing: proportional efforts and results. It is in the non-linear world that infinitesimal inputs may result in macroscopic outputs (or vice versa). To appreciate what I am hinting at: if electronics were linear we would have no fun with transistors and computers; we would have no TV; in fact you would not be reading these lines.

There is also no safety in ignoring such outlandish things as nonstandard analysis, superspace and anticommuting integration, p-adic and ultrametric space. All three have applications in both electrical engineering and physics. Once, complex numbers were equally outlandish, but they frequently proved the shortest path between 'real' results. Similarly, the first two topics named have already provided a number of 'wormhole' paths. There is no telling where all this is leading - fortunately.

Thus the original scope of the series, which for various (sound) reasons now comprises five subseries: white (Japan), yellow (China), red (USSR), blue (Eastern Europe), and green (everything else), still applies. It has been enlarged a bit to include books treating of the tools from one subdiscipline which are used in others. Thus the series still aims at books dealing with:

- a central concept which plays an important role in several different mathematical and/or scientific specialization areas;
- new applications of the results and ideas from one area of scientific endeavour into another;
- influences which the results, problems and concepts of one field of enquiry have, and have had, on the development of another.

Random fields, the 2-D analogues of stochastic processes are tough to work with as there are practically no systematic books on the topic. This in spite of their obvious importance and relevance. Here, apparently (as in analysis) the 'dimensional difficulty gap' is between 1 and 2 rather than between 2 and 3, or 3 and 4, as is more customary in geometry, topology and mathematical physics. There remains a great deal of fundamental work to be done as becomes already clear when one starts thinking about what a 2D-Wiener process should be, an active area of current research.

It is well known how important limit theorems have been in the theory of stochastic processes and how important they are and have been to built up intuition. Thus it is most fortunate and valuable that the authors manage to say much about precisely limit theorems; in addition there is a great deal of material on the estimation of mathematical expectations and correlation functions. Precisely the stuff therefore which is needed for the many potential applications in image processing, statistical physics, meteorology, turbulence and many other fields.

This valuable book - in my opinion - is a substantially updated and expanded translation of its Russian original of some two years ago.

Perusing the present volume is not guaranteed to turn you into an instant expert, but it will help, though perhaps only in the sense of the last quote on the right below.

The shortest path between two truths in the real domain passes through the complex domain.

J. Hadamard

Never lend books, for no one ever returns them; the only books I have in my library are books that other folk have lent me.

Anatole France

La physique ne nous donne pas seulement l'occasion de résoudre des problèmes ... elle nous fait pressentir la solution.

H. Poincaré

The function of an expert is not to be more right than other people, but to be wrong for more sophisticated reasons.

David Butler

Bussum, March 1989

Michiel Hazewinkel

Preface

This book is devoted to an investigation of the basic problems of the statistics of random fields. It should be mentioned at the outset that up until now there have been no books available in which these problems have been tackled. On the whole, the theory of random fields is poorly represented in monographic literature. I am aware of only two books: "Spectral Theory of Random Fields" (1980) by M.I. Yadrenko and "The Geometry of Random Fields" (1981) by R.J. Adler. This meagre output is in spite of the fact that the theory of random fields continues to attract new applications such as in turbulence theory, meteorology, statistical radiophysics, theory of surface roughness, pattern recognition and identification of parameters of complex systems. Moreover, random functions of many variables, namely, random fields, arise naturally in probability theory proper.

The specific nature of random fields manifests itself when studying random functions whose properties are coordinated with the algebraic structure of the space in the same manner as the specific nature of random processes (a function of a single variable viewed as time) are revealed in coordination with the ordering structure (this gives rise to the theory of martingales and Markov processes).

As is well known, limit theorems play an essential role in solving problems in statistics. It is usually assumed in problems of statistics of random processes that a process (as a rule a single realization of it) is observed on a time interval that extends to infinity. Based on these observations:

(1) one constructs estimators of parameters and (2) one tests hypotheses about the distribution of the process or about the form of its basic characteristics and so on.

An analogous approach is utilized in this book for the study of homogeneous isotropic random fields. It is assumed that a field is observed in an expanding domain and based on these observations, problems (1) and (2) are solved for random fields. Limit theorems for functionals (in particular, additive ones) of sample functions of fields are required as a working tool. We note that limit theorems for functionals of random fields are of interest in their own right. The book devotes major attention to an investigation of limit distributions for various specific functionals of a geometric nature for Gaussian random fields possessing strongly and weakly decreasing correlations. For the first time in monographic literature, functionals useful in applications are investigated, such as measures of excess of a Gaussian field above a fixed or moving level and above a spherical surface, and volumes constrained by realizations of Gaussian fields over sets. The results obtained can be applied to the analysis of surface roughness, for example.

Substantial results are obtained in estimation theory of the first two moments of random fields with a continuous parameter, that is, the mathematical expectation and correlation function. The least squares and least moduli estimators are studied for a multi-dimensional parameter of linear and non-linear regressions. These estimators are not always optimal; however, they have an important quality, being determined as the extremum point of simple integral functionals of realization of an observed field. In addition to parametric estimation problems, non-parametric estimation of the correlation function of homogeneous (or homogeneous isotropic) fields having zero mean value is considered. Correlation-type estimators are investigated. Conditions are provided under which measures corresponding to these estimators converge weakly to Gaussian measures. This allows us to solve the important problem of constructing confidence intervals for an unknown correlation function. We note that in practice it is often impossible to obtain a realization of a random field or even measure an instantaneous value of such a field. However, measurements of the correlograms of random fields can be obtained by means of physical devices.

Both specialists in the theory of random fields and scientists working in related areas that utilize this theory will no doubt find a great deal of new and useful information in this book.

A.V. Skorohod
Member of the Academy of Sciences of the Ukrainian SSR

CHAPTER 1

Elements of the Theory of Random Fields

1.1. Basic Concepts and Notation

This section presents basic concepts and statements that have important applications in this book.

1. SETS

Denote by $\mathbf{R}^n$ a real Euclidean space of dimension $n \geq 1$. Let $\langle x, y \rangle = x_1 y_1 + \ldots + x_n y_n$ be the scalar product of vectors $x = (x_1, \ldots, x_n)$, $y = (y_1, \ldots, y_n) \in \mathbf{R}^n$. Then $|x| = \sqrt{\langle x, x \rangle}$ is the length of the vector x, $\rho_{xy} = |x - y|$ is the distance between x and y in $\mathbf{R}^n$. The symbol $dx = dx_1 \ldots dx_n$ denotes an element of the Lebesgue measure in $\mathbf{R}^n$. Denote by

$$\mathbf{R}_+^n = \{ x \in \mathbf{R}^n : x_i \geq 0, \ i = 1, \ldots, n \}$$

the non-negative octant of $\mathbf{R}^n$ and by $\mathbf{Z}^n$, $n \geq 1$, the integer lattice in $\mathbf{R}^n$.

Let

$$\prod [a, b) = \{ x \in \mathbf{R}^n : a_i \leq x_i < b_i, \ i = 1, \ldots, n \}$$

be a parallelepiped in $\mathbf{R}^n$. Define by analogy $\prod(a, b]$, $\prod[a, b]$ and write $\prod[0, b] = \prod(b)$. A ball and a sphere in $\mathbf{R}^n$ of radius r and the centre at the origin are defined as the sets $v(r) = \{ x \in \mathbf{R}^n : |x| < r \}$ and $s(r) = \{ x \in \mathbf{R}^n : |x| = r \}$ respectively. If necessary we write $s_{n-1}(r)$ specifying the dimension of the sphere.

Along with the Cartesian coordinate system in $\mathbf{R}^n$ we shall use the spherical one (r, u), where

$$r \geq 0, \ u = (\phi, \phi_1, \ldots, \phi_{n-2}) \in s_{n-1}(1), \ 0 \leq \phi < 2\pi,$$

1

$$0 \le \phi_j < \pi, \; j = 1, \ldots, n-2, \; n \ge 2$$

which is related to the Cartesian system by the formulas

$$x_1 = r \sin \phi_{n-2} \ldots \sin \phi_1 \sin \phi, \quad x_2 = r \sin \phi_{n-2} \ldots \sin \phi_1 \cos \phi, \ldots$$

$$\ldots, \quad x_{n-1} = r \sin \phi_{n-2} \cos \phi_{n-3}, \quad x_n = r \cos \phi_{n-2}. \tag{1.1.1}$$

An element of the Lebesgue measure on the sphere $s_{n-1}(r)$ will be denoted by

$$dm(x) = r^{n-1} \sin^{n-2} \phi_{n-2} \ldots \sin \phi_1 d\phi_1 \ldots d\phi_{n-2} d\phi. \tag{1.1.2}$$

We denote by $|\Delta|$ the Lebesgue measure of a measurable set $\Delta \subset \mathbf{R}^n$. For example, $|v(1)| = 2\pi^{n/2}/[n\Gamma(n/2)]$ is the volume of a unit ball in $\mathbf{R}^n$. We denote by $|s(1)| = |s_{n-1}(1)| = 2\pi^{n/2}/\Gamma(n/2)$ the area of the surface of the sphere $s_{n-1}(1)$ when there is no danger of misunderstanding.

Suppose, henceforth, that a partially ordered system of bounded measurable sets $\mathfrak{M} = \{\Delta\}$ is given in $\mathbf{R}^n$ with respect to the order relation "$\preceq$". Assume that $\mathfrak{M}$ is directed upwards, that is, for any $\Delta_1, \Delta_2 \in \mathfrak{M}$ there exists $\Delta \in \mathfrak{M}$ such that $\Delta_1 \preceq \Delta$, $\Delta_2 \preceq \Delta$ (Δ is an upper bound of Δ_1 and Δ_2). Let $a \in \mathbf{R}^n$ be a non-zero vector and $\prod_m = \prod[0, a) + m \otimes a$ the shift of $\prod[0, a)$ by the vector $m \otimes a = (m_1 a_1, \ldots, m_n a_n)$, $m \in \mathbf{Z}^n$.

The collection of sets $\{\prod_m\}$, $m \in \mathbf{Z}^n$, generates a partition of $\mathbf{R}^n$. Given $\Delta \in \mathbf{R}^n$, we define $N_a^+(\Delta)$ as the number of such $\prod_m$ for which $\prod_m \cap \Delta \ne \emptyset$ and $N_a^-(\Delta)$ as the number of such $\prod_m$ satisfying $\prod_m \subset \Delta$. We say that the sets $\Delta \in \mathfrak{M}$ approach infinity in the Van Hove sense (written $\Delta \xrightarrow{V.H.} \infty$) if for any $a \ne 0 \lim N_a^-(\Delta) = \infty$, $\lim N_a^-(\Delta)/N_a^+(\Delta) = 1$ as the sets Δ expand. The latter relation, for example, signifies that for any $\epsilon > 0$ there exists $\Delta(\epsilon) \in \mathfrak{M}$ such that for all $\Delta \succeq \Delta(\epsilon)$, $|N_a^-(\Delta)/N_a^+(\Delta) - 1| < \epsilon$. We define the distance between sets Δ_1 and Δ_2 belonging to $\mathbf{R}^n$ by the formula $r(\Delta_1, \Delta_2) = \inf\{|x - y|, x \in \Delta_1, y \in \Delta_2\}$. For a set $\Delta \subset \mathbf{R}^n$ put $\rho(x, \Delta) = \inf\{|x-y|, y \in \Delta_2\}$, $\Delta_\epsilon = \{x \in \mathbf{R}^n : \rho(x, \Delta) < \epsilon\}$, $\Delta_{-\epsilon} = \mathbf{R}^n \backslash (\mathbf{R}^n \backslash \Delta_\epsilon)$, $\Delta[\epsilon] = \Delta_\epsilon \backslash \Delta_{-\epsilon}$, $\epsilon > 0$. The requirement $\Delta \xrightarrow{V.H.} \infty$ is then equivalent to the following: $\lim |\Delta| = \infty$ and $\lim |\Delta[\epsilon]|/|\Delta| = 0$ for any $\epsilon > 0$. Thus, for a system of parallelepipeds $\mathfrak{M} = \{\prod(b)\}$, $\prod(b) \xrightarrow{V.H.} \infty$ if $\min\{b_i, i = 1, \ldots, n\} \to \infty$; for a system of balls $\mathfrak{M} = \{v(r)\}$, an approach to infinity in the Van Hove sense signifies that $r \to \infty$.

The diameter of a set $\Delta \subset \mathbf{R}^n$ is defined as the quantity $d(\Delta) = \sup\{|x - y|, x, y \in \Delta\}$. We say that the sets Δ approach infinity in the sense of Fisher ($\Delta \xrightarrow{F} \infty$) if 1) $\lim |\Delta| = \infty$; 2) $\sup\{|\Delta[\alpha d(\Delta)]|/|\Delta|], \Delta \in \mathfrak{M}\} = o(\alpha)$ as $\alpha \downarrow 0$.

Since $|\Delta[\alpha d(\Delta)]| \geq |v(\alpha d(\Delta))|$, the convergence $\Delta \xrightarrow{F} \infty$ implies that there exists a constant $k > 0$ independent of Δ such that $d(\Delta) \leq k|\Delta|^{1/n}$. In particular, if $0 \in \Delta$, then $\Delta \subseteq v(k|\Delta|^{1/n})$. The latter inclusion restricts the degree of elongation of sets Δ. Below, when considering the convergence in the Fisher sense we shall assume that $0 \in \Delta$.

The requirement that $\Delta \xrightarrow{F} \infty$ is more restrictive than that of $\Delta \xrightarrow{V.H.} \infty$. For example, as $s \to \infty$ and $b = (s, s^2, \ldots, s^n)$, $\prod(b) \xrightarrow{V.H.} \infty$ but $\prod(b) \xrightarrow{F} \infty$ provided $n \geq 2$.

We shall write $\Delta \to \infty$ if the mode of convergence of Δ to infinity is not specified; here we shall assume that $|\Delta| \to \infty$.

Let $\Delta \subset \mathbf{R}^n$ be a bounded measurable set containing the origin. Denote by $\Delta(\lambda)$ the image of Δ under the homothetic transformation with the centre at the point $0 \in \Delta$ and coefficient $\lambda > 0$. Note that $|\Delta(\lambda)| = \lambda^n |\Delta|$.

The closure of a set Δ in $\mathbf{R}^n$ will be denoted by Δ^c.

2. RANDOM VARIABLES

Let $(\Omega, \mathfrak{F}, P)$ be a complete probability space. A random element taking on values in a measurable space $(X, \mathfrak{B})$ is a mapping $\xi : \Omega \to X$ such that $\{\omega : \xi(\omega) \in A\} \in \mathfrak{F}$ for any $A \in \mathfrak{B}$. The measure $P_\xi(A) = P\{\xi \in A\}$ defined on the σ-algebra $\mathfrak{B}$ is called the distribution of ξ. If X is a metric space, the σ-algebra of Borel sets $\mathfrak{B}(X)$ is usually taken as $\mathfrak{B}$. For $X = \mathbf{R}^m$ we denote the Borel σ-algebra by $\mathfrak{B}^m = \mathfrak{B}(\mathbf{R}^m)$. A random element taking on values in $(\mathbf{R}^m, \mathfrak{B}^m)$ is called a random variable if $m = 1$ and a random vector if $m \geq 2$. In either case we shall use the abbreviation "r.v.".

To define $P_\xi(A)$ for a r.v. $\xi = (\xi_1, \ldots, \xi_m) \in \mathbf{R}^m$ it suffices to define the distribution function (d.f.) $F(x) = F_\xi(x_1, \ldots, x_m) = P\{\xi_1 < x_1, \ldots, \xi_m < x_m\}$. A distribution $P_\xi(A)$ is said to be absolutely continuous if $P_\xi(A) = \int_A p(x)dx, A \in \mathfrak{B}^m$. The function $p(x) = p_\xi(x_1, \ldots, x_m)$, $x \in \mathbf{R}^m$, is called the density function of the distribution of r.v. ξ. Clearly, $\int_{\mathbf{R}^m} p(x)dx = 1$ and for almost all $x \in \mathbf{R}^m$,

$$p_\xi(x) \geq 0, \ p_\xi(x) = p_\xi(x_1, \ldots, x_m) = (\partial^m / \partial x_1 \ldots \partial x_m) F_\xi(x_1, \ldots, x_m).$$

Let $g : \mathbf{R}^m \to \mathbf{R}^1$ be a Borel function. The mathematical expectation of the function $g(\xi)$ of the r.v. $\xi \in \mathbf{R}^m$ is computed as the Lebesgue integral with respect to measure P, $Eg(\xi) = \int_\Omega g(\xi(\omega))P(d\omega)$ or as the m-fold Lebesgue-Stieltjes integral $Eg(\xi) = \int_{\mathbf{R}^m} g(x)dF_\xi(x)$ provided this integral exists. In the case when the r.v. has a density function, $Eg(\xi) = \int_{\mathbf{R}^m} g(x)p_\xi(x)dx$. For a r.v. $\xi = (\xi_1, \ldots, \xi_m) \in \mathbf{R}^m$ we set by definition $E\xi = (E\xi_1, \ldots, E\xi_m)$. The symbols $\operatorname{cov}(\xi, \eta) = E\xi\eta - E\xi E\eta$ and $\operatorname{var}\xi = E\xi^2 - (E\xi)^2$ will be used to denote the covariance of the r.v.'s ξ and η and the variance of the r.v. ξ. The matrix $B = (\operatorname{cov}(\xi_i, \xi_j))_{i,j=1,\ldots,m}$ is said to be the correlation matrix of the r.v. $\xi = (\xi_1, \ldots, \xi_m)$. For a r.v. ξ with $E\xi^2 < \infty$ Chebyshev's inequality holds: for any $\epsilon > 0$, $P\{|\xi - E\xi| > \epsilon\} \leq (\operatorname{var}\xi)/\epsilon^2$.

The characteristic function (ch.f.) of a r.v. $\xi \in \mathbf{R}^q$, $q \geq 1$ is defined as the function

$$f_\xi(t) = Ee^{i\langle t, \xi\rangle} = \int_{\mathbf{R}^q} e^{i\langle t, x\rangle} dF_\xi(x), \ t \in \mathbf{R}^q.$$

The r.v. $\xi \in \mathbf{R}^1$ is called Gaussian if its ch.f. has the form $f_\xi(t) = E\exp\{it\xi\} = \exp\{ita - t^2\sigma^2/2\}$, $a \in \mathbf{R}^1, \sigma^2 \geq 0$.

In the non-singular case ($\sigma^2 > 0$) the Gaussian r.v. ξ has the density function

$$\phi_{a,\sigma^2}(u) = \frac{1}{\sqrt{2\pi}\sigma}\exp\left\{-\frac{(u-a)^2}{2\sigma^2}\right\}, \ u \in \mathbf{R}^1 \qquad (1.1.3)$$

with $a = E\xi$, $\sigma^2 = \operatorname{var}\xi$. If $a = 0$,

$$E\xi^{2j} = (2j-1)!!\sigma^{2j}, \ E\xi^{2j+1} = 0, \ j = 1, 2, \ldots.$$

A r.v. $\xi \in \mathbf{R}^q$, $q > 1$ is called Gaussian if its ch.f. is of the form $f_\xi(t) = \exp\{i\langle t, a\rangle - \frac{1}{2}\langle Bt, t\rangle\}$, $t \in \mathbf{R}^q$, where $a \in \mathbf{R}^q$ and $B = (b_{ij})_{i,j=1,\ldots,q}$ is a non-negative definite symmetric matrix. If $B > 0$, the r.v. ξ has the density function

$$\phi_{a,B}(u) = (2\pi)^{-q/2}(\det B)^{-1/2}\exp\{-\frac{1}{2}\langle B^{-1}(u-a), u-a\rangle\} \qquad (1.1.4)$$

with $a = (a_1, \ldots, a_q), a_i = E\xi_i$, $i = 1, \ldots, q$; $b_{ij} = \operatorname{cov}(\xi_i, \xi_j)$, $i, j = 1, \ldots, q$, are the elements of the correlation matrix of the r.v. ξ. In what follows, a Gaussian r.v. $\xi \in \mathbf{R}^q$, $q > 1$ will be denoted by $N_q(a, B)$ and the corresponding d.f. by $\Phi_{a,B}^{(q)}(x)$, $x \in \mathbf{R}^q$; the Gaussian r.v. $\xi \in \mathbf{R}^1$ will be denoted by $N(a, \sigma^2)$ and the corresponding d.f. by $\Phi_{a,\sigma^2}(x)$, $x \in \mathbf{R}^1$. A system of r.v.'s $\xi_\alpha \in \mathbf{R}^1$, $\alpha \in \mathfrak{A}$, where $\mathfrak{A}$ is a set of indices is said to be Gaussian if for any integer $l \geq 1$ and any $\alpha_1, \ldots, \alpha_l \in \mathfrak{A}$ the r.v. $(\xi_{\alpha_1}, \ldots, \xi_{\alpha_l})$ is Gaussian (any linear combination $k_1\xi_{\alpha_1} + \ldots + k_l\xi_{\alpha_l}$, $k_i \in \mathbf{R}^1$, $i = 1, \ldots, l$ is a Gaussian r.v.).

A sequence of r.v.'s ξ_m, $m = 1, 2 \ldots$, converges in probability to the r.v. ξ ($\xi_m \xrightarrow{P} \xi$) if for any $\epsilon > 0$ $P\{|\xi_m - \xi| > \epsilon\} \to 0$ as $m \to \infty$. A sequence of r.v.'s ξ_m, $m = 1, 2, \ldots$, converges almost surely (a.s.) to the r.v. ξ if $P\{\xi_m \xrightarrow[m\to\infty]{} \xi\} = 1$. In order for a sequence of r.v.'s ξ_m to converge to the r.v.

ξ it is sufficient that the series $\sum_{m=1}^{\infty} P\{|\xi_m - \xi| > \epsilon\}$ converge for any $\epsilon > 0$. A.s. convergence and convergence in probability are defined for collections of r.v.'s indexed by directed sets in a similar manner.

R.v.'s $\xi_1, \ldots, \xi_m$ are said to be independent if $P\{\xi_1 \in B_1, \ldots, \xi_m \in B_m\} = P\{\xi_1 \in B_1\} \ldots P\{\xi_m \in B_m\}$ for any $B_i \in \mathbf{B}^1$, $i = 1, \ldots, m$. A sequence of r.v.'s ξ_i, $i = 1, 2, \ldots$, is said to be a sequence of independent r.v.'s if the r.v.'s $\xi_1, \ldots, \xi_m$ are independent for any $m \geq 1$.

The symbol $\chi(A)$ will denote the indicator function of event A, that is,

$$\chi(A) = \chi_\omega(A) = \begin{cases} 1, & \omega \in A, \\ 0, & \omega \notin A. \end{cases}$$

3. RANDOM FIELDS

Let $T \subseteq \mathbf{R}^n$ be a set. A random field is defined as a function $\xi(\omega, x) : \Omega \times T \to \mathbf{R}^m$ such that $\xi(\omega, x)$ is a r.v. for each $x \in T$. A random field will also be denoted as $\xi(x)$, $x \in T$. If $n = 1$, $\xi(x)$ is a random process. For $n > 1$ and $m = 1$ $\xi(x)$ is a scalar random field; for $m > 1$ it is a vector random field.

In particular, if $\xi(x)$, $x \in T$, is a Gaussian system of r.v.'s, the (scalar) field $\xi(x)$, $x \in T$ is said to be Gaussian.

Finite dimensional distributions of the random field $\xi(x)$, $x \in T$, are defined as a set of distributions $P\{\xi(x^{(i)}) \in B_i, i = 1, \ldots, r\}$ where $B_i \in \mathbf{B}^m$, $i = 1, \ldots, r$, $r = 1, 2, \ldots$. We shall write $\xi(x) \overset{d}{=} \eta(x)$, $x \in T$, if finite dimensional distributions of random fields $\xi(x)$, $x \in T$, and $\eta(x)$, $x \in T$, coincide ($\xi(x)$ and $\eta(x)$ are stochastically equivalent).

For a fixed ω, the function $\xi(x)$, $x \in T$, is called a realization of the random field or a sample function.

A random field $\xi(x)$, $x \in T$, having a finite second order moment is said to be mean-square continuous (m.s. continuous) at point x_0 if $E|\xi(x) - \xi(x_0)|^2 \to 0$ as $\rho_{xx_0} \to 0$. If this relation holds for any $x_0 \in T$, the field $\xi(x)$, $x \in T$ is called m.s. continuous on T. In order for a field $\xi(x)$, $x \in T$, having $E\xi^2(x) < \infty$ to be m.s. continuous it is necessary and sufficient that the function $\text{cov}(\xi(x), \xi(y))$ be continuous along the diagonal $\{(x, y) \in T \times T : x = y\}$. Clearly, a m.s. continuous field $\xi(x)$, $x \in T$, is stochastically continuous, that is, for any $x_0 \in T$, $P\{|\xi(x) - \xi(x_0)| > \epsilon\} \to 0$ for any $\epsilon > 0$ as $\rho_{xx_0} \to 0$.

Let $T \in \mathbf{B}^n$. A random field $\xi(\omega, x) : \Omega \times T \to \mathbf{R}^m$ is said to be measurable if for any $A \in \mathbf{B}^m$ $\{(\omega, x) : \xi(\omega, x) \in A\} \in \mathfrak{F} \times \mathbf{B}(T)$. If $\xi(x)$, $x \in T$, is stochastically continuous on T, a measurable random field $\tilde{\xi}(x) \overset{d}{=} \xi(x)$, $x \in T$, exists.

The Fubini-Tonelli theorem will be used throughout this book without specific reference.

Theorem 1.1.1 [30] Let $\xi(\omega, x) : \Omega \times \mathbf{R}^n \to \mathbf{R}^1$ be a measurable random field. The following assertions are valid.

1. Sample functions $\xi(x)$ are a.s. $\mathcal{B}^n$-measurable functions of $x \in \mathbf{R}^n$.

2. If $E\xi(x)$ exists for all $x \in \mathbf{R}^n$, then $m(x) = E\xi(x)$ is a $\mathcal{B}^n$-measurable function of $x \in \mathbf{R}^n$.

3. If $\Delta \in \mathcal{B}^n$ then the field $\xi(x)$ is integrable over $\Omega \times \Delta$ if at least one of the iterated integrals $\int_\Delta \{E|\xi(x)|\}dx$ and $E\{\int_\Delta |\xi(x)|dx\}$ is finite; in this case, the two-fold integral $(\int_\Delta E\xi(x)dx)$ is identical to the iterated ones $\int_\Delta \{E\xi(x)\}dx$ and $E\{\int_\Delta \xi(x)dx\}$.

Let $\Theta \subseteq \mathbf{R}^q$, $\mathfrak{M} = \{\Delta\}$ be a system of bounded measurable subsets of $\mathbf{R}^n$ and $g_\Delta(\cdot, \theta) : \mathfrak{M} \times \mathbf{R}^n \times \Theta \to \mathbf{R}^1$ a family of functions which are $\mathcal{B}^n \times \mathcal{B}(\Theta)$-measurable for each $\Delta \in \mathfrak{M}$.

If for all $\theta \in \Theta$ the function $g_\Delta(\cdot, \theta)$ is integrable over $\Delta \in \mathfrak{M}$ and the measurable scalar field $\xi(x)$, $x \in \mathbf{R}^n$, possesses the property $\sup\{E|\xi(x)|, x \in \mathbf{R}^n\} < \infty$, then by Theorem 1.1.1, the integrals

$$s_\Delta(\theta) = \int_\Delta g_\Delta(x, \theta)\xi(x)dx \qquad (1.1.5)$$

are a.s. finite for any $\Delta \in \mathfrak{M}$. In the same manner integrals of type (1.1.5) may be defined for the scalar field $\xi(x)$, $x \in \mathbf{R}^n$, and the vector function $g_\Delta(x, \theta) : \mathfrak{M} \times \mathbf{R}^n \times \Theta \to \mathbf{R}^m$, $m > 1$.

Theorem 1.1.1 remains valid for vector random fields. Its formulation does not change if the Lebesgue measure on $\Delta \in \mathcal{B}^n$ is replaced, for example, by the Lebesgue measure $m(x)$ on the surface of the sphere $s_{n-1}(r)$.

The field $\xi(\omega, x) : \Omega \times T \to \mathbf{R}^1$ is said to be separable with respect to the set $I \subset T$ if I is countable and dense in T and there exists a set $N \in \mathfrak{F}$, $P(N) = 1$, such that for any ball $v(r) \subset \mathbf{R}^n$

$$\{\omega : \sup_{x \in I \cap v(r)} \xi(x) = \sup_{x \in T \cap v(r)} \xi(x)\} \supset N,$$

$$\{\omega : \inf_{x \in I \cap v(r)} \xi(x) = \inf_{x \in T \cap v(r)} \xi(x)\} \supset N.$$

For any random field $\xi(x)$, $x \in T$, there exists a separable field $\tilde{\xi}(x) \overset{d}{=} \xi(x)$, $x \in T$. Henceforth only measurable separable fields will be considered.

4. WEAK CONVERGENCE OF DISTRIBUTIONS

Let X be a metric space and $C(X)$ the space of all bounded continuous functions defined on X with norm $\|f\|_X = \sup\{|f(x)|, \ x \in X\}$.

A net [13] of probability measures $\{P_\Delta, \ \Delta \in \mathfrak{M}\}$ on $(X, \mathcal{B}(X))$ converges weakly to a probability measure P on $(X, \mathcal{B}(X))$ as $\Delta \to \infty$: $P_\Delta \Rightarrow P$, if for any function $f \in C(X)$

$$\lim_{\Delta \to \infty} \int_X f(x) dP_\Delta(x) = \int_X f(x) dP(x).$$

In what follows we shall deal with limit theorems uniform in $\theta \in \Theta$ for integrals (1.1.5). In this connection the following definition will be required.

A family of distributions $P_{\Delta,\theta}$ in $X = \mathbf{R}^n$, $n \geq 1$, converges weakly to a distribution $P_\theta (P_{\Delta,\theta} \xrightarrow{\mathfrak{D}} P_\theta)$ uniformly in Θ as $\Delta \to \infty$, if for any $f \in C(\mathbf{R}^n)$

$$\lim_{\Delta \to \infty} \int_{\mathbf{R}^n} f(x) dP_{\Delta,\theta}(x) = \int_{\mathbf{R}^n} f(x) dP_\theta(x)$$

uniformly in $\theta \in \Theta$.

In this definition $P_{\Delta,\theta}$, P_θ may be matrix-valued measures (see Ch. 3).

We now formulate two statements to be used below; refer to [53] for the proof.

Theorem 1.1.2. Let distributions $P_{\Delta,\theta}$, P_θ, $\theta \in \Theta$ possess the properties:

1) $\sup\{P_{\Delta,\theta}(x : |x| > A), \ \theta \in \Theta\} \to 0$ as $A \to \infty$,

2) $f_{\Delta,\theta}(t) = \int_{\mathbf{R}^n} e^{i(t,x)} dP_{\Delta,\theta}(x) \to \int_{\mathbf{R}^n} e^{i(t,x)} dP_\theta(x) = f_\theta(t)$ uniformly in $\theta \in \Theta$ as $\Delta \to \infty$.

Then $P_{\Delta,\theta} \xrightarrow{\mathfrak{D}} P_\theta$ uniformly in Θ as $\Delta \to \infty$.

Theorem 1.1.3. Let ξ_r, $r = 1, 2, \ldots$ be independent r.v.'s with d.f. $F_{r,\theta}(x)$ depending on $\theta \in \Theta$ and $B_{r,\theta}^2 = \sum_{j=1}^r \operatorname{var} \xi_j$. If

1) $\sup_{\theta \in \Theta} \ \sup_{j=1,2,\ldots} E_\theta |\xi_j|^{2+\delta} < \infty$ for some $\delta > 0$;

2) $B_{r,\theta}^{-1-\delta/2} \sum_{j=1}^r E_\theta |\xi_j|^{2+\delta} \to 0$ as $r \to \infty$ uniformly in Θ, then

$$\Big[\sum_{j=1}^r (\xi_j - E_\theta \xi_j)\Big] / B_{r,\theta} \xrightarrow{\mathfrak{D}} N(0,1)$$

uniformly in Θ as $r \to \infty$.

Theorem 1.1.3 is a uniform version of the Lyapunov theorem.

A family of probability measures $\{P_\Delta, \ \Delta \in \mathfrak{M}\}$ on a metric space $(X, \mathcal{B}(X))$ is said to be weakly compact if any sequence of measures from this family contains a weakly convergent subsequence.

Let $K \subseteq \mathbf{R}^n$ be compact. Consider an a.s. continuous scalar field $\xi(\omega, x)$, $x \in K$, that is, a field in which almost all sample functions are continuous on K. A distribution $P_\xi(A) = P\{\xi(\omega, \cdot) \in A\}$, $A \in \mathcal{B}(C(K))$, is said to be a measure generated (induced) by the random field $\xi(x)$, $x \in K$ in $C(K)$ or the distribution of the random field (for details, see [30,Ch.5]).

Let $\xi(x), \xi_\Delta(x)$, $x \in K$, $\Delta \in \mathfrak{M}$, be a family of a.s. continuous random fields and P, P_Δ probability measures in $C(K)$ induced by the fields $\xi(x), \xi_\Delta(x)$ respectively.

The symbol $\mathcal{D}$ will denote a set of Boolean vectors $\nu = (\nu_1, \ldots, \nu_n)$, and we set $|\nu| = \nu_1 + \ldots + \nu_n$. Define increments in $\prod[x, y]$ for an arbitrary function $f : \mathbf{R}^n \to \mathbf{R}^1$ as follows:

$$\delta_f^{(n)}[y, x] = \sum_{\nu \in \mathcal{D}} (-1)^{|\nu|} f(y - \nu \otimes (y - x)). \tag{1.1.6}$$

Theorem 1.1.4 [131,132,75]. Let the finite-dimensional distributions of the random fields $\xi_\Delta(x)$, $x \in K = \prod(b)$, $\Delta \in \mathfrak{M}$, converge to finite-dimensional distributions of the field $\xi(x)$, $x \in K$, as $\Delta \to \infty$ and let the following conditions hold:

(1) for all $0 \le y_i^{(1)} < y_i^{(2)} \le b_i$, $i = 1, \ldots, n$, there exists a $k_1 > 0$ such that

$$E|\delta_{\xi_\Delta}^{(n)}[y^{(2)}, y^{(1)}]|^p \le k_1 |\prod[y^{(1)}, y^{(2)}]|^{1+q},$$

where $p > 0$, $q > 0$;

(2) there exist $z_i^{(0)}$, $i = 1, \ldots, n$, such that for all $0 \le y_i^{(1)} \le y_i^{(2)} \le b_i$, $i = 1, \ldots, n$, there exists a $k_2 > 0$ such that

$$E|\xi_\Delta(z_1^{(0)}, \ldots, y_i^{(2)}, \ldots, z_n^{(0)}) - \xi_\Delta(z_1^{(0)}, \ldots, y_i^{(1)}, \ldots, z_n^{(0)})|^p \le$$

$$\le k_2 |y_i^{(2)} - y_i^{(1)}|^{1+q}, \quad i = 1, \ldots, n,$$

where $p > 0$, $q > 0$.

Then $P_\Delta \Rightarrow P$ in $C(\prod(b))$ as $\Delta \to \infty$.

We note that the symbols k, k_i, $k_i(n, \alpha, \Delta)$, $\tilde{k}_i$, etc. denote positive constants whose particular values are usually non-essential. The parameters on which these constants depend will be indicated in parentheses if necessary. The numbering of such constants holds only within the relevant subsection. Symbols c, $c_i(n, \alpha)$, etc. will be used to denote constants essential for the exposition. The numbering of such constants holds within each chapter. Assumptions are numbered by Roman numerals I, II, III, etc. They are valid (without special mention) throughout the relevant chapter.

1.2. Homogeneous and Isotropic Random Fields

We present an overview of basic concepts of the spectral theory of homogeneous isotropic random fields.

Rotations of the Euclidean space $\mathbf{R}^n$ are defined as linear transformations g on this space that do not change its orientation and preserve the distances of the points from the origin: $|gx| = |x|$. The rotations of $\mathbf{R}^n$ generate the group $SO(n)$.

The motions of $\mathbf{R}^n$ are defined as non-homogeneous linear transformations which preserve the distances between points of this space and its orientation. It is well known that any motion in $\mathbf{R}^n$ can be written in the form $x \to gx + \tau$, $g \in SO(n)$, where $\tau \in \mathbf{R}^n$ may be treated as an element of the group $T = \{\tau\}$ of the shifts $x \to x + \tau$ in $\mathbf{R}^n$. The motions of a Euclidean space generate the group $M(n)$.

A random field $\xi : \Omega \times \mathbf{R}^n \to \mathbf{R}^m$ is called homogeneous in the strict sense if all of its finite-dimensional distributions are invariant with respect to the group T of shifts in $\mathbf{R}^n$, that is, $\xi(x) \overset{d}{=} \xi(x + \tau)$, $\tau \in \mathbf{R}^n$.

A random field $\xi : \Omega \times \mathbf{R}^n \to \mathbf{R}^1$ satsifying $E\xi^2(x) < \infty$ is called homogeneous in the wide sense if its mathematical expectation $m(x) = E\xi(x)$ and correlation function (cor.f.) $B(x,y) = E[\xi(x) - m(x)][\xi(y) - m(y)]$ are invariant with respect to the group of shifts τ in $\mathbf{R}^n$, that is, $m(x) = m(x + \tau)$, $B(x,y) = B(x + \tau, y + \tau)$ for any x, y, $\tau \in \mathbf{R}^n$. It means that $E\xi(x) = k =$const, and the cor.f. $B(x,y) = B(x - y)$ depends only on the difference $x - y$. Without loss of generality it will usually be assumed that $k = 0$.

If a field $\xi : \Omega \times \mathbf{R}^n \to \mathbf{R}^1$ that is homogeneous in the strict sense possesses a second order moment, it is homogeneous in the wide sense too. In the case of Gaussian one-dimensional random fields both concepts of homogeneity coincide.

If not specified, homogeneity will be treated in what follows as homogeneity in the wide sense.

The cor.f. $B(x - y)$ of a homogeneous random field is a non-negative definite kernel on $\mathbf{R}^n \times \mathbf{R}^n$, that is, for any $r \geq 1$, $x^{(j)} \in \mathbf{R}^n$, $z_j \in \mathbf{C}^1$, $j = 1, \ldots, r$, $\sum_{i,j=1}^{r} B(x^{(i)} - x^{(j)}) z_i \bar{z}_j \geq 0$. If the function $B(x)$ is continuous at the point $x = 0$, then the field is mean-square (m.s.) continuous at each point $x \in \mathbf{R}^n$ and vice versa.

Let K be a class of functions which can serve as correlation functions of homogeneous random fields. Now if $B_1(x), B_2(x) \in K$ and k_1, k_2 are constants, then $k_1 B_1(x) + k_2 B_2(x) \in K$, $B_1(x) B_2(x) \in K$.

The Bochner-Khinchin theorem implies that a function $B(x)$ is the cor.f. of a m.s. continuous random field $\xi(x)$, $x \in \mathbf{R}^n$, if and only if there exists a finite measure $F(\cdot)$ on $(\mathbf{R}^n, \mathcal{B}^n)$ such that

$$B(x - y) = \int_{\mathbf{R}^n} e^{i\langle \lambda, x-y \rangle} F(d\lambda) = \int_{\mathbf{R}^n} \cos\langle \lambda, x - y \rangle F(d\lambda), \qquad (1.2.1)$$

with $F(\mathbf{R}^n) = B(0) < \infty$.

Representation (1.2.1) is called the spectral decomposition of the correlation function; $F(\cdot)$ is called the spectral measure of the field $\xi(x)$, $x \in \mathbf{R}^n$. The function $F(\lambda) = F(\prod(-\infty, \lambda))$, $\prod(-\infty, \lambda) = \underset{j=1}{\overset{n}{\times}}(-\infty, \lambda_j)$, is called the spectral function of the field $F(\infty, \ldots, \infty) = B(0)$, $F(-\infty, \ldots, -\infty) = 0$.

A spectral measure $F(\cdot)$ admits the Lebesgue decomposition into absolutely continuous, discrete and singular parts. If the last two components are missing, the spectral measure is absolutely continuous: $F(\Delta) = \int_\Delta f(\lambda) d\lambda$, $\Delta \in \mathcal{B}^n$.

The function $f(\lambda)$, $\lambda \in \mathbf{R}^n$, which is integrable over $\mathbf{R}^n$, is called the homogeneous second order spectral density function, or simply the spectral density function of the homogeneous random field.

If the spectral density function exists, then the spectral representation (1.2.1) may be written as

$$B(x) = \int_{\mathbf{R}^n} e^{i\langle \lambda, x \rangle} f(\lambda) d\lambda. \qquad (1.2.2)$$

Hereafter $L_p(T)$, $T \subseteq \mathbf{R}^n$, will be used to denote the space of measurable functions $f : T \to \mathbf{R}^1$ such that $\|f\|_p = \left[\int_T |f(x)|^p dx \right]^{1/p} < \infty$, $1 \le p < \infty$. If $B(x) \in L_1(\mathbf{R}^n)$ then, clearly, $f(\lambda)$ exists.

Due to Karhunen's theorem [30, p.293-294], representation (1.2.1) implies the spectral decomposition of the field $\xi(x)$, $x \in \mathbf{R}^n$, itself, that is, there exists a complex-valued orthogonal random measure $Z(\Delta)$, $\Delta \in \mathcal{B}^n$ such that for every $x \in \mathbf{R}^n$ (P-a.s.)

$$\xi(x) = \int_{\mathbf{R}^n} e^{i\langle \lambda, x \rangle} Z(d\lambda), \qquad (1.2.3)$$

where $E|Z(\Delta)|^2 = F(\Delta)$, and the stochastic integral (1.2.3) is viewed as an integral with respect to the random measure $Z(\cdot)$ with the structure function $F(\cdot)$ [30].

We note the following properties of the random measure $Z(\cdot)$: 1) $EZ(\Delta) = 0$; 2) $Z(\Delta) = \overline{Z(-\Delta)}$, where $-\Delta = \{\lambda : -\lambda \in \Delta\}$; 3) $EZ(\Delta_1)\overline{Z(\Delta_2)} = $

$F(\Delta_1 \cap \Delta_2)$, $\Delta_1, \Delta_2 \in \mathcal{B}^n$; 4) $Z\left(\bigcup_{j=1}^r \Delta_j\right) = \sum_{j=1}^r Z(\Delta_j)$ a.s. for any disjoint $\Delta_1, \ldots, \Delta_r \in \mathcal{B}^n$.

A random field $\xi : \Omega \times \mathbf{R}^n \to \mathbf{R}^1$ is called isotropic in the wide sense if $E\xi^2(x) < \infty$ and the mathematical expectation $m(x)$ along with the correlation function $B(x, y)$ are invariant with respect to the group $SO(n)$, that is, $m(x) = m(gx)$, $B(x, y) = B(gx, gy)$ for any $x, y \in \mathbf{R}^n$, $g \in SO(n)$.

There exist isotropic fields which are not homogeneous. For example, the Lévy multiparametric Brownian motion, that is, a Gaussian random field $V(x)$, $x \in \mathbf{R}^n$, with $EV(x) = 0$, $EV(x)V(y) = \{|x|+|y|-|x-y|\}/2$, $x, y \in \mathbf{R}^n$, is such a field. If $\xi : \Omega \times \mathbf{R}^n \to \mathbf{R}^1$ is a homogeneous isotropic random field, then $E\xi(x) = k =$ const (we assume that $k = 0$) and the correlation function $B(x - y) = B(|x - y|)$ depends only on the Euclidean distance $\rho_{xy} = |x - y|$ between points x and y. This is valid if and only if $F(\Delta) = F(g\Delta)$ for any $g \in SO(n)$, $\Delta \in \mathcal{B}^n$.

For $\nu > -\frac{1}{2}$, we denote by $J_\nu(z) = \sum_{m=0}^\infty (-1)^m (z/2)^{2m+\nu} [m! \Gamma(m + \nu + 1)]^{-1}$ the Bessel function of the first kind of order ν and by $Y_n(z) = 2^{(n-2)/2} \Gamma(n/2) J_{(n-2)/2}(z) z^{(2-n)/2}$ the spherical Bessel function.

A function $B(\rho)$, $\rho \in \mathbf{R}_+^1$, is a correlation function of a homogeneous isotropic m.s. continuous random field $\xi(x)$, $x \in \mathbf{R}^n$, if and only if there exists a finite measure $G(\cdot)$ such that

$$B(\rho) = \int_0^\infty Y_n(u\rho) G(du), \tag{1.2.4}$$

with $G(\mathbf{R}_+^1) = F(\mathbf{R}^n) = B(0) < \infty$.

The bounded non-decreasing function $G(u) = \int_{\{\lambda : |\lambda| < u\}} F(d\lambda)$ is called the spectral function of the homogeneous isotropic random field. Clearly

$$\int_0^\infty dG(u) = F(\mathbf{R}^n) < \infty$$

and the inversion formula

$$G(u) = 2^{(2-n)/2} \Gamma^{-1}(n/2) \int_0^\infty J_{n/2}(ur)(ur)^{n/2} B(r) r^{-1} dr \tag{1.2.5}$$

holds. Thus, for $n = 2$ formula (1.2.4) becomes

$$B(\rho) = \int_0^\infty J_0(u\rho) G(du), \tag{1.2.6}$$

and for $n = 3$

$$B(\rho) = \int_0^\infty \sin\{u\rho\} (u\rho)^{-1} G(du). \tag{1.2.7}$$

Formula (1.2.4) follows from (1.2.1) if we proceed to spherical coordinates (1.1.1) and use the relation

$$\frac{1}{|s_{n-1}(u)|}\int_{s_{n-1}(u)}e^{i\langle\lambda,x-y\rangle}dm(\lambda)=Y_n(u|x-y|), \qquad (1.2.8)$$

where $m(\cdot)$ is the Lebesgue measure on the sphere $s_{n-1}(u)$ (cf. (1.1.2)).

Remark 1.2.1. If $B(0) = 1$, then $G(\cdot)$ is a probability measure on R^1_+. For $n > 1$ consider the symmetric measure on $[-1, 1]$:

$$G_{(n-2)/2}(du) = (1-u^2)^{(n-3)/2}[\sqrt{\pi}\Gamma((n-1)/2)]^{-1}\Gamma(n/2)du, \ u \in [-1,1].$$

Using the Poisson integral

$$J_\nu(z)\Gamma(\nu+\tfrac{1}{2}) = \pi^{-1/2}(z/2)^\nu\int_{-1}^1 e^{itz}(1-t^2)^{\nu-1/2}dt, \quad \mathrm{Re}\,\nu > -\tfrac{1}{2}, \quad (1.2.9)$$

we find that the characteristic function of the distribution $G_{(n-2)/2}$ equals $Y_n(t)$; thus formula (1.2.4) may be written as $B(t) = EY_n(t\eta) = E\exp\{it\eta\zeta\}$ where η and ζ are independent r.v.'s with distributions G and $G_{(n-2)/2}$ respectively. On the other hand, (1.2.4) may be treated as a Hankel type transform

$$H_\nu(G,t) = B(t) = \int_0^\infty Y_\nu(ut)G(du)$$

with the kernel $Y_\nu(t) = \Gamma(\nu+1)J_\nu(t)(t/2)^{-\nu}$ defined for $\nu > -\tfrac{1}{2}$ (or $n > 1$). However, as $\nu \to -\tfrac{1}{2}+0$, $\lim Y_\nu(t) = \cos t$ while the function $\cos t$ is the characteristic function of the probability measure $M = (e_1 + e_{-1})/2$ concentrated at the points 1 and -1 and assigning masses $\tfrac{1}{2}$ to each of them. Therefore, $\cos t$ may be denoted by $Y_{-1/2}(t)$ and the measure M by $G_{-1/2}(\cdot)$; thus one can assume by continuity that $\nu \geq -\tfrac{1}{2}$; this also includes the case dimension $n = 1$.

Remark 1.2.2. Sometimes as an extension of the class of isotropic correlation functions $B(x) = B(|x|)$, $x \in \mathbf{R}^n$ depending on the variable $\rho = |x|$ only, one considers the class of ellipsoidal correlation functions, that is, functions $B(\langle A^{-1}x, x\rangle)$ depending on the value of the positive definite quadratic form

$$x'A^{-1}x = \langle A^{-1}x, x\rangle = \sum_{i,j=1}^n \frac{A_{ij}}{\det A}x_ix_j, \ A = (a_{ij})_{i,j=1,\ldots,n}$$

where x' is the transpose of the vector x, A is an n by n positive definite matrix and A_{ij} is the cofactor of the element a_{ij}. Since any positive definite quadratic form can be diagonalized by an appropriate change of variables, any ellipsoidal correlation function can be reduced to the form depending on the variable

$$\tau = \left\{ \left(\frac{x_1}{a_1}\right)^2 + \ldots + \left(\frac{x_n}{a_n}\right)^2 \right\}^{\frac{1}{2}}$$

where $a_1, \ldots, a_n$ are (constant) scale factors. In turn, a correlation function depending on the variable $\tau \geq 0$ can be made isotropic (1.2.4) by introducing $x_i' = x_i/a_i$, $i = 1, \ldots, n$. Thus from a mathematical point of view, the investigation of a class of ellipsoidal correlation functions can be reduced to that of isotropic correlation functions. However, in the context of physical applications, ellipsoidal correlation functions are of interest since the constants $a_1, \ldots, a_n$ may reflect the isotropy of physical fields in various directions [227,116,105,99,113].

Remark 1.2.3. Sometimes direction-dependent correlation functions are considered. For example, let $n = 2$ and $x = (x_1, x_2) = (t \cos \theta, t \sin \theta)$, $x' = (x_1', x_2') = (t' \cos \theta', t' \sin \theta')$. In general, the correlation function $B(x, x') = E\xi(x)\xi(x')$ of a random field $\xi(x)$ with $E\xi(x) = 0$ depends on t, t', θ, θ'. If the field $\xi(x)$ is homogeneous, then the correlation function $B(x, x') = B(x - x')$ depends on the variables $x - x' = \tau = (\tau_1, \tau_2)$ only, $\tau_i = t_i - t_i'$, $i = 1, 2 : B(\tau_1, \tau_2) = B(t \cos \theta - t' \cos \theta', t \sin \theta - t' \sin \theta')$. We say that a correlation function is direction-dependent if for any pair of points that lie on a line defined by the angle θ, the correlation function $B(\tau_1, \tau_2)$ can be expressed in the form

$$B(\tau, \theta) = B(\tau \cos \theta, \tau \sin \theta)$$

where $\tau = t - t'$ is the distance between points measured along the line defined by the angle θ. If $B(\tau, \theta) = B(\tau)$ does not depend on θ, $B(\tau)$ is an isotropic correlation function. Sometimes it is assumed that $B(\tau, \theta) = B(\tau)l(\theta)$ where, for example, $l(\theta) = [\cos(\theta - \theta_0)]^{2k}$, $k = 1, 2, \ldots$. In the case when the correlation function is ellipsoidal (see Remark 1.2.2), by considering only points lying on a line defined by the angle θ we find that the direction-dependent correlation function $B(\tau, \theta)$ depends on τ and θ through the variable

$$w = \left(\frac{\tau_1}{a_1}\right)^2 + \left(\frac{\tau_2}{a_2}\right)^2 = \frac{(\tau \cos \theta)^2}{a_1^2} + \frac{(\tau \sin \theta)^2}{a_2^2}$$

where a_1, a_2 are constants. Let $1 - e^2 = (a_2/a_1)^2$, then $w = (\tau/a_2)^2(1 - e^2 \cos^2 \theta)$. Thus the function $B(\tau, \theta)$ depends on $\tau\sqrt{1 - e^2 \cos^2 \theta}$ where e measures the eccentricity of the isocorrelation ellipses. The correlation structure becomes isotropic when $e = 0$.

Lemma 1.2.1. If $B(\rho)$, $\rho \geq 0$, is the correlation function of a m.s. continuous homogeneous isotropic random field, then:

1) $B(\rho) \geq -\frac{B(0)}{n}$,

2) for $n \geq 3$ the function $B(\rho)$ as a function in ρ is differentiable $[\frac{n-1}{2}]$ times.

Proof. Let $x^{(j)} \in \mathbf{R}^n$, $j = 1, 2, \ldots, n+1$ with $|x^{(i)} - x^{(j)}| = \rho$, $i \neq j$. Then $E[\sum_{j=1}^{n+1} \xi(x^{(j)})]^2 = (n+1)[B(0) + nB(\rho)]$, whence 1) follows. Relation 2) follows from the fact that (1.2.4) can be differentiated with respect to ρ using the formula $\frac{d}{dz}(\frac{J_\nu(z)}{z^\nu}) = -\frac{J_{\nu+1}(z)}{z^\nu}$ and the integrals obtained by formal differentiation converge uniformly.

Let $n \geq 2$ and (r, u), $u \in s_{n-1}(1)$, $r \geq 0$, be spherical coordinates of the point $x \in \mathbf{R}^n$ (cf. (1.1.1). Denote by $S_m^l(u)$, $u \in s_{n-1}(1)$, the real spherical harmonics of degree m, and let $h(n, m) = (2m+n-2)(m+n-3)!/[(n-2)!m!]$ be the number of such harmonics [145,24,p.431-467].

In view of Karhunen's theorem and the addition theorem for Bessel functions:

$$2^{(n-2)/2}\Gamma(n/2)J_{(n-2)/2}\sqrt{r_1^2 + r_2^2 - 2r_1r_2\cos\gamma}(r_1^2 + r_2^2 - 2r_1r_2\cos\gamma)^{(2-n)/2}$$

$$= c_1^2(n)\sum_{m=0}^{\infty}\sum_{l=1}^{h(n,m)}S_m^l(u_1)S_m^l(u_2)\times$$

$$\times J_{m+(n-2)/2}(r_1)J_{m+(n-2)/2}(r_2)(r_1r_2)^{(2-n)/2},$$

where $x = (r_1, u_1)$, $y = (r_2, u_2)$, $u_j \in s_{n-1}(1)$, $j = 1, 2$, $\cos\gamma = \langle x, y\rangle/[|x| \cdot |y|]$ is the cosine of the angle between x and y, $|x - y| = \sqrt{r_1^2 + r_2^2 - 2r_1r_2\cos\gamma}$, and

$$c_1^2(n) = 2^{n-1}\Gamma(n/2)\pi^{n/2}, \tag{1.2.10}$$

we arrive at the following spectral decomposition for the field $\xi(x)$, $x \in \mathbf{R}^n$:

$$\xi(x) = c_1(n)\sum_{m=0}^{\infty}\sum_{l=1}^{h(n,m)}S_m^l(u)\int_0^{\infty}\frac{J_{m+(n-2)/2}(\mu r)}{(\mu r)^{(n-2)/2}}Z_m^l(d\mu). \tag{1.2.11}$$

The series in formula (1.2.11) converges in the mean square sense and $Z_m^l(\cdot)$ is a family of real-valued random measures on $(\mathbf{R}_+^1, \mathcal{B}(\mathbf{R}_+^1))$ such that relations:
1) $EZ_m^l(\Delta) = 0$; 2) $EZ_m^l(\Delta_1)Z_p^q(\Delta_2) = \delta_m^p\delta_l^q G(\Delta_1 \cap \Delta_2)$, $\Delta_1, \Delta_2 \in \mathcal{B}(\mathbf{R}_+^1)$, hold.*

---

* $\delta_r^m = 1$, $r = m$; 0, $r \neq m$, is the Kronecker symbol.

For $n = 2$, $c_1(n) = \sqrt{2\pi}$, $h(2, m) = 2$, there exist for each $m \in Z^1$ two complex harmonics $\exp\{im\phi\}/\sqrt{2\pi}$ in terms of which, representation (1.2.11) can be written as

$$\xi(\rho, \phi) = \sum_{m=-\infty}^{\infty} \int_0^{\infty} J_m(\mu\rho) Z_m(d\mu), \qquad (1.2.12)$$

where $J_m(z)$ is the Bessel function of the first kind of integral order $m \geq 0$, $J_{-m}(z) = (-1)^m J_m(z)$ and $Z_m(\cdot)$ are complex random measures such that $E Z_m(\Delta_1)\overline{Z_{m'}(\Delta_2)} = \delta_m^{m'} G(\Delta_1 \cap \Delta_2)$, $\Delta_1, \Delta_2 \in \mathbf{B}(\mathbf{R}_+^1)$ and $G(\cdot)$ is defined in (1.2.6).

For $n = 3$ the field $\xi(x)$, $x = (r, \phi, \phi_1) \in \mathbf{R}^3$, $r \geq 0$, $0 \leq \phi_1 < \pi$, $0 \leq \phi < 2\pi$, may be represented by the series

$$\xi(r, \phi, \phi_1) = \pi\sqrt{2} \sum_{m=-\infty}^{\infty} \sum_{l=-m}^{m} S_m^l(\phi, \phi_1) \int_0^{\infty} \frac{J_{m+1/2}(\mu r)}{(\mu r)^{1/2}} Z_m^l(d\mu), \qquad (1.2.13)$$

where $Z_m^l(\cdot)$ are complex random measures having the structure function $G(\cdot)$ given in (1.2.7) and the complex spherical harmonics are of the form:

$$S_m^l(\phi, \phi_1) = e^{il\phi} \Theta_m^l(\cos\phi_1)/\sqrt{2\pi}, \quad -m \leq l \leq m,$$

$$\Theta_m^l(x) = k_1(m, l)(1 - x^2)^{l/2}(d^{l+m}/dx^{l+m})(1 - x^2)^m =$$
$$= k_2(m, l)(1 - x^2)^{-l/2}(d^{m-l}/dx^{m-l})(1 - x^2)^m, \, l \geq 0,$$

$$\Theta_m^{-l}(x) = (-1)^l \Theta_m^l(x), \quad \overline{S}_m^l(\phi, \phi_1) = (-1)^l S_m^{-l}(\phi, \phi_1),$$

$$k_1(m, l) = (-1)^m \sqrt{(2m + 1)(m - l)!/[2(m + l)!]}/[2^m m!],$$

$$k_2(m, l) = (-1)^{m-l} \sqrt{(2m + 1)(m + l)!/[2(m - l)!]}/[2^m m!]. \qquad (1.2.14)$$

The functions $S_m^l(\phi, \phi_1)$ satisfy the orthogonality property:

$$\int_{s_2(1)} S_m^l(\phi, \phi_1)\overline{S_{m'}^{l'}(\phi, \phi_1)} dm_2(\phi, \phi_1) = \delta_l^{l'} \delta_m^{m'}, \qquad (1.2.15)$$

where $dm(\phi, \phi_1) = \sin\phi_1 d\phi_1 d\phi$, $0 \leq \phi_1 < \pi$, $0 \leq \phi < 2\pi$ (see (1.1.2)).

Remark 1.2.4 For $n = 3$, the functions $S_m^l(\phi, \phi_1)$ are expressed in terms of the associated Legendre functions of the first kind or in terms of Gegenbauer polynomials [145].

The field $\xi(x)$, $x \in \mathbf{R}^n$ has an absolutely continuous spectrum, that is, the measure $F(\cdot)$ given in (1.2.1) is absolutely continuous with respect to the Lebesgue measure if and only if

$$\xi(x) = \int_{\mathbf{R}^n} b(u - x)W(du), \qquad (1.2.16)$$

where $b(\cdot) \in L_2(\mathbf{R}^n)$, $W(\cdot)$ is a complex Wiener measure on $(\mathbf{R}^n, \mathbf{B}^n)$, that is, $E|W(\Delta)|^2 = |\Delta|$, $\Delta \in \mathbf{B}^n$. In representation (1.2.16) $b(x)$ is the Fourier-Plancherel transform of the function $\alpha(\lambda) \in L_2(\mathbf{R}^n)$, where $|\alpha(\lambda)|^2 = f(\lambda)$ is the spectral density function of the field $\xi(x)$, $x \in \mathbf{R}^n$.

If the function $b(x)$ in (1.2.16) is radial, $b(x) = b(|x|)$, then the field $\xi(x)$, $x \in \mathbf{R}^n$ is homogeneous and isotropic.

If $\xi(x)$, $x \in \mathbf{R}^n$, is a homogeneous isotropic field possessing an absolutely continuous spectrum, that is, if there exists a function $g(\mu)$ such that

$$G'(\mu) = |s(1)|\mu^{n-1}g(\mu), \qquad \mu^{n-1}g(\mu) \in L_1(\mathbf{R}^1_+), \qquad (1.2.17)$$

then the function $g(\mu)$ is called the isotropic spectral density function of the field $\xi(x)$, $x \in \mathbf{R}^n$.

In this case, representation (1.2.4) may be rewritten as

$$B(\rho) = |s(1)| \int_0^\infty Y_n(\mu\rho)\mu^{n-1}g(\mu)d\mu. \qquad (1.2.18)$$

Accordingly, the spectral representation (1.2.16) becomes

$$\xi(x) = (2\pi)^{n/2} \sum_{m=0}^{\infty} \sum_{l=1}^{h(n,m)} S_m^l(u) \int_0^\infty \frac{\sqrt{\mu r}}{r^{(n-1)/2}} J_{m+(n-2)/2}(\mu r)\sqrt{g(\mu)}W_m^l(d\mu), \qquad (1.2.19)$$

where $\{W_m^l(\cdot)\}$ $l = 1, \ldots, h(n,m)$, $m = 0, 1, \ldots$, is a family of jointly uncorrelated real Wiener measures on $(\mathbf{R}^1_+, \mathbf{B}(\mathbf{R}^1_+))$, that is, $EW_m^l(\Delta_1)W_p^q(\Delta_2) = \delta_m^p \delta_l^q |\Delta_1 \cap \Delta_2|$.

We provide examples of correlation functions for homogeneous isotropic random fields together with spectral representations (1.2.4), (1.2.18). In all the examples the constant $a > 0$.

Example 1.2.1. Let the function $G(u)$, $u \in \mathbf{R}^1_+$ appearing in representation (1.2.4) have the unique point of discontinuity:

$$G(u) = \begin{cases} 0, & u \leq a, \\ 1, & u > a. \end{cases}$$

The correlation function of the homogeneous isotropic random field is then of the form:

$$B(\rho) = Y_n(a\rho) = 1 - \frac{\rho^2}{2n} + \frac{\rho^4}{2 \cdot 4 \cdot n(n+2)} - \frac{\rho^6}{2 \cdot 4 \cdot 6n \cdot (n+2)(n+4)} + \cdots$$

Example 1.2.2. Let the function $G(u)$, $u \in \mathbf{R}_+^1$ appearing in representation (1.2.4) be of the form

$$G(u) = \begin{cases} (\frac{u}{a})^n, & u \le a, \\ 1, & u > a. \end{cases}$$

Then the correlation function of the homogeneous isotropic random field is of the form

$$B(\rho) = n2^{(n-2)/2}\Gamma(\tfrac{n}{2})(a\rho)^{-n/2}J_{n/2}(a\rho) = Y_{n+2}(a\rho).$$

Example 1.2.3. Let the correlation function of a homogeneous random field be of the form

$$B(\rho) = \exp\{-a\rho^2\}.$$

Then relations (1.2.17), (1.2.18) hold with

$$G'(\mu) = \mu^{n-1}\frac{\exp\{-\frac{\mu^2}{4a}\}}{2^{n-1}a^{n/2}\Gamma(\tfrac{n}{2})}.$$

Example 1.2.4. Let the correlation function of a homogeneous isotropic field be of the form

$$B(\rho) = \exp\{-a\rho\}.$$

Then representations (1.2.17), (1.2.18) hold with

$$G'(\mu) = \mu^{n-1}\frac{a(n-1)!}{2^{n-2}\Gamma^2(\tfrac{n}{2})}\frac{1}{(\mu^2+a^2)^{(n+1)/2}}.$$

Example 1.2.5. For $n \ge 3$ let

$$B(\rho) = \frac{a^{n-1}}{(\rho^2+a^2)^{(n-2)/2}}.$$

Then representations (1.2.17), (1.2.18) hold with

$$G'(\mu) = \frac{a^{n-1}}{(n-2)!}\mu^{n-2}e^{-a\mu}.$$

Example 1.2.6. Let
$$B(\rho) = \frac{a^{n+1}}{(\rho^2 + a^2)^{(n+1)/2}}.$$

Then
$$G'(\mu) = \frac{a^{n-1}}{(n-1)!}\mu^{n-1}e^{-a\mu}.$$

Example 1.2.7. The correlation function of a homogeneous isotropic random field is of the form:

$$B(\rho) = \sigma^2\frac{1}{(1+\rho^2)^\alpha}, \quad \sigma^2 = \operatorname{var}\xi(0), \quad 0 < \alpha < \frac{n}{2}.$$

The asymptotics of the spectral density function of such a field are described below (see Lemmas 1.5.4, 2.10.1). Note that the above correlation function occurs in applications [116].

Example 1.2.8. Let

$$B(\rho) = (\tfrac{\pi}{2})^{n/2}(\tfrac{\rho}{a})^p K_p(a\rho)\tfrac{G_0}{2}, \quad G_0 > 0,$$

where $K_p(z)$ is McDonald's function (a Bessel function of the second kind of an imaginary argument) of order $p = 2 - \frac{n}{2}$ defined by the relation

$$K_\nu(z) = \int_0^\infty e^{-z\cosh t}\cosh\{\nu t\}dt.$$

Note that for large z McDonald's function can be expanded as the asymptotic series:

$$K_p(z) \simeq \sqrt{\frac{\pi}{2z}}e^{-z}\Big[1 + \frac{4p^2-1}{1!8z} + \frac{(4p^2-1)(4p^2-3^2)}{2!(8z)^2} + \ldots\Big].$$

Thus for $n = 3$ $(p = \tfrac{1}{2})$ we have

$$B(\rho) = \frac{G_0\pi}{8a}e^{-a\rho};$$

for $n = 2$ $(p = 1)$ we have

$$B(\rho) = \frac{\pi G_0}{4a^2}(a\rho)K_1(a\rho).$$

A random field $\xi(x)$, $x \in \mathbf{R}^n$ satisfying the equation

$$(\nabla^2 - a^2)\xi(x) = W'(x),$$

where $W'(x)$ is the white noise in $\mathbf{R}^n$, $(W(\cdot)$ is the Wiener measure in $\mathbf{R}^n)$ and

$$\nabla^2 = \frac{\partial^2}{\partial x_1^2} + \ldots + \frac{\partial^2}{\partial x_n^2}$$

is the Laplace operator, has correlation function $B(\rho)$.

In this case

$$G'(\mu) = \frac{|s(1)|}{(2\pi)^n} \frac{\mu^{n-1}}{(\mu^2 + a^2)^2}.$$

A generalization of this example is presented in [140].

We also note that for $n = 3$, the correlation functions

$$\overline{B}(\rho) = \frac{a^2}{2^{\nu-1}\Gamma(\nu)} \left(\frac{r}{r_0}\right)^\nu K_\nu\left(\frac{r}{r_0}\right), \quad r_0 > 0, \ \nu > 0$$

are often used (see, for example, [122]). In this case relation (1.2.17) holds for $n = 3$ and

$$G'(\mu) = |s_2(1)|\mu^2 \frac{\Gamma(\nu + 3/2)}{\pi\sqrt{\pi}\Gamma(\nu)} \frac{a^2 r_0^3}{(1 + \mu^2 r_0^2)^{\nu+3/2}}.$$

The function $\overline{B}(\rho)$ is also utilized as the correlation function of a stationary random process (see, for example, [122]).

Example 1.2.9. Let

$$B(r) = arK_1(ar).$$

Then relation (1.2.17) holds with

$$G'(\mu) = 2^{n/2}a^2\Gamma^2(\tfrac{n+2}{2})\frac{\mu^{n-1}}{(\mu^2 + a^2)^{(n+2)/2}}.$$

We emphasize the following points relating to the definition of a vector-valued random field $\xi : \Omega \times \mathbf{R}^n \to \mathbf{R}^m$.

Homogeneous random fields $\xi_i : \Omega \times \mathbf{R}^n \to \mathbf{R}^1$, $i = 1, 2$, (in the wide sense) with zero mean are said to be jointly homogeneous if their cross-correlation function $B_{12}(x, y) = E\xi_1(x)\xi_2(y) = B_{12}(x - y)$ depends only on the difference $x - y$. Jointly homogeneous random fields are jointly isotropic if their cross-correlation function $B_{12}(x-y)$ depends only on the distance $\rho_{xy} = |x-y|$. A collection of m homogeneous (isotropic) fields $\xi(x) = [\xi_1(x), \ldots, \xi_m(x)]$ which are jointly homogeneous (isotropic) is said to be a vector homogeneous (isotropic) random field $\xi : \Omega \times \mathbf{R}^n \to \mathbf{R}^m$.

Let $B_{rs}(x) = E\xi_r(x+y)\xi_s(y)$. Then the matrix $B(x) = (B_{rs}(x))_{r,s=1,\ldots,m}$ is the correlation function of the homogeneous (isotropic) vector field $\xi : \Omega \times$

$\mathbf{R}^n \to \mathbf{R}^m$. Let $B(x)$ be continuous at the point $x = 0$. Then for all r, s in $\{1, \ldots, m\}$ the functions $B_{rs}(x)$ admit the representation (1.2.1) in the case of a homogeneous field and the representation (1.2.4) in the case of a homogeneous isotropic field. We denote by $F_{rs}(\cdot)$, $G_{rs}(\cdot)$ the spectral measures involved in these decompositions and refer to the matrix measures $F = (F_{rs}(\cdot))_{r,s=1,\ldots,m}$ and $G = (G_{rs}(\cdot))_{r,s=1,\ldots,m}$ respectively as the spectral measures of a vector homogeneous and homogeneous isotropic field $\xi : \Omega \times \mathbf{R}^n \to \mathbf{R}^m$. Then the correlation function $B(x)$ of a vector homogeneous (isotropic) random field admits a decomposition (1.2.1) (respectively (1.2.4)), where $F(\cdot)$ and $G(\cdot)$ are the matrix measures defined above. If all of the elements $F_{rs}(\cdot)$ of the matrix F are absolutely continuous with respect to the Lebesgue measure, that is, there exist $f_{rs}(\lambda) = (\partial^n / \partial \lambda_1 \ldots \partial \lambda_n) F_{rs}(\prod(-\infty, \lambda))$, $r, s = 1, \ldots, m$, then the multidimensional field $\xi : \Omega \times \mathbf{R}^n \to \mathbf{R}^m$ is said to possess the spectral density function $f(\lambda) = (f_{rs}(\lambda))_{r,s=1,\ldots,m}$.

In this case the correlation function admits a decomposition (1.2.2) where $f(\lambda)$ is to be viewed as the matrix-valued function defined above. In the same manner vector analogues may be introduced for the other concepts defined above for scalar fields.

1.3. Spectral Properties of Higher Order Moments of Random Fields

We introduce the Fortet-Blanc-Lapierre classes of random fields.

Let $\mathbf{T}^{(m)}$ be a class of random fields $\xi : \Omega \times \mathbf{R}^n \to \mathbf{R}^1$ for which $E|\xi(x)|^m \le k_m < \infty$, where $k_m > 0$ are constants and $m \ge 1$ is an integer.

Denote the moment and cumulant functions of order $r \in \{1, \ldots, m\}$ of the field $\xi(x) \in \mathbf{T}^{(m)}$ by

$$m_r(x^{(1)}, \ldots, x^{(r)}) = E \prod_{j=1}^{r} \xi(x^{(j)}), \quad s_r(x^{(1)}, \ldots, x^{(r)}) =$$

$$= i^{-r} \frac{\partial^r}{\partial \alpha_1 \ldots \partial \alpha_r} \ln E \exp \left\{ i \sum_{j=1}^{r} \alpha_j \xi(x^{(j)}) \right\} \Big|_{\alpha_1 = \ldots = \alpha_r = 0}. \quad (1.3.1)$$

If necessary, a superscript will be used in the notation: $m_r^{(\xi)}(\cdot, \ldots, \cdot)$, $s_r^{(\xi)}(\cdot, \ldots, \cdot)$. Clearly, $m_r(x^{(1)}, \ldots, x^{(r)})$, $s_r(x^{(1)}, \ldots, x^{(r)})$ are symmetric functions.

Let $I = \{1, 2, \ldots, r\}$, $I_p = \{i_1, \ldots, i_{l_p}\} \subseteq I$, $m(I_p) = m_{l_p}(x^{(i_1)}, \ldots, x^{(i_{l_p})})$, $s(I_p) = s_{l_p}(x^{(i_1)}, \ldots, x^{(i_{l_p})})$.

Then the following Leonov-Shiryaev formulas relating moments to cumulants are valid:

$$m(I) = \sum_{A_q} \prod_{p=1}^{q} s(I_p); \quad s(I) = \sum_{A_q} (-1)^{q-1}(q-1)! \prod_{p=1}^{q} m(I_p), \qquad (1.3.2)$$

where $\sum_{A_q} \ldots$ denotes summation over all unordered partitions $A_q = \left\{ I = \bigcup_{p=1}^{q} I_p \right\}$ of the set I into sets $I_1, \ldots, I_q$ such that $I = \bigcup_{p=1}^{q} I_p$, $I_i \cap I_j \neq \emptyset$, $i \neq j$.

In particular, if $\xi(x) \in \mathbf{T}^{(4)}$ and $E\xi(x) = 0$ for $I = \{1, 2, 3, 4\}$, then

$$m_4(x^{(1)}, x^{(2)}, x^{(3)}, x^{(4)}) = s_4(x^{(1)}, x^{(2)}, x^{(3)}, x^{(4)}) + m_2(x^{(1)}, x^{(2)}) \times$$

$$\times m_2(x^{(3)}, x^{(4)}) + m_2(x^{(1)}, x^{(3)}) m_2(x^{(2)}, x^{(4)}) + m_2(x^{(1)}, x^{(4)}) \times$$

$$\times m_2(x^{(2)}, x^{(3)}). \qquad (1.3.3)$$

Consider the groups of motions $M(n)$, τ, $SO(n)$ defined in §1.2.

If $\xi(x) \in \mathbf{T}^{(m)}$ and for all $1 \leq r \leq m$, $\tau \in \mathbf{R}^n$

$$m_r(x^{(1)}, \ldots, x^{(r)}) = m_r(x^{(1)} + \tau, \ldots, x^{(r)} + \tau), \qquad (1.3.4)$$

we shall say that $\xi(x) \in \mathbf{S}_1^{(m)}$.

If for all $g_i \in SO(n)$, $1 \leq r \leq m$, the relation

$$m_r(x^{(1)}, \ldots, x^{(r)}) = m_r(g_1 x^{(1)}, \ldots, g_r x^{(r)}) \qquad (1.3.5)$$

holds, we shall say that $\xi(x) \in \mathbf{S}_2^{(m)}$. Instead of $\xi(x) \in \mathbf{S}_1^{(m)} \cap \mathbf{S}_2^{(2)}$ we write $\xi(x) \in \mathbf{S}^{(m)}$. In this case all moments and cumulants up to order m are invariant with respect to the group $M(n)$.

In view of (1.3.2), the cumulants $s_r(\cdot, \ldots, \cdot)$ may be used in (1.3.4), (1.3.5) in place of the moments $m_r(\cdot, \ldots, \cdot)$. If $\xi(x) \in \mathbf{T}^{(m)}$ and for all $1 \leq r \leq m$ there exists a totally finite complex measure $F(\cdot)$ on $(\mathbf{R}^{rn}, \mathbf{B}^{rn})$ such that for all $1 \leq j \leq r$

$$s_r(x^{(1)}, \ldots, x^{(r)}) = \int_{\mathbf{R}^{rn}} \exp\left\{ i \sum_{j=1}^{r} \langle x^{(j)}, \lambda^{(j)} \rangle \right\} F(d\lambda^{(1)}, \ldots, d\lambda^{(r)}),$$

then $\xi(x) \in \Phi^{(m)}$. If the field $\xi(x) \in \Phi^{(m)}$, then $\xi(x) \in \mathbf{S}_1^{(m)}$ if and only if for $1 \leq r \leq m$ the measures $F(\cdot)$ are supported by the subspace $L^{(r)}(0)$ defined by the equation $\lambda^{(1)} + \ldots + \lambda^{(r)} = 0$, that is,

$$F(\Lambda) = F(\Lambda \cap L^{(r)}(0)), \quad \Lambda \in \mathbf{B}^{rn}, \quad 1 \leq r \leq m.$$

We say that a field $\xi(x) \in \Xi_1^{(m)}$ if $\xi(x) \in \Phi^{(m)} \cap S_1^{(m)}$ and for each $1 \leq r \leq m$, the measure $F(\Lambda)$, $\Lambda \in \mathcal{B}^{rn}$, is absolutely continuous with respect to the $(r-1)n$-dimensional Lebesgue measure on the subspace $L^{(r)}(0)$. In this case functions $f_r(\lambda^{(1)}, \ldots, \lambda^{(r)})$, called spectral density functions of order $r \geq 2$, exist such that $\lambda^{(r)} = -\sum_{j=1}^{r-1} \lambda^{(j)}$ and

$$\int_{\mathbf{R}^{(r-1)n}} |f_r(\lambda^{(1)}, \ldots, \lambda^{(r)})| d\lambda^{(1)} \ldots d\lambda^{(r-1)} < \infty. \qquad (1.3.6)$$

Note that the functions $f_r(\lambda^{(1)}, \ldots, \lambda^{(r)})$ are symmetric and generally complex-valued. In fact the variable $\lambda^{(r)} = -\lambda^{(1)} - \ldots - \lambda^{(r-1)}$ is a dummy, therefore the notation $f(\lambda^{(1)}, \ldots, \lambda^{(r-1)})$, $r \in \{2, \ldots, m\}$, will also be used. The second order spectral density function $f_2(\lambda^{(1)}, -\lambda^{(1)}) = f(\lambda^{(1)})$ (see (1.2.2)) is the most often used along with the fourth order spectral density function $f_4(\lambda^{(1)}, \lambda^{(2)}, \lambda^{(3)})$. For a Gaussian field, all the spectral density functions $f_m(\cdot, \ldots, \cdot)$ vanish for $m > 2$. Thus if $\xi(x) \in \Xi_1^{(m)}$, then for all $r \in \{2, \ldots, m\}$, homogeneous spectral density functions $f_r(\lambda^{(1)}, \ldots, \lambda^{(r-1)})$ exist such that (1.3.6) holds and

$$s_r(x^{(1)}, \ldots, x^{(r)}) = s_r(x^{(1)} - x^{(r)}, \ldots, x^{(r-1)} - x^{(r)}) =$$

$$= \int_{R^{(r-1)n}} \exp\left\{i \sum_{j=1}^{r-1} \langle \lambda^{(j)}, x^{(j)} - x^{(r)} \rangle \right\} \times$$

$$\times f_r(\lambda^{(1)}, \ldots, \lambda^{(r-1)}) d\lambda^{(1)} \ldots d\lambda^{(r-1)}. \qquad (1.3.7)$$

Representation (1.3.7) becomes (1.2.2) for $r = 2$. As is well known, for representation (1.3.7) to be valid, it suffices that, for example,

$$\int_{R^{(r-1)n}} |s_r(x^{(1)}, \ldots, x^{(r-1)}, 0)| dx^{(1)} \ldots dx^{(r-1)} < \infty. \qquad (1.3.8)$$

For a Gaussian field, (1.3.8) signifies that the correlation function of the field is absolutely integrable. If the field $\xi(x) \in \Phi^{(m)} \cap S_2^{(m)}$, then for any $g_i \in SO(n)$, $1 \leq r \leq m$,

$$F(\lambda^{(1)}, \ldots, \lambda^{(r)}) = F(g_1 \lambda^{(1)}, \ldots, g_r \lambda^{(r)}). \qquad (1.3.9)$$

In particular, if $\xi(x) \in \Phi^{(m)} \cap S_1^{(m)}$, relation (1.3.9) holds on $L^{(m)}(0)$. Indeed, if $g_j \in SO(n)$, $u^{(j)} = g_j^{-1}\lambda^{(j)}$, $1 \le j \le r \le m$, then

$$s_r(x^{(1)}, \ldots, x^{(r)}) = s_r(g_1 x^{(1)}, \ldots, g_r x^{(r)}) =$$

$$= \int_{\mathbf{R}^{rn}} \exp\left\{i\sum_{j=1}^{r} \langle \lambda^{(j)}, g_j x^{(j)} \rangle\right\} F(d\lambda^{(1)}, \ldots, d\lambda^{(r)}) =$$

$$= \int_{\mathbf{R}^{rn}} \exp\left\{i\sum_{j=1}^{r} \langle g_j^{-1}\lambda^{(j)}, x^{(j)} \rangle\right\} F(d\lambda^{(1)}, \ldots, d\lambda^{(r)}) =$$

$$= \int_{\mathbf{R}^{rn}} \exp\left\{i\sum_{j=1}^{r} \langle u^{(j)}, x^{(j)} \rangle\right\} F(d(g_1 u^{(1)}) \ldots, d(g_r u^{(r)})).$$

Suppose that $\xi(x) \in \Phi^{(m)} \cap \mathbf{S}^{(m)}$. Then the rth order cumulants ($2 \le r \le m$) depend only on $\rho_j = |x^{(j)}|$ and admit the representation

$$s_r(\rho_1, \ldots, \rho_{r-1}, 0) = \int_{\mathbf{R}_+^{r-1}} \prod_{j=1}^{r-1} Y_n(\mu_j \rho_j) dG(\mu_1, \ldots, \mu_{r-1}), \qquad (1.3.10)$$

where $G(\mu_1, \ldots, \mu_{r-1}) = \int_{\{|\lambda^{(j)}| \le \mu_j, \, j=1,\ldots,r-1\}} F(d\lambda^{(1)}, \ldots, d\lambda^{(r-1)})$.

To prove (1.3.10), we integrate both sides of the formula

$$s_r(x^{(1)}, \ldots, x^{(r-1)}, 0) = \int_{\mathbf{R}^{n(r-1)}} \exp\left\{i\sum_{j=1}^{r-1} \langle \lambda^{(j)}, x^{(j)} \rangle\right\} F(d\lambda^{(1)}, \ldots, d\lambda^{(r-1)})$$

$r-1$ times over the surfaces of the spheres $s(\rho_1), \ldots, s(\rho_{r-1})$ and use relations (1.2.8) and (1.3.9).

Representation (1.3.10) is similar to (1.2.4). We say that a field $\xi(x) \in \Xi^{(m)}$ if $\xi(x) \in \Phi^{(m)} \cap \mathbf{S}^{(m)}$ and for all $2 \le r \le m$ there exist the derivatives

$$\frac{\partial^{r-1}}{\partial\mu_1 \ldots \partial\mu_{r-1}} G(\mu_1, \ldots, \mu_{r-1}) = |s(1)|^{r-1} g_r(\mu_1, \ldots, \mu_{r-1}) \prod_{j=1}^{r-1} \mu_j^{n-1}.$$

The functions $g_r(\mu_1, \ldots, \mu_{r-1})$ will be called isotropic spectral density functions of order $r \in \{2, \ldots, m\}$.

In the case $r = 2$ we obtain (1.2.17). Thus, if $\xi(x) \in \Xi^{(2)}$, (1.2.18) is valid.

If $\xi(x) \in \Xi^{(4)}$, the fourth order cumulant admits the representation

$$s_4(\rho_1, \rho_2, \rho_3, 0) = |s(1)|^3 \int_{\mathbf{R}_+^3} \left\{\prod_{j=1}^{3} Y_n(\mu_j \rho_j) \mu_j^{n-1}\right\} g_4(\mu_1, \mu_2, \mu_3) d\mu_1 d\mu_2 d\mu_3.$$

$$(1.3.11)$$

This representation will be used in Ch. 4.

It is quite natural to refer to the classes $\mathbf{T}^{(m)}$, $\mathbf{S}^{(m)}$, $\Phi^{(m)}$ introduced above as the Fortet-Blanc-Lapierre classes of random fields.

1.4. Some Properties of the Uniform Distribution

This section presents properties of the uniform distribution on convex sets and related topics of geometric probabilities.

A set $\Delta \subset \mathbf{R}^n$ is called convex if for any $x, y \in \Delta$, the points $z = \mu x + (1 - \mu)y \in \Delta$ for all $0 \le \mu \le 1$.

Let $\Delta \subset \mathbf{R}^n$ be a convex bounded measurable set with $|\Delta| > 0$. Consider the uniform distribution on Δ, that is, the distribution having the density function

$$p_\Delta(u) = \begin{cases} 1/|\Delta|, & u \in \Delta, \\ 0, & u \notin \Delta. \end{cases} \qquad (1.4.1)$$

The characteristic function of the uniform distribution (1.4.1) is of the form

$$K(y, \Delta) = \int_\Delta e^{i(y,x)} p_\Delta(x) dx = |\Delta|^{-1} \int_\Delta e^{i(y,x)} dx. \qquad (1.4.2)$$

Lemma 1.4.1. Let Δ be a bounded convex set such that $0 \in \Delta$. Then as $\lambda \to \infty$ the function

$$\Psi_{\Delta(\lambda)}(y) = |K(y, \Delta(\lambda))|^2 \lambda^n |\Delta| (2\pi)^{-n} \qquad (1.4.3)$$

possesses the kernel property: for any $\epsilon > 0$

$$\int_{\mathbf{R}^n} \Psi_{\Delta(\lambda)}(y) dy = 1; \qquad \lim_{\lambda \to \infty} \int_{\mathbf{R}^n \backslash \Delta(\epsilon)} \Psi_{\Delta(\lambda)}(y) dy = 0. \qquad (1.4.4)$$

The first of the relations (1.4.4) is implied by the Plancherel theorem. Theorem 2 in [177] (see also Theorem 2 in [117]) yields the following assertion: if Δ is a convex set and $\mathbf{mes}^{(n-1)}\{\Delta\}$ is its surface area, then for $\epsilon > 0$

$$\int_{|y|>\epsilon} \left| \int_\Delta e^{i(x,y)} dx \right|^2 dy \le 8\epsilon^{-1} \mathbf{mes}^{(n-1)}\{\Delta\} \left[\int_0^\pi \sin^n \alpha \, d\alpha \right]^{-1}$$

is valid. This inequality and homothety properties yield the second relation in (1.4.4). ∎

If Δ is a ball $v(r)$, the well known formula (see, for example, [30])

$$\int_{v(r)} e^{i(y,x)} dx = (2\pi r/|y|)^{n/2} J_{n/2}(r|y|) \qquad (1.4.5)$$

yields an explicit expression for the function

$$K(y, v(r)) = |v(r)|^{-1} \left(\frac{2\pi r}{|y|} \right)^{n/2} J_{n/2}(r|y|). \qquad (1.4.6)$$

If $\Delta = \prod[a, b)$, it is obvious that

$$K(y, \Pi[a, b)) = \prod_{j=1}^{n} (e^{ib_j y_j} - e^{ia_j y_j})/[iy_j (b_j - a_j)]. \qquad (1.4.7)$$

For the function $\Psi_{v(r)}(y)$ relations (1.4.4) are implied by (1.4.6). Indeed, for any $\epsilon > 0$, $r \to \infty$

$$\int_{\mathbf{R}^n \setminus v(\epsilon)} |K(y, v(r))|^2 r^n (2\pi)^{-n} |v(1)| dy =$$

$$= 2\Gamma(\tfrac{n}{2} + 1)\Gamma^{-1}(\tfrac{n}{2}) \int_{r\epsilon}^{\infty} J_{n/2}^2(\mu) \frac{d\mu}{\mu} \to 0.$$

If $\Delta = \Pi(b)$. (1.4.7) yields $\Psi_{\Pi(b)}(y) = (2\pi)^{-n} |\Pi(b)| \; |K(y, \Pi(b))|^2 = \prod_{j=1}^{n} \Phi_{b_j}(y_j)$, where $\Phi_T(z) = [\sin\{\tfrac{zT}{2}\}/(\tfrac{z}{2})]^2 (2\pi T)^{-1}$ is the Fejér kernel. In this case it is easy to derive properties of the kernel $\Psi_{\Pi(b)}(y)$ of the type (1.4.4) as $\min\{b_j, \; j = 1, \ldots, n\} \to \infty$.

In what follows, considerable importance will be attached to the distribution function $F_\Delta(z) = F_{\rho_{\alpha\beta}}(z)$ of the distance $\rho_{\alpha\beta} = |\alpha - \beta|$ between two independent random points α and β selected in accordance with the uniform law on the set Δ.

Denote the incomplete beta-function by

$$I_x(p, q) = \frac{\Gamma(p + q)}{\Gamma(p)\Gamma(q)} \int_0^x t^{p-1}(1 - t)^{q-1} dt, \; x \in [0, 1], \; p > 0, \; q > 0. \qquad (1.4.8)$$

Lemma 1.4.2 [186]. The density function $\psi_r(u)$ of the distance between two independent and uniformly distributed points inside the ball $v(r)$ is

$$\psi_r(u) = nu^{n-1}r^{-n} I_{1-u^2/(4r^2)}(\tfrac{n+1}{2}, \tfrac{1}{2}), \; 0 \le u \le 2r. \qquad (1.4.9)$$

Remark 1.4.1. For $n = 1, 2, 3$ relation (1.4.9) yields respectively the expressions: $(1 - u/2r)/r$, $8(u/2r)[\cos^{-1}(u/2r) - (u/2r)\sqrt{1 - (u/2r)^2}]/\pi r$, $6(u/2r)^2(1 - u/2r)^2(2 + u/2r)/r$.

Lemma 1.4.3. Let $v(r_1)$ and $v(r_2)$ be two concentric spheres in $\mathbf{R}^n (r_1 < r_2)$, α a random point selected from the first ball according to the uniform law, and β a random point selected from the second ball $v(r_2)$ according to the uniform law independently of α. Then the distribution density function of the distance $\rho_{\alpha\beta}$ between the points is of the form

$$\Psi_{r_1, r_2}(u) = 2^{(n-2)/2}\Gamma(\tfrac{n}{2})n^2 u^{n/2}(r_1 r_2)^{n/2} \times$$

$$\times \int_0^\infty \frac{J_{(n-2)/2}(u\rho)J_{n/2}(r_1\rho)J_{n/2}(r_2\rho)d\rho}{\rho^{n/2}}, \quad u \in [0, r_1 + r_2]. \qquad (1.4.10)$$

Lemma 1.4.3 may be proved using the characteristic functions method, only note that for $r_1 = r_2 = r$, (1.4.10) becomes (1.4.9).

Lemma 1.4.4. Let $s(r)$ be a sphere in $\mathbf{R}^n$, $n \geq 2$. If α and β are independent random points selected on $s(r)$ according to the uniform law, that is, $P\{\alpha \in \Delta\} = P\{\beta \in \Delta\} = \int_{\Delta \cap s(r)} dm(x)/[r^{n-1}|s(1)|]$, $\Delta \in s(r)$, then the density function of the distance $\rho_{\alpha\beta}$ between α and β is

$$f_{r_1,r_2}(u) = \tfrac{1}{\sqrt{\pi}}\Gamma(\tfrac{n}{2})\Gamma^{-1}(\tfrac{n-1}{2})r^{1-n}u^{n-2}\times$$

$$\times \left(1 - \frac{u^2}{4r^2}\right)^{(n-3)/2}, \quad 0 < u < 2r. \qquad (1.4.11)$$

Lemma 1.4.5. Let $s(r_1)$ and $s(r_2)$ be two concentric spheres in $\mathbf{R}^n$, $n \geq 2$, with radii r_1 and r_2 ($r_1 < r_2$), let α and β be independent points uniformly distributed on the corresponding spheres. Then the density function of the distance $\rho_{\alpha\beta}$ between these points equals 0 for $u < r_2 - r_1$, $u > r_1 + r_2$ and $f_{r_1,r_2}(u)$ for $u \in [r_2 - r_1, r_1 + r_2]$, where

$$f_{r_1,r_2}(u) = \tfrac{1}{\sqrt{\pi}}\Gamma(\tfrac{n}{2})\Gamma^{-1}(\tfrac{n-1}{2})2^{3-n}(r_1r_2)^{2-n}\times$$

$$\times[(r_1 + r_2)^2 - u^2]^{(n-3)/2}[u^2 - (r_1 - r_2)^2]^{(n-3)/2}. \qquad (1.4.12)$$

The proofs of Lemmas 1.4.4 and 1.4.5 utilize geometric probabilities. We shall outline the proof of Lemma 1.4.5 only.

Proof of Lemma 1.4.5. Assume that one of the points is fixed and is attached to the "North Pole" of the larger sphere, $r_2 - r_1 < u < r_2 + r_1$. Changing to spherical coordinates (1.1.1), we find that the probability of the random event $\{\rho_{\alpha\beta} < u\}$ equals the ratio of the area of surface of the segment $\{(\phi, \phi_1, \ldots, \phi_{n-2} : 0 \leq \phi_1 \leq A\}$ (where $A = \cos^{-1}\{(r_1^2 + r_2^2 - u^2)/(2r_1r_2)\}$ which equals $2\pi^{(n-1)/2}r_1^{(n-1)/2}\Gamma^{-1}(\tfrac{n-1}{2})\int_0^A \sin^{n-2}\phi_1 d\phi_1$) to the area of surface of the sphere $s(r_1)$. Hence

$$P\{\rho_{\alpha\beta} < u\} = \tfrac{1}{\sqrt{\pi}}\Gamma(\tfrac{n}{2})\Gamma^{-1}(\tfrac{n-1}{2})\int_0^A \sin^{n-2}\phi_1 d\phi_1.$$

Thus (1.4.12) follows. ∎

Various statements to be derived in what follows by applying Lemmas 1.4.2–1.4.5 may be extended to different sets whenever the distribution function $F_\Delta(z)$ is known. For example, for $n = 2$ the following assertion holds, which may be proved by means of a geometric probabilities argument or by direct integration.

Lemma 1.4.6 [40]. Let $\Delta = [0, a] \times [0, a] \subset \mathbf{R}^2$ be a square. Then the distribution function $F_\Delta(u)$ of the distance between random points α and β selected independently from Δ according to the uniform law, equals 0 for $u \leq 0$, $f_1(u)$ for $0 < u \leq a$, $f_2(u)$ for $a < u < a\sqrt{2}$ and 1 for $u > a\sqrt{2}$, where

$$f_1(u) = \pi u^2/a^2 - 8u^3/(3a^3) + u^4/(2a^4), \quad f_2(u) = \frac{1}{3} + 4\sqrt{u^2 - a^2}/a -$$

$$-(2 + \pi)u^2/a + 4u^2 \sin^{-1}\{a/u\}/a^2 + 8(u^2 - a^2)^{3/2}/(3a^3) + u^4/(2a^4).$$

1.5. Variances of Integrals of Random Fields

We shall investigate the asymptotic behaviour of variances of integrals of random fields in correlation and spectral terms.

Let $\Delta(\lambda)$ be the homothetic image of a bounded measurable convex set Δ with $|\Delta| > 0$ containing the origin, $\xi : \Omega \times \mathbf{R}^n \to \mathbf{R}^1$ be a m.s. continuous measurable homogeneous isotropic random field satisfying $E\xi(x) = 0$. We set $\eta(\lambda) = \int_{\Delta(\lambda)} \xi(x)dx$.

By randomization of the correlation function we obtain in the notation of §1.4:

$$b(\lambda) = \operatorname{var} \eta(\lambda) = \int_{\Delta(\lambda)} \int_{\Delta(\lambda)} B(|x - y|)dxdy = |\Delta(\lambda)|^2 EB(\rho_{\alpha\beta}) =$$

$$= |\Delta|^2 \lambda^{2n} \int_0^{d(\lambda)} B(z)dF_{\Delta(\lambda)}(z), \tag{1.5.1}$$

where $\rho_{\alpha\beta}$ is the distance between the independent points α and β selected according to the uniform law on the set $\Delta(\lambda)$, while $d(\lambda)$ is the diameter of the set $\Delta(\lambda)$.

Using (1.4.9), we obtain from (1.5.1) for balls $v(r) \subset \mathbf{R}^n$:

$$b(r) = \operatorname{var} \left[\int_{v(r)} \xi(x)dx \right] =$$

$$= c_2(n)r^n \int_0^{2r} z^{n-1}B(z)I_{1-z^2/(2r)^2}(\tfrac{n+1}{2}, \tfrac{1}{2})dz, \qquad (1.5.2)$$

where

$$c_2(n) = 4\pi^n/[n\Gamma^2(\tfrac{n}{2})] \qquad (1.5.3)$$

and $I_x(p,q)$ is defined by (1.4.8).

An asymptotic analysis of formula (1.5.2) allows us to state the following lemma:

Lemma 1.5.1. If

$$\int_0^\infty z^{n-1}|B(z)|dz < \infty, \quad \beta_1 = \int_0^\infty z^{n-1}B(z)dz \neq 0, \qquad (1.5.4)$$

then $b(r) = c_2(n)\beta_1 r^n(1 + o(1))$ as $r \to \infty$.

Similarly, using (1.4.10) we find that for $r_1 < r_2$,

$$E\int_{v(r_1)} \xi(x)dx \int_{v(r_2)} \xi(y)dy = \int_{v(r_1)} \int_{v(r_2)} B(|x - y|)dxdy =$$

$$= (r_1 r_2)^n |v(1)|^2 EB(\rho_{\alpha\beta}) = c_3(n)(r_1 r_2)^{n/2} \int_0^{r_1+r_2} B(r)r^{n/2} \times$$

$$\times \left[\int_0^\infty \rho^{-n/2} J_{(n-2)/2}(r\rho) J_{n/2}(r_1\rho) J_{n/2}(r_2\rho)d\rho \right] dr,$$

where $c_3(n) = 2^{n/2}\pi^n/\Gamma(\tfrac{n}{2})$ and $\rho_{\alpha\beta}$ is the distance between random points α and β selected in $v(r_1)$ and $v(r_2)$ respectively according to the uniform law. The last relation together with (1.5.2) implies that

$$b(r_1, r_2) = \text{var}\left[\int_{r_1 \le |x| < r_2} \xi(x)dx \right] = c_2(n) \int_0^{r_1+r_2} z^{n-1}B(z) \times$$

$$\times \left\{ \int_0^\infty Y_n(z\rho)[r_2^{n/2} J_{n/2}(r_2\rho) - r_1^{n/2} J_{n/2}(r_1\rho)]^2 \rho^{-1}d\rho \right\}dz. \qquad (1.5.5)$$

Using relation 6.571.1 from [37] one can prove that

$$\int_0^\infty \frac{[(\lambda r_2)^{n/2} J_{n/2}(\lambda r_2) - (\lambda r_1)^{n/2} J_{n/2}(\lambda r_1)]^2}{\lambda^{n+1}}d\lambda = \frac{r_2^n - r_1^n}{n} \qquad (1.5.6)$$

The function $Y_n(z)$ is continuous and $|Y_n(z)| \le 1$. Therefore (1.5.5) and (1.5.6) yield the following statement.

Lemma 1.5.2. Under conditions (1.5.4) for $r_1 + r_2 \to \infty$ the asymptotic formula $b(r_1, r_2) = c_2(n)\beta_1(r_2^n - r_1^n)(1 + o(1))$ is valid.

Similarly, from (1.4.11) we derive for $n \geq 2$ the representation

$$l(r) = \mathrm{var}\left[\int_{s(r)} \xi(x)dm(x)\right] =$$

$$= c_4(n)r^{n-1}\int_0^{2r} z^{n-2}\left(1 - \frac{z^2}{4r^2}\right)^{(n-3)/2} B(z)dz, \qquad (1.5.7)$$

where

$$c_4(n) = 4\pi^{n-1/2} \big/ [\Gamma(\tfrac{n}{2})\Gamma(\tfrac{n-1}{2})] = 2^n\pi^{n-1}/(n-2)! \qquad (1.5.8)$$

By studying the asymptotic behaviour of the integral (1.5.7) we obtain a statement similar to that of Lemma 1.5.1.

Lemma 1.5.3. Let $n \geq 2$ and let

$$\int_0^\infty z^{n-2}|B(z)|dz < \infty, \quad \beta_2 = \int_0^\infty z^{n-2}B(z)dz \neq 0. \qquad (1.5.9)$$

Then $l(r) = c_4(n)\beta_2 r^{n-1}(1 + o(1))$ as $r \to \infty$.

To state the analogous assertions in spectral terms, we present a simple Tauberian-type assertion for the integral

$$S(r) = \int_0^\infty K(\lambda r)\lambda^\alpha g(\lambda)d\lambda.$$

Lemma 1.5.4. Suppose that: a) the function $g(\lambda)$ is continuous in a neighbourhood of zero, $g(0) \neq 0$ and $g(\lambda)$ is bounded on $[0, \infty)$; b) $\int_0^\infty K(z)z^\alpha dz = \gamma(\alpha)$, $\int_0^\infty |K(z)|z^\alpha dz = \tilde{\gamma}(\alpha) < \infty$. Then $S(r) = r^{-\alpha-1}g(0)\gamma(\alpha)(1 + o(1))$ as $r \to \infty$.

Proof. Note that $r^{-\alpha-1}\gamma(\alpha) = \int_0^\infty K(\lambda r)\lambda^\alpha g(\lambda)d\lambda$. Choose $\beta(r) \to 0$, $r\beta(r) \to \infty$ as $r \to \infty$. Note the identity $S(r) = r^{-\alpha-1}g(0)\gamma(\alpha) + f(r)$, where $f(r) = \int_0^\infty K(\lambda r)[g(\lambda) - g(0)]\lambda^\alpha d\lambda$. We shall prove that under the conditions a), b), $f(r) = o(r^{-\alpha-1})$ as $r \to \infty$. Indeed,

$$|f(r)| \leq \int_0^{\beta(r)} |K(\lambda r)|\,|g(\lambda) - g(0)|\lambda^\alpha d\lambda + \int_{\beta(r)}^\infty |K(\lambda r)|\times$$

$$\times|g(\lambda) - g(0)|\lambda^\alpha d\lambda \leq \max_{0\leq\lambda\leq\beta(r)} |g(\lambda) - g(0)|\tilde{\gamma}(\alpha)r^{-\alpha-1}+$$

$$+2 \sup_{0 \le \lambda < \infty} |g(\lambda)| r^{-\alpha-1} \int_{r\beta(r)}^{\infty} |K(z)| z^{\alpha} dz.$$

Hence, $f(r) = o(r^{-\alpha-1})$, $r \to \infty$. ∎

If $\xi(x)$, $x \in \mathbf{R}^n$, is a homogeneous random field with $E\xi(x) = 0$, (1.2.1) and (1.4.5) imply that

$$b(r) = \text{var}\left[\int_{v(r)} \xi(x)dx\right] = (2\pi r)^n \int_{\mathbf{R}^n} \frac{J_{n/2}^2(|\lambda|r)}{|\lambda|^n} F(d\lambda), \qquad (1.5.10)$$

where $F(\cdot)$ is as given in (1.2.1).

If furthermore the field is isotropic, then

$$b(r) = (2\pi)^n r^{2n} \int_0^{\infty} \frac{J_{n/2}^2(\lambda r)}{(\lambda r)^n} G(d\lambda), \qquad (1.5.11)$$

where $G(\cdot)$ is as in (1.2.4).

Similarly, using (1.2.1) and (1.2.8) we obtain for $n \ge 2$

$$l(r) = (2\pi)^n r^{2(n-1)} \int_0^{\infty} \frac{J_{(n-2)/2}^2(\lambda r)}{(\lambda r)^{n-2}} G(d\lambda). \qquad (1.5.12)$$

We now introduce the following assumption.

I. The spectral function $G(\lambda)$, $\lambda \in \mathbf{R}_+^1$, of a m.s. continuous homogeneous isotropic random field is absolutely continuous and

$$G'(\lambda) = |s(1)| \lambda^{\alpha} g(\lambda), \ \alpha \in (-1, n), \qquad (1.5.13)$$

where $g(\lambda)$ is a continuous function in a neighbourhood of zero, $g(0) \ne 0$ and $g(\lambda)$ is bounded on $[0, \infty)$.

Lemma 1.5.5. Let I hold and $r \to \infty$. Then

$$b(r) = (2\pi)^n |s(1)| c_5(n, \alpha) g(0) r^{2n-\alpha-1}(1 + o(1)), \qquad (1.5.14)$$

where $c_5(n, \alpha) = \Gamma(n - \alpha)\Gamma(\frac{\alpha+1}{2})/[2^{n-\alpha}\Gamma^2(\frac{n-\alpha+1}{2})\Gamma(\frac{2n-\alpha+1}{2})]$. If in assumption I, $\alpha \in (-1, n-2)$, then for $r \to \infty$,

$$l(r) = (2\pi)^n g(0)|s(1)| c_5(n - 2, \alpha) r^{2n-\alpha-3}(1 + o(1)). \qquad (1.5.15)$$

Proof. We substitute (1.5.13) into (1.5.11) and set in Lemma 1.5.4 $K(z) = J_{n/2}^2(z) z^{-n}$. Utilizing relation 6.574.2 from [37]:

$$\int_0^{\infty} J_{\nu}(\alpha t) J_{\mu}(\alpha t) t^{-\lambda} dt = \alpha^{\lambda-1} \Gamma(\lambda) \Gamma(\frac{\nu+\mu-\lambda+1}{2}) \times$$

$$\times [2^\lambda \Gamma(\tfrac{\mu-\nu+\lambda+1}{2})\Gamma(\tfrac{\nu+\mu+\lambda+1}{2})\Gamma(\tfrac{\nu-\mu+\lambda+1}{2})]^{-1},$$

$$\alpha > 0, \quad \mathrm{Re}(\nu+\mu+1) > \mathrm{Re}\,\lambda > 0, \tag{1.5.16}$$

for $-1 < \alpha < n$ we set $\int_0^\infty K(z)z^\alpha dz = c_5(n,\alpha)$. Then (1.5.14) follows from Lemma 1.5.4. Similarly, on substituting (1.5.13) into (1.5.12) and choosing in Lemma 1.5.4 $K(z) = J^2_{(n-2)/2}(z)z^{2-n}$ for $n \geq 2$ we obtain (1.5.15). For such a choice of $K(z)$, to ensure convergence of the integral $\int_0^\infty K(z)z^\alpha dz$ it is necessary to assume that $\alpha \in (-1,\, n-2)$. ■

Lemma 1.5.6. If I holds for $\alpha = n-1$ and $r_1 + r_2 \to \infty$, then

$$b(r_1,r_2) = \mathrm{var}\left[\int_{r_1 \leq |x| < r_2} \xi(x)dx\right] = c_6(n)(r_2^n - r_1^n)(1 + o(1))$$

where $c_6(n) = |s(1)|(2\pi)^n g(0)/n$.

Proof. The assertion of the lemma follows from the representation

$$b(r_1,r_2) = (2\pi)^n \int_0^\infty \frac{[(\lambda r_2)^{n/2} J_{n/2}(\lambda r_2) - (\lambda r_1)^{n/2} J_{n/2}(\lambda r_1)]^2}{\lambda^{2n}} G(d\lambda)$$

using (1.5.6). ■

Remark 1.5.1. For $n = 2$, (1.5.1) may also be investigated when Δ is a square. The constants will be different but they can be obtained with the aid of Lemma 1.4.6.

If $\xi(x)$, $x \in \mathbf{R}^n$, is a homogeneous random field, $\Pi = \Pi[a,b]$ is a parallelepiped such that $|\Pi| \to \infty$ and

$$\int_{\mathbf{R}^n} |B(x)|dx < \infty, \quad \beta_3 = \int_{\mathbf{R}^n} B(x)dx \neq 0, \tag{1.5.17}$$

it is then easy to show that $\mathrm{var}[\int_\Pi \xi(x)dx] = \beta_3|\prod|(1 + o(1))$.

If the field $\xi(x)$ has a continuous and bounded spectral density function $f(\lambda) : f(0) \neq 0$ (its existence, continuity and boundedness follow, for example, from (1.5.17)), then for $|\Pi| \to \infty$, $\mathrm{var}[\int_\Pi \xi(x)dx] = (2\pi)^n f(0)|\Pi|(1 + o(1))$.

Remark 1.5.2. One can show that if for any parallelepiped $\Pi = \Pi[a,b]$ such that $|\Pi| \to \infty$, $\mathrm{var}[\int_\Pi \xi(x)dx] \asymp |\Pi|$, then $\mathrm{var}[\int_\Delta \xi(x)dx] \asymp |\Delta|$ as $\Delta \xrightarrow{V.H.} \infty$.*

---

* $f(x) \asymp g(x)$, if for large enough x there exist constants $0 < k_1 < k_2 < \infty$ such that $k_1 f(x) \leq g(x) \leq k_2 f(x)$.

1.6. Weak Dependence Conditions for Random Fields

In this section mixing conditions for random fields are defined and bounds for covariance in terms of mixing rate are provided. Recall that a r.v. η is called the conditional mathematical expectation of a r.v. ξ with respect to a σ-algebra $\mathcal{F}_0 \subset \mathcal{F}$ if: 1) η is measurable, 2) for all $A \in \mathcal{F}_0$ $\int_A \xi(\omega) P(d\omega) = \int_A \eta(\omega) P(d\omega)$.

The conditional mathematical expectation of a r.v. ξ with respect to $\mathcal{F}_0$ will be denoted by $E(\xi/\mathcal{F}_0)$, which exists whenever $E|\xi| < \infty$. We note the following properties of conditional mathematical expectations: 1) $E(\xi\zeta/\mathcal{F}_0) = \xi E(\zeta/\mathcal{F}_0)$, if $E\xi\zeta$ exists and ξ is $\mathcal{F}_0$-measurable; 2) $E\xi = EE(\xi/\mathcal{F}_0)$ if $E|\xi| < \infty$.

Let $\mathfrak{N}_1, \mathfrak{N}_2$ be σ-algebras of $\mathcal{F}$ and let

$$\alpha(\mathfrak{N}_1, \mathfrak{N}_2) = \sup\{|P(AB) - P(A)P(B)|, \ A \in \mathfrak{N}_1, \ B \in \mathfrak{N}_2\}$$

be the Rosenblatt dependence rate, $\operatorname{cov}(\xi_1, \xi_2) = E\xi_1\xi_2 - E\xi_1 E\xi_2$, $\|\xi\|_p = \{E|\xi|^p\}^{1/p}$, $p > 1$.

Lemma 1.6.1 [25]. If ξ_1, ξ_2 are complex r.v.'s measurable with respect to σ-algebras $\mathfrak{N}_1$ and $\mathfrak{N}_2$ respectively and such that $|\xi_1| \leq k_1$, $|\xi_2| \leq k_2$, a.s., then $|\operatorname{cov}(\xi_1, \xi_2)| \leq 16 k_1 k_2 \alpha(\mathfrak{N}_1, \mathfrak{N}_2)$. In the case of real-valued r.v.'s ξ_1, ξ_2 the coefficient 16 can be replaced by 4.

Proof. Let ξ_1, ξ_2 be real-valued r.v.'s. Utilizing properties of conditional mathematical expectations we write $|\operatorname{cov}(\xi_1, \xi_2)| = |E\xi_1[E(\xi_2/\mathfrak{N}_1) - E\xi_2]| \leq k_1 E\xi_1'[E(\xi_2/\mathfrak{N}_1) - E\xi_2] = k_1 \operatorname{cov}(\xi_1', \xi_2)$ where $\xi_1' = \operatorname{sgn}[E(\xi_2/\mathfrak{N}_1) - E\xi_2]$ is a variable measurable with respect to $\mathfrak{N}_1$. By the same method applied to ξ_2 we obtain the bound $|\operatorname{cov}(\xi_1, \xi_2)| \leq k_1 k_2 \sup \operatorname{cov}(\xi_1', \xi_2')$ where the supremum is taken over all ξ_1', ξ_2' measurable with respect to $\mathfrak{N}_1$ and $\mathfrak{N}_2$ respectively and is equal to 1 or -1. Let $A = \{\xi_1' = 1\}$, $B = \{\xi_2' = 1\}$. Then $|\operatorname{cov}(\xi_1', \xi_2')| = |P(AB) + P(\overline{AB}) - P(\overline{A}B) - P(A\overline{B}) - P(A)P(B) - P(\overline{A})P(\overline{B}) + P(\overline{A})P(B) + P(A)P(\overline{B})| \leq 4\alpha(\mathfrak{N}_1, \mathfrak{N}_2)$. If the variables ξ_1 and ξ_2 are complex, separating real and imaginary parts we obtain the required inequality (the coefficient 4 is to be replaced by 16) ∎

Lemma 1.6.2 [41]. Let ξ_1, ξ_2 be real-valued r.v.'s measurable with respect to the σ-algebras $\mathfrak{N}_1$ and $\mathfrak{N}_2$ respectively such that $E|\xi_1|^p < \infty$, $E|\xi_2|^q < \infty$ for some $p > 1$, $q > 1$. Then $|\operatorname{cov}(\xi_1, \xi_2)| \leq 10\|\xi_1\|_p\|\xi_2\|_q\{\alpha(\mathfrak{N}_1, \mathfrak{N}_2)\}^{1/s}$, where $s > 1$ is such that $p^{-1} + q^{-1} + s^{-1} = 1$.

Proof. Assume that $\alpha(\mathfrak{N}_1, \mathfrak{N}_2) > 0$, otherwise the inequality is trivial. Let

$$c_1 = \|\xi_1\|_p / \{\alpha(\mathfrak{N}_1, \mathfrak{N}_2)\}^{1/p}, \quad c_2 = \|\xi_2\|_q / \{\alpha(\mathfrak{N}_1, \mathfrak{N}_2)\}^{1/q}. \tag{1.6.1}$$

Take the truncated r.v.'s $\xi_1(c_1) = \xi_1 \chi(|\xi_1| \leq c_1)$, $\xi_2(c_2) = \xi_2 \chi(|\xi_2| \leq c_2)$. Let $\bar{\xi}_1(c_1) = \xi_1 - \xi_1(c_1)$, $\bar{\xi}_2(c_2) = \xi_2 - \xi_2(c_2)$. Then

$$| \operatorname{cov}(\xi_1, \xi_2)| \leq | \operatorname{cov}(\xi_1(c_1), \xi_2(c_2))| + | \operatorname{cov}(\xi_1(c_1), \bar{\xi}_2(c_2))| +$$
$$+ | \operatorname{cov}(\bar{\xi}_1(c_1), \xi_2(c_2))| + | \operatorname{cov}(\bar{\xi}_1(c_1), \bar{\xi}_2(c_2))|. \tag{1.6.2}$$

By Lemma 1.6.1 and taking (1.6.1) into account we obtain:

$$| \operatorname{cov}(\xi_1(c_1), \xi_2(c_2))| \leq 4 c_1 c_2 \alpha(\mathfrak{N}_1, \mathfrak{N}_2) \leq 4 \|\xi_1\|_p \|\xi_2\|_q \{\alpha(\mathfrak{N}_1, \mathfrak{N}_2)\}^{1/s}. \tag{1.6.3}$$

Since $E|\bar{\xi}_2(c_2)| \leq c_2^{1-q} E|\xi_2|^q$, we have

$$| \operatorname{cov}(\xi_1(c_1), \bar{\xi}_2(c_2))| \leq 2 c_1 E|\bar{\xi}_2(c_2)| \leq 2 \|\xi_1\|_p \|\xi_2\|_q \{\alpha(\mathfrak{N}_1, \mathfrak{N}_2)\}^{1/s}. \tag{1.6.4}$$

By symmetry

$$| \operatorname{cov}(\bar{\xi}_1(c_1), \xi_2(c_2))| \leq 2 \|\xi_1\|_p \|\xi_2\|_q \{\alpha(\mathfrak{N}_1, \mathfrak{N}_2)\}^{1/s}. \tag{1.6.5}$$

Furthermore,

$$| \operatorname{cov}(\bar{\xi}_1(c_1), \bar{\xi}_2(c_2))| \leq E|\bar{\xi}_1(c_1) \bar{\xi}_2(c_2)| + E|\bar{\xi}_1(c_1)| E|\bar{\xi}_2(c_2)|.$$

Using (1.6.1) and the Hölder inequality we find that

$$E|\bar{\xi}_1(c_1) \bar{\xi}_2(c_2)| \leq (c_1 c_2)^{1-pq/(p+q)} E(|\xi_1||\xi_2|)^{pq/(p+q)} \leq$$

$$\leq (c_1 c_2)^{1-pq/(p+q)} [\|\xi_1\|_p \|\xi_2\|_q]^{pq/(p+q)} \leq \|\xi_1\|_p \|\xi_2\|_q \{\alpha(\mathfrak{N}_1, \mathfrak{N}_2)\}^{1/s}. \tag{1.6.6}$$

Similarly, we have

$$E|\bar{\xi}_1(c_1)| E|\bar{\xi}_2(c_2)| \leq \|\xi_1\|_p \|\xi_2\|_q \{\alpha(\mathfrak{N}_1, \mathfrak{N}_2)\}^{1/s}. \tag{1.6.7}$$

Substituting (1.6.3)–(1.6.7) in (1.6.2) we complete the proof. ∎

There exists a great variety of ways of introducing measures of dependence of σ-algebras $\mathfrak{N}_1$ and $\mathfrak{N}_2$ from $\mathcal{F}$ [52, 206, 154, 155]. If $f(\mathfrak{N}_1, \mathfrak{N}_2)$ is a non-negative function equal to zero when the σ-algebras $\mathfrak{N}_1$ and $\mathfrak{N}_2$ are independent, $f(\mathfrak{N}_1, \mathfrak{N}_2)$ may be treated as a measure of dependence. The following types of dependence measure are most frequently used:

$$f_{p,q}(\mathfrak{N}_1, \mathfrak{N}_2) = \sup\Big\{|P(AB) - P(A)P(B)|/\{[P(A)]^p[P(B)]^q\} :$$

$$A \in \mathfrak{N}_1, \ B \in \mathfrak{N}_2, \ P(A) > 0, \ P(B) > 0\Big\}, \ p, q \in [0, 1]. \qquad (1.6.8)$$

(The measure of dependence $f_{0,0}(\mathfrak{N}_1, \mathfrak{N}_2)$ is due to M. Rosenblatt [209], the measure $f_{1,0}(\mathfrak{N}_1, \mathfrak{N}_2)$ is due to I.A. Ibragimov [51], while $f_{1,1}(\mathfrak{N}_1, \mathfrak{N}_2)$ is due to R. Serfling [216].)

We introduce certain mixing conditions for the random fields $\xi : \Omega \times \mathbf{R}^n \to \mathbf{R}^m$, $n \geq 1$, $m \geq 1$. Let a system $\mathfrak{M}$ of measurable sets be chosen in $\mathbf{R}^n$ and associate with each $\Delta \in \mathfrak{M}$ the smallest σ-algebra $\mathfrak{N}(\Delta)$ generated by the collection of r.v.'s $\{\xi(x), x \in \Delta\}$. Define mixing rates $\alpha(r) = \sup \alpha(\mathfrak{N}(\Delta_1), \mathfrak{N}(\Delta_2))$, where the supremum is taken over all pairs of sets $\Delta_1, \Delta_2 \in \mathfrak{M}$ such that the distance between them is at least r; $\alpha(r, d) = \sup \alpha(\mathfrak{N}(\Delta_1), \mathfrak{N}(\Delta_2))$, where the supremum is taken over all pairs of sets $\Delta_1, \Delta_2 \in \mathfrak{M}$ having a diameter less than d and lying at a distance at least r. If no restriction is imposed on the diameters, the rate $\alpha(r, \infty)$ becomes $\alpha(r)$.

If $\mathfrak{M}$ is a system of parallelepipeds, the rates $\alpha(r)$ and $\alpha(r, d)$ will be denoted by $\alpha_*(r)$ and $\alpha_*(r, d)$ respectively. In this case the supremum is taken over all parallelepipeds that: 1) have a diameter of less than d; 2) lie at a distance exceeding r.

The mixing rate $\alpha(r)$ was introduced by R.L. Dobrushin [43], the rates $\alpha(r, d)$ and $\alpha_*(r, d)$ are due to A.V. Bulinsky and I.G. Zhurbenko [20] (see also [48]). The meaning of the latter is that the dependence of the σ-algebras $\mathfrak{N}(\Delta_1)$ and $\mathfrak{N}(\Delta_2)$ may increase as the sets Δ_1 and Δ_2 become larger if the distance between them is preserved; and this dependence will diminish as the sets Δ_1 and Δ_2 become more distant and their diameters do not exceed a given value d.

Remark 1.6.1. Let the homogeneous random field $\xi : \Omega \times \mathbf{R}^n \to \mathbf{R}^1$ have moment of order $2 + \delta$, $\delta > 0$ (or $|\xi(0)| \leq k_3$ a.s.) so that as $r \to \infty$, the mixing rate $\alpha(r) \leq k_4 r^{-n-\epsilon}$, $\epsilon\delta > 2n$ (or $\alpha(r) \leq k_5 r^{-n-\epsilon}$, $\epsilon > 0$).

Under these assumptions the correlation function $B(x)$, $x \in \mathbf{R}^n$, admits a bound $|B(x)| \leq k_6(1 + |x|)^{-n-k_7}$ in view of Lemma 1.6.2 (or Lemma 1.6.1), that is,

$$\int_{\mathbf{R}^n} |B(x)|dx < \infty. \tag{1.6.9}$$

Hence there exists a continuous bounded spectral density function $f(\lambda)$, $\lambda \in \mathbf{R}^n$. If, furthermore, the field is isotropic, (1.6.9) is equivalent to the relation

$$\int_0^\infty z^{n-1}|B(z)|dz < \infty. \tag{1.6.10}$$

As follows from (1.2.18), the function $\rho^{(n-2)/2}B(\rho)$ is the Hankel transform of order $(n-2)/2$ of the function $g(\nu)(2\pi)^{n/2}\nu^{(n-2)/2}$. Therefore

$$g(\nu) = (2\pi)^{-n/2} \int_0^\infty J_{(n-2)/2}(\nu\rho)(\nu\rho)^{(2-n)/2}\rho^{n-1}B(\rho)d\rho. \tag{1.6.11}$$

Thus, (1.6.10) implies the existence of an isotropic spectral density $g(\nu)$, $\nu \in \mathbf{R}_+^1$ and (1.6.11) implies continuity and boundedness of this function.

1.7. A Central Limit Theorem

We shall now provide asymptotic normality conditions for weighted integrals of random fields in terms of constraints on the field moments, mixing rate and the weighting function.

Let $\mathfrak{M}$ be a system of measurable sets in $\mathbf{R}^n$, $g_\Delta(x, \theta) : \mathfrak{M} \times \mathbf{R}^n \times \Theta \to \mathbf{R}^1$ a collection of measurable functions and $\xi : \Omega \times \mathbf{R}^n \to \mathbf{R}^1$ a random field with $E\xi(x) = 0$. Consider the integrals:

$$s_\Delta(\theta) = \int_\Delta g_\Delta(x, \theta)\xi(x)dx. \tag{1.7.1}$$

We say that for $s_\Delta(\theta)$, $\theta \in \Theta$, $\Delta \in \mathfrak{M}$, a uniform central limit theorem (u.c.l.t.) holds as $\Delta \to \infty$ if there exists a function $\sigma^2(\theta) : \Theta \to (0, \infty)$ such that uniformly in Θ

$$P\{s_\Delta(\theta) < a\} \to \Phi_{0,\sigma^2(\theta)}(a) = \frac{1}{\sqrt{2\pi}\sigma(\theta)} \int_{-\infty}^a e^{-t^2/(2\sigma^2(\theta))}dt. \tag{1.7.2}$$

If there is no dependence on $\theta \in \Theta$ we shall write c.l.t.

We introduce the following assumptions.

II. There exists a $\delta > 0$ such that $E|\xi(x)|^{2+\delta} \le k_1 < \infty$ for all $x \in \mathbf{R}^n$.

III. $|\xi(x)| \le k_2 < \infty$ a.s. for all $x \in \mathbf{R}^n$.

IV. Uniformly in $\theta \in \Theta$ there exists $\lim\limits_{\Delta \to \infty} E s_\Delta^2(\theta) = \sigma^2(\theta) > 0$.

V. For $\Delta \to \infty$, $\Delta \in \mathfrak{M}$ there exists a $k_3 > 0$ such that

$$\sup_{\theta \in \Theta} \sup_{x \in \Delta} |g_\Delta(x, \theta)| \le k_3 / \sqrt{|\Delta|}.$$

First assume that $\mathfrak{M}$ is a set of parallelepipeds. Let $a^0 = (a_1^0, \ldots, a_n^0) \in \mathbf{R}^n$, be a fixed vector having non-zero components $\Pi_m^0 = \Pi[0, a^0] + m \otimes a^0$, $m \in \mathbf{Z}^n$. Suppose that $\mathfrak{M}$ is a system of parallelepipeds generated by Π_m^0, $|\Pi[0, a^0]| = \prod_{j=1}^n a_j^0$. Consider a parallelepiped $\Pi = \underset{j=1}{\overset{n}{\times}} [a_j, b_j] \in \mathfrak{M}$. Write $\Pi \to \infty$ to denote that $\min\{l_j, \ j = 1, \ldots, n\} \to \infty$, where $l_j = b_j - a_j$.

VI. Π holds and $\alpha_*(r, d) < k_4(1 + d)^\lambda r^{-n-\epsilon}$ where $\epsilon > 2n/\delta$, $\lambda < \delta(\epsilon\delta - 2n)/\{2n\delta(1 + \delta)\}$.

VII.   III holds and $\alpha_*(r) \le k_5 r^{-n-\epsilon}$, where $\epsilon > 0$.

Theorem 1.7.1. If assumptions IV, V, VI hold, then (1.7.2) is valid for $s_\Pi(\theta)$ as $\Pi \to \infty$.

Proof.   The proof utilizes a sectioning technique [12] adapted for the multidimensional case.

Let $v = |\Pi|$, $\tau = v^{\gamma/n}$, $\tau' = v^{\gamma'/n}$, $0 < \gamma' < \gamma < \delta/[2(1 + \delta)]$. Set $\tau_j = [\tau/a_j^0] a_j^0$, $\tau_j' = [\tau'/a_j^0] a_j^0$, $j \in I = \{1, 2, \ldots, n\}$. We shall call segments τ_j "large" and segments τ_j' "small". If $J = \{j \in I : l_j < \tau_j\} \neq \emptyset$, then for $j = J$ set $\tilde{\tau}_j = l_j$ and for $j \in I \backslash J = \bar{J}$ introduce new "large" segments $\tilde{\tau}_j = [\tilde{\tau}/a_j^0] a_j^0 \ge \tau_j$, where $\tilde{\tau} = \left(v^{-\gamma} \prod_{j \in J} l_j \right)^{-1/|\bar{J}|}$. If $J' = \{j \in \bar{J} : l_j < \tilde{\tau}_j\} \neq \emptyset$, then proceed as above with $J \cup J'$ and so on. In what follows τ_j and $\tilde{\tau}_j$ will not be distinguished.

For $l_j > 3\tau_j$, starting from point a_j, split each edge $[a_j, b_j)$ of the parallelepiped $\Pi[a, b]$ into disjoint segments: $[a_j, b_j) = \Delta_{j_1}^0 \cup \Delta_{j_1}^1 \cup \Delta_{j_2}^0 \cup \ldots \cup \Delta_{j_{s_j}}^0$ where $\Delta_{j_1}^0, \ldots, \Delta_{j_{s_j}-1}^0$ have length τ_j ($\Delta_{j_{s_j}}^0$ has length $\tilde{\tau}_j \le \tau_j$) and $\Delta_{j_1}^1, \ldots, \Delta_{j_{s_j}-1}^1$ have length τ_j'.

Call segments Δ^0 "long" and Δ^1 "short". Draw hyperplanes through the points of partition of each edge perpendicular to this edge. Then the whole parallelepiped will be partitioned into at most 2^n types of parallelepipeds (the type is determined by indicating the axes along which "short" or "long" segments are taken). Each of the disjoint parallelepipeds is of the form $\underbrace{\Delta_{1r_1}^{\epsilon_1} \times \ldots \times \Delta_{nr_n}^{\epsilon_n}}_{n-1}$, where ϵ_j equals 0 or 1 and $r_j \in \{1, \ldots, s_j\}$ if $\epsilon_j = 0$ and $r_j \in \{1, \ldots, s_j - 1\}$ if $\epsilon_j = 1$. It follows that the integral in (1.7.1) is partitioned into the sum

$$s_\Pi(\theta) = \sum_s \sum_{r=1}^{m_s} \zeta(\Pi_r^s) = \sum_{r=1}^{m_1} \zeta(\Pi_r^1) + \sum_{s>1} \sum_{r=1}^{m_s} \zeta(\Pi_r^s) = S_1 + S_2, \qquad (1.7.3)$$

where $\zeta(\Pi^s_r) = \int_{\Pi^s_r} g_\Pi(x,\theta)\xi(x)dx$, $m_s = m_s(\Pi) \asymp v^{1-\gamma}$ is the number of parallelepipeds of type s contained in Π. Below we shall show that $S_2 \to 0$ in
probability uniformly in $\theta \in \Theta$, that is, the main contribution to the distribution of $s_\Pi(\theta)$ is due to the sum S_1 generated by parallelepipeds Π^1_r all having
"long" edges.

Note that for $s = 1$ the volume $v_1 = |\Pi^1_r| \asymp v^\gamma$, while for $s > 1$ the volume
$v_s|\Pi^s_r| = v^{\gamma\gamma_s}$, where $\gamma_s < 1$, $1 \le r \le m_s$, so that as $v \to \infty$, $v_s = o(v_1)$
for all $s > 1$. The metric $\rho^*(x,y) = |x - y|_0 = \max\{|x_j - y_j|, j = 1,\dots,n\}$
is equivalent to the Euclidean one $|x - y|$. Let $\rho^*(\Delta_1, \Delta_2) = \inf\{\rho^*(x,y),$
$x \in \Delta_1, y \in \Delta_2\}$. Parallelepiped Π^s_r contains $N^s_r \asymp v_s$ parallelepipeds Π^0_m.
In Lemma 1.6.2 set $p = q = 2 + \delta$ (δ is given in assumption II). By virtue of
assumption V we have

$$\text{cov}(\zeta(\Pi^s_r), \zeta(\Pi^j_t))| \le \int_{\Pi^s_r}\int_{\Pi^j_t} \sup_{\theta\in\Theta}\sup_{x\in\Pi}|g_\Pi(x,\theta)|\sup_{y\in\Pi}|g(y,\theta)|\times$$

$$\times|\text{cov}(\xi(x),\xi(y))|dxdy \le k_5 v_s v_p v^{-1}\alpha_*^{\delta/(2+\delta)}(\rho^*(\Pi^s_r,\Pi^j_t),d) \le$$

$$\le k_6 v^{\gamma_s+\gamma_p-1}(1+d)^{\lambda\delta/(2+\delta)}[\rho^*(\Pi^s_r,\Pi^j_t)]^{-(n+\epsilon)\delta/(2+\delta)}. \qquad (1.7.4)$$

Let $I_s(r,l)$ be the set of indices t for which at least one edge of Π^s_r is situated
at a distance l from the edge in Π^s_t of the same type, that is, it is separated
from the edge in Π^s_t of the same type by exactly $l-1$ segments of the same
type (for $l = 1$ they are separated by a segment of a different type) and the
remaining edges are situated at a distance not exceeding $l-1$ segments.

Let $C^s_{rt} = \text{cov}(\zeta(\Pi^s_r), \zeta(\Pi^s_t))$. Then (1.7.4) and inequality $d = d(\Pi) \le k_7 v$
imply that

$$|\sum_{r\ne t} C^s_{rt}| \le \sum_{r=1}^{m_s}\sum_{l=1}^{m_s}\sum_{t\in I_s(r,l)}|C^s_{rt}| \le k_8 v^{-1}\sum_{l=1}^{m_s} v^{1-\gamma}l^{n-1}v^{2\gamma\gamma_s}v^{\lambda\delta/(2+\delta)}\times$$

$$\times\{(l\tau')^{(n+\epsilon)\delta/(2+\delta)}\}^{-1} \le k_9 v^\nu \sum_{l=1}^\infty l^{-\mu} \to 0 \qquad (1.7.5)$$

as $v \to \infty$ since in view of VI,

$$\mu = (n+\epsilon)\delta/(2+\delta) - (n-1) > 1,$$

$$\nu = \gamma(2\gamma_s - 1) + \lambda\delta/(2+\delta) - \gamma'(n+\epsilon)\delta/[n(2+\delta)] < 0. \qquad (1.7.6)$$

For $s > 1$, $\left|\sum_{r=1}^{m_s} C_{rr}^s\right| \le k_{10} v_s v^{-\gamma} \to 0$ as $v \to \infty$. Consequently in (1.7.3), $S_2 \xrightarrow{P} 0$ uniformly in $\theta \in \Theta$ and it does not affect the asymptotic distribution, that is, the quantities $s_\Pi(\theta)$ have the same distributions as $S_1 = \sum_{r=1}^{m_1} \zeta(\Pi_r^1)$.

By induction on m_1 along each of the n coordinate axes, we easily derive from Lemma 1.6.1 the inequality

$$\left| E \exp\left\{ it \sum_{r=1}^{m_1} \zeta(\Pi_r^1) \right\} - \prod_{r=1}^{m_1} E \exp\{ it\zeta(\Pi_r^1) \} \right| \le$$

$$\le 16n(m_1 - 1)\alpha_*(\tau_0, d) \le k_{11} v^\nu, \tag{1.7.7}$$

where

$$\nu = 1 + \lambda - \gamma - \gamma'(n + \epsilon)/n < 0, \tag{1.7.8}$$

and $\tau_0' = \min\{\tau_j, \ j = 1, \ldots, n\}$.

We also denote $\min\{\tau_j, \ j \in \mathbf{R}(\Pi)\}$ by τ_0, where $\mathbf{R}(\Pi)$ is the set of numbers j of the axes that were partitioned into alternating segments (see the definition of $\tilde{\tau}_j$ above).

By virtue of (1.7.7) and (1.7.8), as $v \to \infty$ the summands in S_1 may be viewed as independent r.v.'s depending on the parameter $\theta \in \Theta$ and having distributions identical to those of the summands in the sum S_1. We verify that they satisfy the premises of Theorem 1.1.3.

In view of IV, for all $\theta \in \Theta$ we have

$$\lim_{\Pi \to \infty} E s_\Pi^2(\theta) = \lim_{\Pi \to \infty} E S_1^2 = \sigma^2(\theta) > 0, \tag{1.7.9}$$

since it is easy to show utilizing (1.7.4) that

$$v^{-1} \left| \sum_{r,t} \operatorname{cov}(\zeta(\Pi_r^s), \zeta(\Pi_t^p)) \right| \to 0, \ v \to \infty, \ s \ne p.$$

The proof of the last relation is analogous to that of (1.7.5) where the case $s = p$ was considered. The only difference is that the case $l = 0$ may occur (parallelepipeds of different types are adjacent to each other). The covariance of these terms admits an upper bound $k_{12}\sqrt{v_s v_p} v^{-\gamma}$ which tends to zero as $v \to \infty$. In view of V and the Minkowski inequality, we obtain

$$E\left[\int_{\Pi_r^1} g_\Pi(x, \theta)\xi(x)dx \right]^{2+\delta} \le k_{12} \left\{ \sup_{\theta \in \Theta} \sup_{x \in \Pi} |g_\Pi(x, \theta)| \right\}^{2+\delta} \times$$

$$\times E\left[\int_{\Pi_r^1} \xi(x)dx \right]^{2+\delta} \le k_{14} |\Pi_r^1|^{2+\delta}/|\Pi|^{(2+\delta)/2} \le k_{15} v^\nu,$$

$$\nu = \gamma(2 + \delta) - (2 + \delta)/2.$$

Uniformly in $\theta \in \Theta$,

$$\sum_{r=1}^{m_1} E|\zeta_r^1|^{2+\delta} \leq v^\mu, \qquad (1.7.10)$$

where for $\gamma < \delta/[2(1 + \delta)]$

$$\mu = 1 - \gamma + \gamma(2 + \delta) - (2 + \delta)/2 < 0. \qquad (1.7.11)$$

Formulas (1.7.9), (1.7.10), (1.7.11) imply that condition 2) of Theorem 1.1.3 is fulfilled for summands of the sum S_1. Therefore Theorem 1.1.2, whose condition 1) is a corollary of assumption IV and the Chebyshev inequality, implies the asymptotic normality of $s_\Pi(\Theta)$. ∎

Theorem 1.7.2. Under assumptions IV, V, VII and $\Pi \to \infty$, (1.7.2) holds for $s_\Pi(\theta)$.

The proof of Theorem 1.7.2 is analogous to that of Theorem 1.7.1, where Lemma 1.6.2 is replaced by Lemma 1.6.1.

Remark 1.7.1. The proof of Theorem 1.7.1 shows that it is sufficient to require $|\Pi| \to \infty$ only (the requirement that all of the edges increase is not needed).

Remark 1.7.2. In Theorem 1.7.1, assumption VI can be replaced by the following: Π holds, $\sum_{r=1}^\infty r^{n-1} \alpha_*^{\delta/(2+\delta)}(r, d_0) < \infty$, where d_0 is the diameter of $\Pi(a^0)$, and there exists $r(d) \to \infty$, $r(d) = o(d^{1/n})$ as $d \to \infty$ such that $\alpha_*(r, d) = o(d^{-(2+\delta)/2})$.

Remark 1.7.3. If one utilizes the mixing rate $\alpha_*(r) = \alpha_*(r, \infty)$ introduced in §1.6, assumption VI in Theorem 1.7.1 (or VII in Theorem 1.7.2) ensures the absolute integrability of the cor.f. over $\mathbf{R}^n$ in the case of homogeneous random fields (see Remark 1.6.1).

Assume now that $\mathfrak{M}$ consists of all possible finite unions of parallelepipeds Π_m^0. Let $\Delta \in \mathfrak{M}$, $|\Delta| \to \infty$. Suppose that $\Delta \xrightarrow{F} \infty$.

VIII. Π holds with $\alpha_*(r, d) \leq k_{16}(1 + d)^\lambda r^{-n-\epsilon}$, where $\epsilon > 2n/\delta$, $\lambda < \delta(\epsilon\delta - 2n)/\{2\delta(1 + \delta)\}$.

Theorem 1.7.3. Under assumptions IV, V, VIII and $\Delta \xrightarrow{F} \infty$, (1.7.2) holds for $s_\Delta(\theta)$.

Proof. Let $\tau = |\Delta|^{\gamma/n}$, $\tau' = |\Delta|^{\gamma'/n}$, $0 < \gamma' < \gamma < \delta/[2(1 + \delta)]$. Define τ_j, τ_j' in the same manner as in the proof of Theorem 1.7.1.

In general, if $\Delta = \bigcup_{i=1}^{s} A_i$, where $A_i \cap A_j = \emptyset$ with $i \neq j$, there exists an i such that $|\Delta \cap A_i| \leq |\Delta|/s$.

This means that the method used to prove Theorem 1.7.1 permits us to partition the set Δ by hyperplanes into a system of parallelepipeds so that a subset $\Delta_0 \subset \Delta$ which falls into separate slots of width $\phi(\Delta) \asymp \tau'$ would have volume $|\Delta_0| \leq k_{17}\tau'\tau^{-1}|\Delta|$ and $\Delta \backslash \Delta_0$ would consist of at most $[d(\Delta)/\tau]$ parallelepipeds. As $\Delta \xrightarrow{F} \infty$, $d(\Delta) \leq k_{18}|\Delta|^{1/n}$, hence the required assertion can be derived by suitably modifying formulas (1.7.6), (1.7.8) and (1.7.11) and utilizing the concluding part of the proof of Theorem 1.7.1. ∎

Remark 1.7.4. If one utilizes the mixing rate $\alpha_*(r)$, assumption VII in Theorem 1.7.3 should be replaced by the following: II holds and $\alpha_*(r) \leq k_{19}r^{-n-\epsilon}$, $\epsilon\delta > 2n$.

Now let $\Delta \xrightarrow{V.H.} \infty$ and let the values $\tau_j, \tau_j', \tau_0, \tau_0'$ be defined as in Theorems 1.7.1 and 1.7.3. Assume that $\mathfrak{M}$ consists of all possible finite unions of Π_m^0 obtained by shifting a fixed parallelepiped $\Pi[0, a^0)$.

Lemma 1.7.1 [20]. The set Δ can be subdivided into the parts having the volumes $\psi(\Delta) \asymp |\Delta|^\gamma$ (constants involved in the symbol $\asymp$ are independent of Δ) which belong to $\mathfrak{M}$ and lie at a distance equal to τ_0^1 at least, and into the part Δ_0 falling in separate slots such that $|\Delta_0| = o(|\Delta|)$ as $\Delta \to \infty$.

We introduce an additional assumption.

IX. II holds and $\alpha(r) \leq k_{20}r^{-n-\epsilon}$, $\epsilon\delta > 2n$.

Theorem 1.7.4. If assumptions IV, V, IX hold and $\Delta \xrightarrow{V.H.} \infty$, then $s_\Delta(\theta)$ satisfies (1.7.2).

Proof. Consider a partition of $\mathbf{R}^n$ into the parallelepipeds $\Pi_m = \Pi[0, a) + m \otimes a$, $m \in \mathbf{Z}^n$, where $a \in \mathbf{R}^n$ is a vector having non-zero coordinates. Let Δ_a^- and Δ_a^+ consist of Π_m such that $\Pi_m \cap \Delta \neq \emptyset$ and $\Pi_m \subset \Pi$ respectively. Then $\Delta \backslash \Delta_a^- \subset \Delta_a^+ \backslash \Delta_a^-$ and, provided $\Delta \to \infty$

$$|\Delta_a^+ \backslash \Delta_a^-|/|\Delta| \to 0 \tag{1.7.12}$$

for any $a \in \mathbf{R}^n$. Set

$$s_\Delta(\theta) = \int_{\Delta_a^-} g_\Delta(x, \theta)\xi(x)dx + \int_{\Delta\backslash\Delta_a^-} g_\Delta(x, \theta)\xi(x)dx = S_3 + S_4.$$

By virtue of V, IX and (1.7.12) $S_4 \xrightarrow{P} 0$ uniformly in $\theta \in \Theta$. By splitting the integral over Δ_a^- into integrals over the sets stipulated in Lemma 1.7.1 and using the concluding part of the proof of Theorem 1.7.1 we obtain the required assertion. ∎

Consider now a column vector $s_\Delta(\theta) = [s_\Delta^{(1)}(\theta), \ldots, s_\Delta^{(r)}(\theta)]'$, with components defined by formula (1.7.1), where $g_\Delta(x, \theta) = [g_\Delta^{(1)}(x, \theta), \ldots, g_\Delta^{(r)}(x, \theta)]'$ is a column vector of functions satisfying the following assumptions.

X. As $\Delta \to \infty$ there exists a function $\mathfrak{S}^2(\theta)$, taking on values in a set of positive definite $r \times r$ matrices such that

$$\lim_{\Delta \to \infty} [Es_\Delta(\theta)s'_\Delta(\theta) - \mathfrak{S}^2(\theta)] = 0$$

uniformly in $\theta \in \Theta$.

XI. For any $i \in \{1, \ldots, r\}$ the function $g_\Delta^{(i)}(x, \theta)$ satisfies assumption V with a constant $k_3^{(i)} > 0$ (possibly depending on i).

The following assertion can be proved.

Theorem 1.7.5. If assumptions IX–XI hold, then as $\Delta \xrightarrow{V.H.} \infty$ the u.c.l.t. is valid, that is, the random vector $s_\Delta(\theta)$ has uniformly in Θ an asymptotically multidimensional normal distribution $N_r(0, \mathfrak{S}^2(\theta))$ with zero mean vector and covariance matrix $\mathfrak{S}^2(\theta)$.

Theorem 1.7.6. Let $\xi(x) = [\xi^{(1)}(x), \ldots, \xi^{(r)}(x)]'$ be a m.s. continuous weakly homogeneous random field with $E\xi(x) = 0$ and satisfying the assumption:

XII. $E|\xi(0)|^{2+\delta} < \infty$, $\alpha(r) \leq k_{21}r^{-n-\epsilon}$ for some $\delta > 0$, $\epsilon > 2n/\delta$.

Then a bounded spectral density function $f(\lambda) = (f_{ij}(\lambda))_{i,j=1,\ldots,r}$, $\lambda \in \mathbf{R}^n$ exists which is continuous at zero. If moreover $f(0)$ is a non-singular matrix, then as $\Delta \xrightarrow{V.H.} \infty$ the following asymptotic formula holds for the vector $s_\Delta = \int_\Delta \xi(x) dx$:

$$Es_\Delta s'_\Delta = (2\pi)^n |\Delta| f(0)(1 + o(1)), \qquad (1.7.13)$$

and the vector $|\Delta|^{-1/2} s_\Delta$ has the asymptotically multivariate normal distribution $N_r(0, (2\pi)^n f(0))$.

Theorem 1.7.6 is proved by the sectioning technique using Remark 1.6.1 and Plancherel's theorem for the derivation of (1.7.13).

Remark 1.7.5. In Theorems 1.7.5 and 1.7.6 condition $\Delta \xrightarrow{V.H.} \infty$ can be replaced by $\Delta \xrightarrow{F} \infty$. In this case IX and XII may be replaced by VIII and in the case of parallelepipeds by VI as well.

Remark 1.7.6. As $r = 1$, $\Delta \xrightarrow{V.H.} \infty$ under the conditions of Theorem 1.7.6, $\sigma_1^2(n) = (2\pi)^n f(0) > 0$ and the convergence of distributions: $|\Delta|^{-1/2} s_\Delta \xrightarrow{D} N(0, \sigma_1^2(n))$ holds.

If $\sigma_1^2(n) = 0$ the last relation is equivalent to the following: $|\Delta|^{-1/2} s_\Delta \xrightarrow{P} 0$. This follows from (1.7.13) and the Chebyshev inequality.

Theorem 1.7.7. Let $\xi(x)$ be a m.s. continuous homogeneous isotropic random field satisfying assumption XII. Then there exists a continuous bounded isotropic spectral density function $g(\mu)$, $\mu \in \mathbf{R}_+^1$. If $g(0) \neq 0$, then as $r \to \infty$

$$\frac{1}{r^{n/2}} \int_{v(r)} \xi(x) dx \xrightarrow{D} N(0, \sigma_2^2(n)), \quad \sigma_2^2(n) = (2\pi)^n |s(1)| g(0)/n.$$

As $r_1 + r_2 \to \infty$ $(r_1 < r_2)$,

$$\cdot \frac{1}{(r_2^n - r_1^n)^{1/2}} \int_{r_1 \leq |x| < r_2} \xi(x) dx \xrightarrow{D} N(0, \sigma_2^2(n)).$$

The assertion of the theorem follows from Theorem 1.7.4, if one takes $\Delta = v(r)$ (or $\Delta = v(r_2) \backslash v(r_1)$), $g_\Delta(x, \theta) = r^{-n/2}$ (or $\left(\frac{1}{r_2^n - r_1^n}\right)^{1/2}$) and utilizes Lemma 1.5.4 with $\alpha = n - 1$ and Remark 1.6.1.

Remark 1.7.7. The constant $\sigma_2^2(n)$ in Theorem 1.7.7 may be replaced by $c_1(n)\beta_1$ as given in Lemma 1.5.1.

We now state a c.l.t. for spheres which may be proved by the sectioning technique following Yu.A. Rozanov's method [112] where a sphere is partitioned into "large" and "small" spherical layers by hyperplanes perpendicular to the diameter of the sphere. In this case the integrals over spherical layers are additive functions of sets $\Delta \subset \mathbf{R}^1$.

Theorem 1.7.8. Let $n \geq 2$ and suppose that the assumption of Theorem 1.7.7 holds. If $\beta_2 \neq 0$ (see (1.5.9)), then as $r \to \infty$, $r^{(1-n)/2} \int_{s(r)} \xi(x) dm(x) \xrightarrow{D} N(0, \sigma_3^2(n))$, $\sigma_3^2(n) = \beta_2 c_4(n)$, where $c_4(n)$ is defined by formula (1.5.8).

Remark 1.7.8. In the statement of Theorem 1.7.8 the mixing rate $\alpha(r)$ may be defined as in §6 with the least upper bound taken only over the sets Δ_1 and Δ_2 lying on the surfaces of spheres.

Theorems 1.7.1–1.7.4 allow us to derive a c.l.t. for non-standard normalizing factors (different from the normalizing factor $A_\Delta = |\Delta|^{-\frac{1}{2}}$). Refer to §3.1 for more details.

1.8. Moment Inequalities

When studying approximations to distributions of integrals of random fields by means of the normal law, moment inequalities for these integrals play an important role.

Lemma 1.8.1. Let $\xi : \Omega \times \mathbf{R}^n \to \mathbf{R}^1$ be a random field with $E\xi(x) = 0$ and satisfying the following conditions:

1) for some integer $m \geq 1$ and some $\delta > 0$, $E|\xi(x)|^{2m+\delta} \leq k^{(1)} < \infty$, $x \in \mathbf{R}^n$;

2) the mixing condition holds with the mixing rate $\alpha(r)$ such that $\alpha_m = \int_0^\infty \rho^{nm-1}\alpha^{\delta/(2m+\delta)}(\rho)d\rho < \infty$;

3) $\mathbf{B}^n \ni \Delta \to \infty$ so that $\Delta \subseteq v(k^{(2)}|\Delta|^{1/n})$, $k^{(2)} > 0$.

Then a constant $0 \leq c < \infty$ exists independent of Δ such that

$$\mu(m) = \frac{1}{|\Delta|^m} E\left[\int_\Delta \xi(x)dx\right]^{2m} \leq c. \tag{1.8.1}$$

Proof. We subdivide the proof into several parts.

1. In the notation of §1.3

$$\mu(m) \leq \frac{1}{|\Delta|^m} \int_{\Delta^{2m}} |m_{2m}(x^{(1)}, \ldots, x^{(2m)})| \prod_{i=1}^{2m} dx^{(i)}.$$

Let $\rho_{ij} = |x^{(i)} - x^{(j)}|$, $i, j = 1, \ldots, 2m$. Represent the set $\Delta^{2m} \subset \mathbf{R}^{2nm}$ as the union of $(C_{2m}^2)!$ sets (strings) having the property that all of the distances corresponding to each one of these sets are arranged in non-decreasing order. Choose an arbitrary set of this kind Λ^* and subdivide the corresponding string of inequalities connecting ρ_{ij} into two segments. Classify the first m distances ρ_{ij} as the left segment Λ_l^* and the remaining $C_{2m}^2 - m = 2m(m-1)$ distances as the right one Λ_r^*. Two cases are possible: a) an index from the set $J_{2m} = \{1, \ldots, 2m\}$ occurs in Λ_r^* $2m-1$ times; b) no index from J_{2m} occurs $2m-1$ times.

2. Consider case a). Let $1 \leq k \leq m-1$ be the number of indices occurring in Λ_r^* $2m-1$ times and $N_k \subset I_{2m}$ be the set of these indices. Find a value ρ_{ij}^*

in Λ_r^* for which the last index from N_k occurs for the first time as we move along Λ_r^* from left to right. Without loss of generality assume $1 \in N_k$ and $\rho_{ij}^* = \rho_{12}$. Denote by $\Lambda_r^*(N_k)$ the fragment of Λ_r^* from the beginning up to ρ_{12} inclusive. Separate the variables into two sets $\{x^{(1)}\}$ and $\{x^{(2)}, \ldots, x^{(2m)}\}$. By Lemma 1.6.2, $|m_{2m}(x^{(1)}, \ldots, x^{(2m)})| \le k^{(3)} \alpha^{\delta/(2m+\delta)}(|x^{(1)} - x^{(2)}|)$; thus

$$\mu(m, \Lambda^*) = |\Delta|^{-m} \int_{\Delta^{2m} \cap \Lambda^*} |m_{2m}(x^{(1)}, \ldots, x^{(2m)})| \prod_{i=1}^{2m} dx^{(i)} \le$$

$$\le k^{(3)} |\Delta|^{-m} \int_{\Delta^*} \alpha^{\delta/(2m+\delta)}(|x^{(1)} - x^{(2)}|) \prod_{i=1}^{2m} dx^{(i)},$$

$$\Delta^* = \Delta^{2m} \cap (\Lambda_l^* \Lambda_r^*(N_k)). \tag{1.8.2}$$

3. Suppose now that $2 \in N_k$, $k \ge 2$. Consider a graph Γ_l whose vertices are variables $x^{(i)}$, $i \in J_{2m} \backslash N_k$ and whose edges connecting the vertices $x^{(i)}$ and $x^{(j)}$ correspond to values ρ_{ij} from Λ_l^*. Thus Γ_l is a graph possessing $2m - k$ vertices, none of which is isolated, and m edges. In general, Γ_l consists of several disconnected components $\Gamma_l^{(1)}, \ldots, \Gamma_l^{(m_1)}$. Choose a spanning tree in each one of these components to obtain a forest $T_l^{(1)}, \ldots, T_l^{(m_1)}$. Retain in Λ_l^* only those values ρ_{ij} that belong to this forest. Select a vertex $x^{(k_1)}$ in $T_l^{(1)}$. Let $x^{(k_1)}$ be connected to $x^{(k_2)}, \ldots, x^{(k_s)}$ from $T_l^{(1)}$. Perform the change of variables $x^{(k_1)} = t^{(k_1)}$, $x^{(k_2)} - x^{(k_1)} = t^{(k_2)}, \ldots, x^{(k_s)} - x^{(k_1)} = t^{(k_s)}$. If some vertex among the $x^{(k_2)}, \ldots, x^{(k_s)}$ (for example, $x^{(k_s)}$) is connected to the vertices $x^{(r_1)}, \ldots, x^{(r_u)}$ from $T_l^{(1)}$, write $x^{(r_1)} - x^{(k_s)} = t^{(r_1)}, \ldots, x^{(r_u)} - x^{(k_s)} = t^{(r_u)}$. Traverse the whole tree $T_l^{(1)}$ in this manner. Perform an analogous change of variables in $T_l^{(2)}, \ldots, T_l^{(m_1)}$. Note that the number of independent changes of variables of type $x^{(k_1)} = t^{(k_1)}$ equals the number m_1 of disconnected components of the graph Γ_l.

4. Consider $\Lambda_r^*(N_k)$ and the graph Γ_r, corresponding to this segment of the string whose vertices are variables belonging to $\Lambda_r^*(N_k)$ and the edges are those ρ_{ij} belonging to $\Lambda_r^*(N_k)$. It is convenient to describe the subsequent change of variables in (1.8.2) in terms of operations on the graph Γ_r. If ρ_{ij} is an edge of Γ_r with $i, j \notin N_k$, remove ρ_{ij} from Γ_r. Having removed all such edges from Γ_r we obtain a new graph $_1\Gamma_r$. Subdivide $_1\Gamma_r$ into disconnected components $_1\Gamma_r^{(1)}, \ldots, _1\Gamma_r^{(s)}$ and find a spanning tree in each one of them. Obtain a forest $_1T_r^{(1)}, \ldots, _1T_r^{(s)}$. Suppose that a suspended vertex $x^{(1)}$ and thus, vertex $x^{(2)}$ belongs to $_1T_r^{(1)}$. Perform the change of variables $x^{(1)} = t^{(1)}$, $x^{(2)} - x^{(1)} = t^{(2)}$. We shall traverse the tree $_1T_r^{(1)}$ changing the variables: if $x^{(2)}$ is connected with $x^{(i)}$, $i \in N_k$, set $x^{(i)} - x^{(2)} = t^{(i)}$ and move on passing along the tree

and performing linear changes if vertices $x^{(j)}$, $j \in N_k$, occur. After a number of steps we shall arrive at either a suspended vertex of $_1T_r^{(1)}$ with the index from N_k or at a vertex $x^{(j)}$, $j \notin N_k$. The latter, in particular, may happen at the first step of moving from $x^{(2)}$. If we reach $x^{(j)}$, $j \notin N_k$, remove the edge connecting $x^{(j)}$ with the preceding vertex. If $x^{(j)}$ becomes an isolated vertex, it is also removed. If $x^{(j)}$ remains non-isolated, find a tree $T_l^{(q)} \ni x^{(j)}$, $q \in \{1, \ldots, m_1\}$, and proceed traversing the fragment of the tree $_1T_r^{(1)}$ in the above manner beginning from $x^{(j)} \in T_l^{(q)}$ till vertex $x^{(i)}$, $i \notin N_k$ is attained. Next repeat the process again until the whole tree $_1T_r^{(1)}$ is traversed. Find among $_1T_r^{(2)}, \ldots, _1T_r^{(s)}$ the trees containing at least one vertex $x^{(i)}$, $i \notin N_k$. Let these be $_1T_r^{(1)}, \ldots, _1T_r^{(iv)}$. Glue them together by the vertices $x^{(i)}$ to the corresponding trees $T_l^{(1)}, \ldots, T_l^{(m_1)}$. Then traverse the trees $_1T_r^{(i_1)}, \ldots, _1T_r^{(iv)}$ in the same manner as the tree $_1T_r^{(1)}$, that is, performing linear changes of variables $x^{(j)}$, $j \in N_k$, removing the superfluous edges and vertices, gluing together the fragments of trees $_1T_r^{(i_1)}, \ldots, _1T_r^{(iv)}$ and the trees $T_l^{(1)}, \ldots, T_l^{(m_1)}$. It is important to point out that changing the variables in the trees $_1T_r^{(i_1)}, \ldots, _1T_r^{(iv)}$, $_1T_r^{(1)}$ yields only one independent variable $t^{(1)}$. Consider the remaining trees in the forest $_1T_r^{(2)}, \ldots, _1T_r^{(s)}$ (if any). These contain vertices having indices only from N_k. Changing the variables in these trees will produce independent variables whose number is equal to the number $m_2 \geq 0$ of trees of this kind.

5. Changes of variables in paragraphs 3 and 4 with Jacobians not exceeding a constant in absolute value permit us to evaluate integral (1.8.2). Let $q = 2m - m_1 - m_2 - 2$, $\Delta - t = \{y : y = x - t, \ x \in \Delta\}$, $\Delta - \Delta = \bigcup_{t \in \Delta}(\Delta - t)$, $A_q = \{(t^{(i_j)}, \ j = 1, \ldots, q) : |t^{(i_1)}| \leq \ldots \leq |t^{(i_q)}| \leq |t^{(2)}|\}$. Integral (1.8.2) is then bounded by

$$k^{(3)}|\Delta|^{-m+m_1+m_2} \int_{A_q} \prod_{j=1}^{q} dt^{(i_j)} \int_\Delta dt^{(1)} \int_{\Delta - t^{(1)}} \alpha^{\delta/(2m+\delta)}(|t^{(2)}|)dt^{(2)} \leq$$

$$\leq k^{(3)}n^{-q}(q!)^{-1}|s(1)|^q|\Delta|^{-m+m_1+m_2} \int_\Delta dt^{(1)} \times$$

$$\times \int_{\Delta - t^{(1)}} |t^{(2)}|^{qn}\alpha^{\delta/(2m+\delta)}(|t^{(2)}|)dt^{(2)}. \tag{1.8.3}$$

In view of condition 3), $\Delta - \Delta \subseteq v(3k^{(2)}|\Delta|^{1/n})$. Proceeding in (1.8.3) to polar coordinates (1.1.1), we obtain

$$\int_\Delta dt^{(1)} \int_{\Delta - t^{(1)}} |t^{(2)}|^{qn}\alpha^{\delta/(2m+\delta)}(|t^{(2)}|)dt^{(2)} \leq$$

$$\leq |s(1)||\Delta| \int_0^{3k^{(2)}|\Delta|^{1/n}} \rho^{qn+n-1}\alpha^{\delta/(2m+\delta)}(\rho)d\rho \leq$$

$$\leq |s(1)|(3k^{(2)})^{m-m_1-m_2-1}|\Delta|^{m-m_1-m_2}\alpha_m, \qquad (1.8.4)$$

where α_m is as in condition 2). Bounds (1.8.2)–(1.8.4) show that the integral $\mu(m,\Lambda^*)$ in (1.8.2) is bounded uniformly in Δ.

6. The implications in point 5 are valid provided

$$m - m_1 - m_2 - 1 \geq 0. \qquad (1.8.5)$$

We shall show that (1.8.5) is valid. To do this, we find $\max m_1$ and $\max m_2$. The maximal number of trees in the forest $_1T_r^{(1)}, \ldots, _1T_r^{(s)}$ containing the vertices $x^{(i)}$, $i \in N_k\backslash\{1,2\}$, only equals the maximal number of matchings containing $k-2$ vertices at the most. Therefore $m_2 \leq [(k-2)/2]$. We now determine $\max m_1$.

Let a graph Γ have R vertices and r edges, $2r \geq R$. We determine the maximal number of disconnected components in Γ provided it has no isolated vertices. Evidently, one has to determine the maximal number p^* of matchings in Γ. The maximal number of disconnected components is then $p^* + 1$. The value p^* may be obtained by the following arguments. If the graph Γ has p^* matchings, the residual subgraph $\tilde{\Gamma}$ has $R - 2p^*$ vertices and $r - p^*$ edges. The number of edges in $\tilde{\Gamma}$ cannot exceed the number of edges in the complete graph having $R - 2p^*$ vertices, that is,

$$r - p^* \leq C_{R-2p^*}^2. \qquad (1.8.6)$$

On the other hand, increasing p^* by one implies that the number of edges in the graph $\tilde{\Gamma}$ cannot exceed the number of edges in the complete graph having $R - 2(p^* + 1)$ vertices, that is

$$r - p^* - 1 > C_{R-2(p^*+1)}^2. \qquad (1.8.7)$$

The solution of the system of inequalities (1.8.6) and (1.8.7) for p^* with respect to the graph Γ_l with $R = 2m - k$ and $r = m$ yields a bound $m_1 \leq m - [(k+1+(k+1)^{1/2})/2]$. Therefore $1 + m_1 + m_2 \leq 1 + ((k+1+(k+1)^{1/2})/2+1) + (k-2)/2 = m - ((k+1)^{1/2} - 1)/2 \leq m - (3^{1/2} - 1)/2 \simeq m - 0.35$, whence (1.8.5) follows.

7. We return now to the beginning of paragraph 3 and suppose that $2 \notin N_k$, $k \geq 1$. We shall trace how the bound of integral (1.8.2) will change.

Now the number of independent variables obtained while carrying out changes in Γ_l equals $m_1 - 1$ since one of the trees $T_l^{(1)}, \ldots, T_l^{(m_1)}$ containing $x^{(2)}$ has no independent variable. On the other hand, $m_2 \leq [(k-1)/2]$. Therefore one may obtain a bound similar to (1.8.3) and (1.8.4) provided the inequality $m_1 + m_2 \leq m$ analogous to (1.8.5) holds. However, now $m_1 + m_2 < m - \sqrt{k+1}/2 \leq m - \sqrt{2}/2 \simeq m - 0.7$.

8. We return to the end of paragraph 1 and consider case b). Each of the $2m$ indices occurs in Λ_r^* $2m - 2$ times exactly. Since the distances ρ_{1i} occur in Λ_r^* $2m - 2$ times, there exists an index j such that ρ_{1j} does not belong to Λ_r^*. Without loss of generality suppose that $j = 2$ and $\rho_{1i_0} = \min\{\rho_{1i}, \ i \in I_{2m}\backslash\{2\}\} \leq \min\{\rho_{2i}, \ i \in I_{2m}\backslash\{1\}\}$. The variables may then be subdivided into two sets $\{x^{(1)}, x^{(2)}\}$ and $\{x^{(3)}, \ldots, x^{(2m)}\}$. Applying Lemma 1.6.2 once more, we obtain

$$|E\prod_{i=1}^{2m}\xi(x^{(i)}) - E\xi(x^{(1)})\xi(x^{(2)})E\prod_{i=3}^{2m}\xi(x^{(i)})| \leq$$

$$\leq k^{(4)}\alpha^{\delta/(2m+\delta)}(|x^{(1)} - x^{(i_0)}|). \tag{1.8.8}$$

From (1.8.8) it follows that

$$\mu(m, \Lambda^*) \leq k^{(4)}|\Delta|^{-m}\int_{\Delta^{2m}\cap\Lambda^*}\alpha^{\delta/(2m+\delta)}(|x^{(1)} - x^{(i_0)}|)\times$$

$$\times\prod_{i=1}^{2m}dx^{(i)} + \mu(1)\mu(m-1, \tilde{\Lambda}^*), \tag{1.8.9}$$

where the string $\tilde{\Lambda}^*$ is obtained from the string Λ^* by removing all the distances ρ_{1i} and ρ_{2j}. Evaluate the first integral in the right-hand side of (1.8.9). Note that in the case under consideration, the graph Γ_l is a perfect matching and the vertices $x^{(1)}$ and $x^{(2)}$ belong to the same matching.

Set $x^{(1)} = t^{(1)}$, $x^{(i_0)} - x^{(1)} = t^{(i_0)}$, $x^{(2)} - x^{(1)} = t^{(2)}$. Remove from Γ_l the matching which contains vertex $x^{(i_0)}$. Perform $m - 2$ independent changes of variables in the remaining $m - 2$ matchings. Denoting $A_{m-2} = \{|t^{(2)}| \leq |t^{(i_1)}| \leq \ldots \leq |t^{(i_{m-2})}| \leq |t^{(i_0)}|\}$ we obtain

$$\frac{1}{|\Delta|^m}\int_{\Delta^{2m}\cap\Lambda^*}\alpha^{\delta/(2m+\delta)}(|x^{(1)} - x^{(i_0)}|)\prod_{i=1}^{2m}dx^{(i)} \leq$$

$$\leq |\Delta|^{-1}\int_{A_{m-2}}dt^{(2)}\prod_{j=1}^{m-2}dt^{(i_j)}\int_{\Delta}dt^{(1)}\int_{\Delta-t^{(1)}}\alpha^{\delta/(2m+\delta)}(|t^{(i_0)}|)dt^{(i_0)} \leq$$

$$\leq |s(1)|^{m-1} \frac{1}{|\Delta|(m-1)!n^{m-1}} \int_\Delta dt^{(1)} \int_{\Delta-t^{(1)}} |t^{(i_0)}|^{(m-1)n} \times$$

$$\times \alpha^{\delta/(2m+\delta)}(|t^{(i_0)}|)dt^{(i_0)} \leq |s(1)|^m[(m-1)!n^{(m-1)}]^{-1}\alpha_m,$$

where α_m is defined in conditions 2).

Since $\mu(1) < \infty$, the evaluation of the second summand in the right-hand side of (1.8.9) is reduced to the evaluation of integral $\mu(m-1,\tilde\Lambda^*)$ which is bounded above in the same manner as $\mu(m,\Lambda^*)$. Thus, to prove inequality (1.8.1), one has to perform at most $m-2$ steps. ∎

Lemma 1.8.2. Let $\xi(x)$, $x \in R^n$ be a homogeneous random field with zero mean, $E|\xi(0)|^{2m+\delta} < \infty$ for an integer $m \geq 1$ and some $\delta > 0$. If conditions 2) and 3) of Lemma 1.8.1 hold, then (1.8.1) is valid.

Lemma 1.8.3. Let $\xi(x)$, $x \in R^n$ be a random field with $E\xi(x) = 0$ and suppose that for each $x \in R^n$, $|\xi(x)| \leq k^{(5)}$ a.s., $\int_0^\infty \rho^{nm-1}\alpha(\rho)d\rho < \infty$ for some $m \geq 1$. If condition 3) of Lemma 1.8.1. holds, then (1.8.1) is valid.

Lemma 1.8.3 is proved in the same manner as Lemma 1.8.1, utilizing Lemma 1.6.1 instead of Lemma 1.6.2.

Corollary 1.8.1. Let the conditions of Lemma 1.8.1 hold and let

$$|\Delta|^{1/2} \operatorname{ess\,sup}_{x \in \Delta} |g(x)| \asymp \left(\int_\Delta |g(x)|^2 dx\right)^{1/2}, \quad \Delta \to \infty, \qquad (1.8.10)$$

for a locally square integrable function $g(x)$, $x \in R^n$. Then

$$E\left[\int_\Delta g(x)\xi(x)dx\right]^{2m} \leq k^{(6)}\left[\int_\Delta |g(x)|^2 dx\right]^m, \qquad (1.8.11)$$

where the constant $0 < k^{(6)} < \infty$ is independent of Δ.

Inequality (1.8.11) is easily derived without the additional requirement (1.8.10) under different assumptions on the field $\xi(x)$, $x \in R^n$.

Lemma 1.8.4. If a random field $\xi(x) \in \Xi^{(2m)}$, $m \geq 1$ (see §1.3) and

$$\beta_r = \sup_{\lambda^{(1)} \in R^n} \int_{R^{(r-2)n}} |f_r(\lambda^{(1)},\dots,\lambda^{(r)})|^2 d\lambda^{(2)}\dots d\lambda^{(r-1)} < \infty,$$

$r = 2,\dots,2m$, then (1.8.11) holds for any locally square integrable function $g(x)$, $x \in R^n$.

Proof. By formulas (1.3.2) the moment

$$E\Big[\int_\Delta g(x)\xi(x)dx\Big]^{2m} = \int_{\Delta^{2m}} \prod_{i=1}^{2m} g(x^{(i)}) m_{2m}(x^{(1)},\dots,x^{(2m)}) \prod_{i=1}^{2m} dx^{(i)}$$

may be represented as the sum of products of integrals of cumulants of the field $\xi(x)$ of order $r = 2,\dots,2m$ weighted by the function $g(x)$. Using (1.3.7) with fixed $2 \le r \le 2m$, we evaluate the integral

$$\int_{\Delta^r} \prod_{i=1}^r g(x^{(i)}) s_r(x^{(1)},\dots,x^{(r)}) dx^{(1)} \dots dx^{(r)} = \int_{R^{(r-1)n}} f_r(\lambda^{(1)},\dots,\lambda^{(r)}) \times$$

$$\times \Big[\prod_{j=1}^r \int_\Delta e^{i\langle\lambda^{(j)},x^{(j)}\rangle} g(x^{(j)})dx^{(j)}\Big] d\lambda^{(1)} \dots d\lambda^{(r-1)} \le$$

$$\le \Big[\int_{R^{(r-1)n}} |f_r(\lambda^{(1)},\dots,\lambda^{(r)})|^2\Big] \int_\Delta e^{i\langle\lambda^{(r)},x^{(r)}\rangle} \times$$

$$\times g(x^{(r)})dx^{(r)}\Big|^2 d\lambda^{(1)} \dots d\lambda^{(r-1)} \Big[\int_\Delta |g(x)|^2 dx\Big]^{(r-1)/2}. \qquad (1.8.12)$$

Carrying out the change of variables $-\lambda^{(1)} - \dots - \lambda^{(r-1)} = \lambda^{(r)} = \lambda_*^{(1)}, \lambda^{(2)} = \lambda_*^{(2)},\dots,\lambda^{(r-1)} = \lambda_*^{(r-1)}$ in the integral on the right-hand side of (1.8.12) we obtain in view of the symmetry of the density function $f_r(\cdot,\dots,\cdot)$,

$$\Big[\int_{R^{(r-1)n}} |f_r(\lambda^{(1)},\dots,\lambda^{(r)})|^2\Big]\Big|\int_\Delta e^{i\langle\lambda^{(r)},x^{(r)}\rangle} \times$$

$$\times g(x^{(r)})dx^{(r)}\Big|^2 d\lambda^{(1)} \dots d\lambda^{(r-1)}\Big]^{1/2} =$$

$$= \Big[\int_{R^{(r-1)n}} |f_r(-\lambda_*^{(1)} - \dots - \lambda_*^{(r-1)}, \lambda_*^{(2)},\dots,\lambda_*^{(1)})|^2 \times$$

$$\times \Big|\int_\Delta e^{i\langle\lambda_*^{(1)},x^{(r)}\rangle} g(x^{(r)})dx^{(r)}\Big|^2 d\lambda_*^{(1)} \dots d\lambda_*^{(r-1)}\Big]^{1/2} \le$$

$$\le \beta_r^{1/2}\Big[\int_\Delta |g(x)|^2 dx\Big]^{1/2}. \qquad \blacksquare$$

1.9. Invariance Principle

We provide the conditions for weak convergence of measures generated by integrals of random fields to a measure generated by a Brownian motion process.

Let $C[0,1]$ be the space of functions continuous on the segment $[0,1]$ equipped with the uniform topology, let $\xi : \Omega \times \mathbf{R}^n \to \mathbf{R}^1$ be a random field, $E\xi(x) = 0$, $T(r) = \int_{v(r)} \xi(x)dx$. Denote by P_r the probability measures in $C[0,1]$ generated by elements $X_r(t) = r^{-n/2}T(t^{1/n}r)$, $t \in [0,1]$; let W_b be the probability measure in $C[0,1]$ generated by a Brownian motion process $w_b(t)$, $t \in [0,1]$, that is by an a.s. continuous Gaussian process with $Ew_b(t) = 0$, $Ew_b(t)w_b(s) = b\min\{t,s\}$, $w_b(0) = 0$ a.s.

Theorem 1.9.1. If $\xi(x)$ is a m.s. continuous homogeneous isotropic random field with $|\xi(0)| < \infty$ a.s. and

$$\alpha(r) \le k_1 r^{-p}, \quad p > 2n, \tag{1.9.1}$$

then there exists a continuous bounded isotropic spectral density function $g(\mu)$, $\mu \in \mathbf{R}^1_+$. If $g(0) \ne 0$, then $P_r \Rightarrow W_b$ in $C[0,1]$ as $r \to \infty$, where $b = (2\pi)^n g(0)/n$.

Theorem 1.9.2. If condition (1.9.1) in Theorem 1.9.1 is replaced by the condition

$$E|\xi(0)|^{4+\delta} < \infty, \quad \alpha(r) \le k_2 r^{-p}, \quad p > 2n\left(1 + \frac{4}{\delta}\right), \quad \delta > 0, \tag{1.9.2}$$

then $P_r \Rightarrow W_b$ in $C[0,1]$ as $r \to \infty$.

The proofs of Theorems 1.9.1 and 1.9.2 are analogous.

Lemma 1.9.1. Let the conditions of Theorem 1.9.2 hold but in (1.9.2) only $p > n(1 + \frac{2}{\delta})$ is required. Then the finite-dimensional distributions of the processes $X_r(t)$, $t \in [0,1]$ converge to finite-dimensional distributions of the process $w_b(t)$, $t \in [0,1]$ as $r \to \infty$.

Proof. The existence of the spectral density function $g(\mu)$ follows from Remark 1.6.1. For $s \ge 1$ consider the points $0 = t_0 < t_1 < \ldots < t_s \le 1$. Let $\lambda_j \in \mathbf{R}^1$, $j = 1, \ldots, s$,

$$\zeta_r = \sum_{j=1}^{s} \lambda_j X_r(t_j) = \sum_{j=1}^{s} a_j (X_r(t_j) - X_r(t_{j-1})),$$

where $a_j = \lambda_1 + \ldots + \lambda_s$. Define $m_j = m_j(r)$, $j = 1, \ldots, s$ so that $m_j < t_j^{1/n} r < m_{j+1}$, with $r^{-1} m_j(r) \uparrow t_j$ as $r \to \infty$. Set $p_j = p_j(r) = m_j - \mu(r)$ for $1 \le j < s$ and $p_s = m_s$, where $\mu(r)$ is such that $\mu(r) \to \infty$, $\mu(r)/r \to 0$ as $r \to \infty$.

Let $\zeta_r^{(j)} = a_j r^{-n/2}(T(p_j) - T(m_{j-1})) = a_j r^{-n/2}\xi_r^{(j)}$, $j = 1,\dots,s$. Then $\zeta_r = \sum_{j=1}^{s} \zeta_r^{(j)} + Q_r$. From Lemma 1.5.6 we easily obtain that $\lim_{r \to \infty} \mathrm{var}\, Q_r = 0$. Thus $Q_r \xrightarrow{P} 0$ as $r \to \infty$ and it does not affect the asymptotic behaviour of ζ_r. However,

$$\zeta_r^{(j)} = \left\{a_j \xi_r^{(j)} \big/ \sqrt{\mathrm{var}\, \xi_r^{(j)}}\right\}\left\{r^{-n/2}\sqrt{\mathrm{var}\, \xi_r^{(j)}}\right\}.$$

From Lemma 1.5.6, $\lim_{r \to \infty}\left\{r^{-n/2} \big/ \sqrt{\mathrm{var}\, \xi_r^{(j)}}\right\} = \sqrt{t_j - t_{j-1}}$. From Theorem 1.7.7, the quantity $\zeta_r^{(j)}$ has asymptotically normal distribution with zero mean and variance $ba_j^2(t_j - t_{j-1})$ as $r \to \infty$, that is,

$$E\exp\{i\tau\zeta_r^{(j)}\} \to \exp\{-\tau ba_j^2(t_j - t_{j-1})/2\}. \tag{1.9.3}$$

From Lemma 1.6.1,

$$\left|E\prod_{j=1}^{s}\exp\{i\tau\zeta_r^{(j)}\} - \prod_{j=1}^{s}E\exp\{i\tau\zeta_r^{(j)}\}\right| \le 16s\alpha(\mu(r));$$

hence as $r \to \infty$, the values $\zeta_r^{(j)}$ may be considered independent, as in Theorem 1.7.1. If $f_r(\tau) = E\exp\{i\tau\zeta_r\}$, then by (1.9.3), as $r \to \infty$,

$$\lim f_r(\tau) = \lim E\exp\left\{i\tau\sum_{j=1}^{s}\zeta_r^{(j)}\right\} = \exp\{-\tau\rho^2/2\},$$

where $\rho^2 = b\sum_{j=1}^{s} a_j^2(t_j - t_{j-1})$. However, $f_r(1)$ is the ch.f. of the vector $(X_r(t_1),\dots,X_r(t_s))$, and $\exp\{-\rho^2/2\}$ is the ch.f. of $(w_b(t_1),\dots,w_b(t_s))$. Thus the assertion of the lemma follows from the continuity theorem for ch.f.'s. ∎

Lemma 1.9.2. *If in the conditions of Theorem 1.9.1 only the condition $p > n$ is required (cf. (1.9.1)), then finite-dimensional distributions of the process $X_r(t)$, $t \in [0,1]$, converge to finite-dimensional distributions of the process $w_b(t)$, $t \in [0,1]$ as $r \to \infty$.*

Lemma 1.9.3. *If the conditions of Theorem 1.9.1 (or those of 1.9.2) hold, then the set of measures P_r is weakly compact in $C[0,1]$.*

Proof. From Lemma 1.8.1 (or 1.8.3) for $m = 2$ inequality $E(X_r(t) - X_r(s))^4 \le k_5(t-s)^2$ holds for any $t, s \in [0,1]$. Theorem 1.1.4 with n=1 now implies the assertion of the lemma. ∎

The assertions of Theorems 1.9.1 and 1.9.2 follow from Lemmas 1.9.1–1.9.3 and Theorem 1.1.4.

Remark 1.9.1. The conditions of Theorems 1.9.1 and 1.9.2 can be weakened. For example, one may consider:

$$T(\Delta) = \int_\Delta \xi(x)dx,$$

where $\Delta \in \mathcal{B}^n$ is a bounded convex set containing the origin (not necessarily a ball) and the measures P_λ in $C[0,1]$ induced by the elements

$$X_\lambda(t) = |\Delta|^{-1/2}\lambda^{-n/2}T(\lambda t^{1/n}\Delta), \ t \in [0,1].$$

The isotropy of the field and even the homogeneity can be discarded by replacing (1.9.2) with the conditions

$$E|\xi(x)|^{4+\delta} \le k_6 < \infty \quad \text{for all } x \in \mathbf{R}^n,$$

$$\alpha(r) \le k_7 r^{-p}, \ p > 2n(1+4/\delta), \ \delta > 0$$

and an additional condition on the behaviour of the variance similar to IV in §1.7 ensuring the existence of a constant $\sigma^2(\Delta) > 0$ such that, as $\lambda \to \infty$,

$$\operatorname{var} T(\lambda\Delta) = \sigma^2(\Delta)\lambda^n(1 + o(1)).$$

Then as $\lambda \to \infty$, the measures $P_\lambda \Rightarrow W_{\sigma^2(\Delta)}$ in $C[0,1]$. One may discard the convexity of the Δ and the requirement: $0 \in \Delta$. The limiting process can then be defined by means of the stochastic integral

$$w_\Delta(t) = \frac{\sigma^2(\Delta)}{|\Delta|} \int_{t^{1/n}\Delta} dw(s_1,\ldots,s_n), \ t \in [0,1] \tag{1.9.4}$$

where $w(x)$, $x \in \mathbf{R}^n$ is a Wiener random field according to N.N. Chentsov [131, 132], that is, an a.s. continuous Gaussian field such that
1) $w(x) = 0$ a.s. if $x_i = 0$ for at least one $i \in \{1,\ldots,n\}$;
2) $Ew(x) = 0$;
3) $Ew(x)w(y) = \prod_{i=1}^n \min\{x_i, y_i\}$, $x, y \in \mathbf{R}^n$.
The correlation function of the process (1.9.4) is

$$Ew_\Delta(t)w_\Delta(s) = \frac{\sigma^2(\Delta)}{|\Delta|}|s^{1/n}\Delta \cap t^{1/n}\Delta|.$$

If the set Δ is not convex, the process $w(t)$, $t \in [0,1]$ may not be a Wiener process.

To ensure the weak convergence of the measures P_λ to a limiting Gaussian measure W_Δ in $C[0,1]$, an additional requirement should be imposed which is that the process $w_\Delta(t)$, $t \in [0,1]$, induces a measure W_Δ in $C[0,1]$.

The truncation technique permits us to relax the requirement on the existence of the moment of order $4 + \delta$, $\delta > 0$, to that of the existence of the moment of order $2 + \delta$, $\delta > 0$.

For a more detailed treatment of the topics covered in this section the reader is referred to [35, 171, 172]. The weakest constraints in the invariance principle for random sequences are presented in reference [204].

CHAPTER 2

Limit Theorems for Functionals of Gaussian Fields

2.1. Variances of Integrals of Local Gaussian Functionals

In Chapter 2, mainly limit theorems for random fields having a slowly decreasing correlation are discussed. Only non-linear transformations of Gaussian fields are treated herein.

We examine the asymptotic behaviour of the variances of integrals of non-linear transforms of Gaussian fields.

Let $\phi(u) = \exp\{-u^2/2\}/\sqrt{2\pi}$, $u \in \mathbf{R}^1$, be the density function of a Gaussian random variable having parameters $(0,1)$, (see (1.1.3)), let $L_2(\mathbf{R}^1, \phi(u)du)$ be the Hilbert space of equivalence classes of Lebesgue measurable functions $G : \mathbf{R}^1 \to \mathbf{R}^1$ satisfying the inequality $\int_{\mathbf{R}^1} G^2(u)\phi(u)du < \infty$.

It is well known that the Chebyshev-Hermite polynomials with the leading coefficient equal to 1 form a complete orthonormal system in the space $L_2(\mathbf{R}^1, \phi(u)du)$, that is

$$\int_{\mathbf{R}^1} H_m(u)H_q(u)\phi(u)du = \delta_m^q m! . \qquad (2.1.1)$$

Using the representation $H_m(u) = (-1)^m \exp\{u^2/2\}\frac{d^m}{du^m}\exp\{-u^2/2\}$, $m = 0, 1, \ldots$, one can derive expressions for the first few polynomials:

$$H_0(u) = 1, \ H_1(u) = u, \ H_2(u) = u^2 - 1, \ H_3(u) = u^3 - 3u, \ H_4(u) = u^4 - 6u^2 + 3,$$

and so on.

We shall assume that:

I. $\xi : \Omega \times \mathbf{R}^n \to \mathbf{R}^1$ is a measurable m.s. continuous homogeneous Gaussian isotropic random field with $E\xi(x) = 0$, $E\xi^2(x) = 1$ and cor. f. $B(|x|) = E\xi(0)\xi(x) \to 0$ as $|x| \to \infty$.

54

II.   The function $G : \mathbf{R}^1 \to \mathbf{R}^1$ is such that $EG(\xi(0)) = C_0 < \infty$, $EG^2(\xi(0)) < \infty$.

Under assumption II, $G(u)$ may be expanded in the series:

$$G(u) = \sum_{q=0}^{\infty} C_q H_q(u)/q!, \quad C_q = \int_{\mathbf{R}^1} G(u) H_q(u) \phi(u) du, \quad q = 0, 1, \ldots,$$

which converges in the space $L_2(\mathbf{R}^1, \phi(u)du)$, and in view of the Parseval equality,

$$\sum_{q=0}^{\infty} C_q^2/q! = EG^2(\xi(0)) < \infty. \tag{2.1.2}$$

III.   The function $G(u)$ satisfies assumption II and an integer $m \geq 1$ exists such that $C_1 = \ldots = C_{m-1} = 0$, $C_m \neq 0$.

If III holds, m is called the rank of G, denoted as $m = \text{rank}G$. For example, the function $G(u) = H_m(u)$, $m \geq 1$, is of rank m.

Below, the density function $\phi(u, v, \rho)$ of a bivariate Gaussian vector (ξ, η) with $E\xi = E\eta = 0$, $E\xi^2 = E\eta^2 = 1$, $E\xi\eta = \rho$ will be used. It is of the form

$$\phi(u, v, \rho) = \frac{1}{2\pi\sqrt{1 - \rho^2}} \exp\left\{ -\frac{u^2 + v^2 - 2uv\rho}{2(1 - \rho^2)} \right\}. \tag{2.1.3}$$

Lemma 2.1.1. If (ξ, η) is a Gaussian vector having the density function (2.1.3), then for all $m \geq 1$, $q \geq 1$

$$EH_m(\xi)H_q(\eta) = \delta_m^q \rho^m m! \tag{2.1.4}$$

Proof. The following expansion of function (2.1.3) is well-known [112,p.251]:

$$\phi(u, v, \rho) = \frac{1}{2\pi} e^{-\frac{u^2+v^2}{2}} \sum_{\nu=0}^{\infty} \frac{\rho^\nu}{\nu!} H_\nu(u) H_\nu(v) =$$

$$= \sum_{\nu=0}^{\infty} \frac{\rho^\nu}{\nu!} \Phi^{(\nu+1)}(u) \Phi^{(\nu+1)}(v). \tag{2.1.5}$$

In view of (2.1.1), (2.1.5), we obtain

$$EH_m(\xi)H_q(\eta) = \int_{\mathbf{R}^2} H_m(u) H_q(v) \phi(u, v, \rho) du dv =$$

$$= \sum_{\nu=0}^{\infty} \frac{\rho^\nu}{\nu!} \int_{\mathbf{R}^1} H_m(u) H_\nu(u) \phi(u) du \int_{\mathbf{R}^1} H_q(v) H_\nu(v) \phi(v) dv = \delta_m^q \rho^m m! \quad \blacksquare$$

We shall use the notation of §§1.4 and 1.5. By Lemma 2.1.1 we obtain for $m \geq 1$ similarly to (1.5.1):

$$\sigma_m^2(\lambda) = E \int_{\Delta(\lambda)} H_m(\xi(x))dx \int_{\Delta(\lambda)} H_q(\xi(y))dy =$$

$$= m! \delta_m^q \int_{\Delta(\lambda)} \int_{\Delta(\lambda)} B^m(|x - y|)dxdy =$$

$$= m! \delta_m^q |\Delta(\lambda)|^2 E B^m(\rho_{\alpha\beta}) = m! \delta_m^q |\Delta|^2 \lambda^{2n} \int_0^{d(\lambda)} B^m(z)dF_{\Delta(\lambda)}(z).$$

We shall write $\sigma_m^2(r)$ instead of $\sigma_m^2(\lambda)$ when $\Delta(\lambda)$ is a ball $v(r)$.

In view of Lemma 1.4.2,

$$\sigma_m^2(r) = c_1 r^n \int_0^{2r} z^{n-1} B^m(z) I_{1-z^2/(2r)^2}(\tfrac{n+1}{2}, \tfrac{1}{2})dz, \qquad (2.1.6)$$

where

$$c_1 = c_1(n, m) = 4m! \pi^n \Gamma^{-2}(\frac{n}{2}) \frac{1}{n} \qquad (2.1.7)$$

and $I_x(p, q)$ is defined in (1.4.8).

Recall that the function $L : (0, \infty) \to (0, \infty)$ is called slowly varying at infinity if for all $\lambda > 0$, $\lim_{t \to \infty} L(\lambda t)/L(t) = 1$. For example, the functions $\ln^a t$, $\ln \ln^a(1 + t)$ vary slowly at infinity for any fixed $a \in \mathbf{R}^1$.

We shall state several well-known properties of slowly varying functions.

Theorem 2.2.1 [51, 104]. Let L be a function which is slowly varying at infinity and integrable over any finite interval. Then: 1) relation $L(\lambda t)/L(t) \to 1$ holds uniformly in $\lambda \in [a, b]$, $0 < a < b < \infty$ as $t \to \infty$; 2) for any $\delta > 0$, $t^\delta L(t) \to \infty$, $t^{-\delta} L(t) \to 0$ as $t \to \infty$; 3) for any $0 < a < b < \infty$

$$\lim_{t \to \infty} \int_a^b \frac{L(ts)}{L(t)} ds = \int_a^b \lim_{t \to \infty} \frac{L(ts)}{L(t)} ds = b - a;$$

4) for all $0 < a < b < \infty$, $\lim_{t \to \infty} \sup_{at < s < bt} |\frac{L(s)}{L(t)} - 1| = 0$; 5) if L is a function integrable in the Riemann or Lebesgue sense (in the latter case the measurability of the function is assumed in its definition) over any finite interval $[a, b]$, then L admits the canonical representation: $L(t) = c(t) \exp\{\int_a^t \epsilon(s)s^{-1}ds\}$ where $c(t)$, $\epsilon(t)$ are functions such that $c(t) \to c > 0$, $\epsilon(t) \to 0$ as $t \to \infty$.

Let $\mathcal{L}$ be the class of functions that are slowly varying at infinity and bounded on each finite interval.

IV. For some $\alpha > 0$, $B(|x|) = |x|^{-\alpha}L(|x|)$, where $|x| > 0$, $|x| \to \infty$ and $L \in \mathcal{L}$.

Lemma 2.1.2. If assumptions I, IV hold for $\alpha \in (0, n/m)$, $n \geq 1$, $m \geq 1$, then there exists a constant

$$c_2 = c_2(n, m, \alpha, \Delta) = m! \int_\Delta \int_\Delta |x - y|^{-\alpha m} dx\, dy =$$

$$= m!|\Delta|^2 \int_0^{d(\Delta)} z^{-\alpha m} dF_\Delta(z), \qquad (2.1.8)$$

such that

$$\sigma_m^2(\lambda) = c_2 \lambda^{2n - m\alpha} L^m(\lambda)(1 + o(1)) \qquad (2.1.9)$$

as $\lambda \to \infty$.

Proof. Transform the variable $u = z/\lambda$ in (2.1.5). Using the consistency of the uniform distribution with a homothety transformation and relation 3) of Theorem 2.1.1, we obtain (2.1.9). ■

If an explicit expression for the distribution function $F_\Delta(z) = F_{\rho_{\alpha\beta}}(z)$ (cf. Lemmas 1.4.2, 1.4.6) is available, one can determine the constant. For example, if $\Delta = v(1)$ is a ball, the change of variables $u = z/zr$ in (2.1.6) and an asymptotic analysis of this formula together with Theorem 2.1.1 yield the following assertion.

Lemma 2.1.3. Let I, IV hold and $r \to \infty$. Then for $\alpha \in (0, n/m)$

$$\sigma_m^2(r) = c_2 r^{2n - m\alpha} L^m(r)(1 + o(1)), \qquad (2.1.10)$$

where $c_2 = c_2(n, m, \alpha, v(1)) = m! 2^{n - m\alpha + 1} \pi^{n - 1/2} \Gamma(\frac{n - m\alpha + 1}{2}) / [(n - m\alpha)\Gamma(\frac{n}{2}) \times \Gamma(\frac{2n - m\alpha + 2}{2})]$. For $\alpha = \frac{n}{m}$,

$$\sigma_m^2(r) = c_1 r^n L^m(r) \ln(2r)(1 + o(1)). \qquad (2.1.11)$$

For $\alpha > \frac{n}{m}$,

$$c_3(n, m) = \int_0^\infty z^{n-1} |B(z)|^m dz < \infty \qquad (2.1.12)$$

and if $c_4 = c_4(n, m) = \int_0^\infty z^{n-1} B^m(z) dz \neq 0$, then

$$\sigma_m^2(r) = c_4 c_1 r^n (1 + o(1)). \qquad (2.1.13)$$

Lemma 2.1.4. Let assumptions I, III, IV hold with $\alpha \in (0, n/m)$, $n \geq 1$, $m \geq 1$. Then

$$\text{var}\left[\int_{\Delta(\lambda)} G(\xi(x))dx\right] = \frac{C_m^2}{(m!)^2}c_2(n, m, \alpha, \Delta) \times$$

$$\times \lambda^{2n-m\alpha} L^m(\lambda)(1 + o(1))$$

as $\lambda \to \infty$.

Proof. From Lemma 2.1.1 and assumption III we represent the variance as

$$\text{var}\left[\int_{\Delta(\lambda)} G(\xi(x))dx\right] = S_1(\lambda) + S_2(\lambda),$$

where in view of Lemma 2.1.2,

$$S_1(\lambda) = \frac{C_m^2}{(m!)^2}\sigma_m^2(\lambda) = \frac{C_m^2}{(m!)^2}c_2\lambda^{2n-m\alpha} L^m(\lambda)(1 + o(1))$$

as $\lambda \to \infty$ and the expression

$$S_2(\lambda) = \sum_{q=m+1}^{\infty} \frac{C_q^2}{(q!)^2}\sigma_q^2(\lambda) \leq$$

$$\leq \frac{\sigma_{m+1}^2(\lambda)}{(m+1)!} \sum_{q=m+1}^{\infty} \frac{C_q^2}{q!}$$

divided by $\lambda^{2n-m\alpha}$ tends to zero as $\lambda \to \infty$. In fact, in view of (2.1.2), $k_1 = \sum_{q=m+1}^{\infty} C_q^2/q! < \infty$. By assumption I, for any $\epsilon > 0$ there exists $A_0 > 0$ such that $|B(|x - y|)| < \epsilon$ for $|x - y| > A_0$. Let

$$\Delta_1 = \{(x, y) \in \Delta(\lambda) \times \Delta(\lambda) : |x - y| \leq A_0\},$$

$$\Delta_2 = \{(x, y) \in \Delta(\lambda) \times \Delta(\lambda) : |x - y| > A_0\};$$

then

$$\frac{\sigma_{m+1}^2(\lambda)}{(m+1)!} = \left\{\int_{\Delta_1} + \int_{\Delta_2}\right\} B^{m+1}(|x - y|)dxdy =$$

$$= S_3(\lambda) + S_4(\lambda).$$

Using the bound $|B(\cdot)|^{m+1} \leq 1$ on the set Δ_1 and the bound $|B(\cdot)|^{m+1} < \epsilon|B(\cdot)|^m$ on the set Δ_2, we obtain

$$|S_3(\lambda)| = \left|\int_{\Delta_1} B^{m+1}(|x - y|)dxdy\right| \leq k_2\lambda^n;$$

$$|S_4(\lambda)| = \left|\int_{\Delta_2} B^{m+1}(|x - y|)dxdy\right| \leq \epsilon k_3 \lambda^{2n-m\alpha},$$

where $k_2 = k_2(n, m, \Delta, A_0) > 0$, $k_3 = k_3(n, m, \alpha, \Delta) > 0$. Since $\epsilon > 0$ is arbitrary, the expression $S_2(\lambda) \leq k_1(k_2\lambda^n + \epsilon\lambda^{2n-m\alpha}k_3)$ divided by $\lambda^{2n-m\alpha}$ tends to zero for $\alpha \in (0, n/m)$ as $\lambda \to \infty$. ∎

Lemma 2.1.5. Let assumptions I, III hold and let (2.1.10) hold for some $\alpha \in (0, \frac{n}{m})$ as $r \to \infty$, where $L \in \mathcal{L}$. Then

$$\text{var}\left[\int_{v(r)} G(\xi(x))dx\right] = \frac{C_m^2}{(m!)^2}c_2(n, m, \alpha, v(1)) \times$$

$$\times r^{2n-m\alpha}L^m(r)(1 + o(1)) \qquad (2.1.14)$$

as $r \to \infty$.

Proof. Relation

$$\text{var}\left[\int_{v(r)} G(\xi(x))dx\right] = \sum_{q=m}^{\infty} \frac{C_q^2}{(q!)^2}\sigma_q^2(r) \qquad (2.1.15)$$

is handled in the same manner as in the proof of Theorem 2.1.4 using (2.1.6) and (2.1.10). ∎

Let $r_1 < r_2$. Utilizing Lemmas 1.4.3, 2.1.1 we obtain for $m \geq 1$, $q \geq 1$,

$$\sigma_m^2(r_1, r_2) = E\int_{v(r_1)} H_m(\xi(x))dx \int_{v(r_2)} H_q(\xi(y))dy =$$

$$= \delta_m^q m!2^{n/2}\pi^n\Gamma^{-1}\left(\frac{n}{2}\right)(r_1r_2)^{n/2}\int_0^{r_1+r_2} B^m(z)z^{n/2} \times$$

$$\times \left[\int_0^{\infty} \rho^{-n/2}J_{(n-2)/2}(z\rho)J_{n/2}(r_1\rho)J_{n/2}(r_2\rho)d\rho\right]dz. \qquad (2.1.16)$$

V. There exists an integer $m \geq 1$ such that

$$c_3(n, m) = \int_0^{\infty} z^{n-1}|B(z)|^m dz < \infty.$$

Lemma 2.1.6. Let assumptions I, III, V hold. Then for all $q \geq m$ (m is given in assumptions III, V) there exist limits $c_5(n, q) = \lim_{r\to\infty}\{\sigma_q^2(r)/r^n\}$ and

$$c_6(n, m) = \sum_{q=m}^{\infty} C_q^2 c_5(n, q)/(q!)^2 < \infty. \qquad (2.1.17)$$

If $c_6(n, m) \neq 0$, then

$$\text{var}\left[\int_{v(r)} G(\xi(x))dx\right] = c_6(n, m)r^n(1 + o(1)) \tag{2.1.18}$$

as $r \to \infty$. For $r_1 < r_2$, $r_1 + r_2 \to \infty$

$$\text{var}\left[\int_{v(r_2)\backslash v(r_1)} G(\xi(x))dx\right] = c_6(n, m)(r_2^n - r_1^n)(1 + o(1)). \tag{2.1.19}$$

Proof. For $q \geq m$ we obtain from I, IV and (2.1.6) that

$$\lim_{r \to \infty} r^{-n}\sigma_q^2(r) = \lim_{r \to \infty} c_1(n, q) \int_0^{2r} z^{n-1}B^q(z)(1 + o(1))dz = c_5(n, q).$$

Since $c_5(n, m) \leq c_5(n, q)$, (2.1.15) and (2.1.2) imply (2.1.18). Relation (2.1.19) is proved using (1.2.16) in the same manner as in Lemma 1.5.1. ∎

VI. There exists an integer $m \geq 1$ such that the function $L_1(r) = r^{-n}\sigma_m^2(r)$ lies in $\mathcal{L}$ and the limits $\lim_{r \to \infty} [\sigma_q^2(r)/\sigma_m^2(r)] = c_7(n, q)$ exist for all $q \geq m$.

Assumption VI holds, for example, when IV holds with $\alpha = n/m$ (cf. (2.1.11)).

Lemma 2.1.7. Under assumptions I, III, VI,

$$c_8(n, m) = \sum_{q=m}^{\infty} C_q^2 c_7(n, q)/(q!)^2 < \infty.$$

If $c_8(n, q) \neq 0$, we have for $r \to \infty$

$$\text{var}\left[\int_{v(r)} G(\xi(x))dx\right] = c_8(n, m)r^n L_1(r)(1 + o(1)).$$

The proof of Lemma 2.1.7 is analogous to that of Lemma 2.1.6.

All the facts presented above for averages over balls become simpler in the case of spherical averages. We state the analogues to Lemmas 2.1.3 and 2.1.4 only. Let $dm(x)$ be defined according to (1.1.2). Using Lemma 1.4.4 we derive for $n \geq 2$

$$\tilde{\sigma}_m^2(r) = E\int_{s(r)} H_m(\xi(x))dm(x)\int_{s(r)} H_q(\xi(y))dm(y) =$$

$$\delta_m^q c_9(n, m)r^{n-1}\int_0^{2r} z^{n-2}\left(1 - \frac{z^2}{4r^2}\right)^{(n-3)/2} B^m(z)dz, \tag{2.1.20}$$

where $c_9(n,m) = m!2^n \pi^{n-1}/(n-2)!$.

Lemma 2.1.8. Let $n \geq 2$, $m \geq 1$, $r \to \infty$ and suppose that assumptions I, IV hold. Then for $\alpha \in (0, (n-1)/m)$,

$$\tilde{\sigma}_m^2(r) = c_{10}(n,m)r^{2n-2-m\alpha}L^m(r)(1+o(1)),$$

where $c_{10}(n,m) = c_9(n,m)2^{n-2-m\alpha}B(\frac{n-1-m\alpha}{2}, \frac{n-1}{2})$. For $\alpha = (n-1)/m$,

$$\tilde{\sigma}_m^2(r) = [c_9(n,m)/4]r^{n-1}L^m(r)\ln(2r)(1+o(1)).$$

For $\alpha > (n-1)/m$,

$$\tilde{\sigma}_m^2(r) = c_9(n,m)c_{11}(n,m)r^{n-1}(1+o(1)),$$

where $c_{11}(n,m) = \int_0^\infty z^{n-2}B^m(z)dz \neq 0$.
Proof. It suffices to change the variable $u = z/2r$ in (2.1.20) and carry out the asymptotic analysis of this integral using Theorem 2.1.1. ∎

Lemma 2.1.9. Let assumptions I, III, IV hold with $\alpha \in (0, (n-1)/m)$, $n \geq 2$, $m \geq 1$. Then as $r \to \infty$,

$$\text{var}\left[\int_{s(r)} G(\xi(x))dm(x)\right] = C_m^2 c_{10}(n,m)(m!)^{-2}r^{2n-2-m\alpha}L^m(r)(1+o(1)).$$

The proof of the lemma is analogous to that of Lemma 2.1.5.

2.2. Reduction Conditions for Strongly Dependent Random Fields

In this section conditions are provided under which the limiting distributions of integrals $\int_{v(r)} G(\xi(x))dx$ for a sufficiently wide class of functions $G(\cdot)$ coincide with the corresponding limiting distributions of the integrals $\int_{v(r)} H_m(\xi(x))dx$ where $H_m(u)$ is the mth Chebyshev-Hermite polynomial and $\xi(x)$ is a homogeneous isotropic Gaussian field such that $\int_{\mathbb{R}^n} |B(x)|dx = \infty$.

Consider a random field $\xi : \Omega \times \mathbb{R}^n \to \mathbb{R}^1$ whose cor.f. satisfies the assumption:

VII. There exist $\delta \in (0,1)$ and an integer $m \geq 1$ such that

$$\lim_{0 \to \infty} \frac{1}{r^{n\delta}} \int_z^{2r} z^{n-1}B^m(z)I_{1-z^2/(2r)^2}\left(\frac{n+1}{2}, \frac{1}{2}\right)dz = \infty$$

with $B(z) \downarrow 0$ as $z \to \infty$.

The function $I_x(p, q)$ in assumption VII is defined in (1.4.8).

Theorem 2.2.1. Let assumptions I, III and VII hold, with $m \geq 1$ given in assumption III. Then the limiting distributions of the r.v.'s.

$$X_r(1) = \left\{ \int_{v(r)} G(\xi(x))dx - C_0 r^n |v(1)| \right\} / [|C_m| \sigma_m(r)/m!]$$

and that of the r.v.'s.

$$X_{m,r}(1) = \operatorname{sgn}\{C_m\} \int_{v(r)} H_m(\xi(x))dx / \sigma_m(r) \qquad (2.2.1)$$

coincide as $r \to \infty$ ($\sigma_m(r)$ is defined in (2.1.6)).*

Proof. We write

$$S(r) = \int_{v(r)} G(\xi(x)dx - C_0 r^n |v(1)| =$$

$$= \frac{C_m}{m!} \int_{v(r)} H_m(\xi(x))dx + \sum_{q=m+1}^{\infty} \frac{C_q}{q!} \int_{v(r)} H_q(\xi(x))dx = S_1(r) + S_2(r).$$

The expansion is in the sense of convergence in the Hilbert space $L_2(\Omega)$ of the random variables ξ with $E\xi^2 < \infty$ and the scalar product $\langle \xi, \eta \rangle = \operatorname{cov}(\xi, \eta)$.

From Lemma 2.1.1, $\operatorname{var} S(r) = \operatorname{var} S_1(r) + \operatorname{var} S_2(r)$. Let $\delta \in (0, 1)$ be as given in VII. Utilizing (2.1.2), the bound $B(z) \leq 1$ for $0 \leq z \leq (2r)^\delta$ and the relation

$$\int_0^{(2r)^\delta} z^{n-1} I_{1-z^2/(2r)^2} \left(\frac{n+1}{2}, \frac{1}{2} \right) dz \leq k_1(n) r^{n\delta}$$

(the integral is computed by changing the variables $u = z/2r$), we obtain

$$\operatorname{var} S_2(r) \leq \sum_{q=m+1}^{\infty} \frac{C_q^2}{q!} \sigma_q^2(r) \leq \max_{q \geq m+1} \left\{ \int_{v(r)} \int_{v(r)} B^q(|x - y|)dxdy \right\} \times$$

$$\times \sum_{q=m+1}^{\infty} \frac{C_q^2}{q!} \leq k_2(n) \max_{q \geq m+1} \left\{ r^n \left[\int_0^{(2r)^\delta} + \int_{(2r)^\delta}^{2r} \right] z^{n-1} B^q(z) \times$$

---

* In statements of theorems of this type, the coincidence of the limiting distributions means the following: if the limiting distribution of one collection of r.v.'s exists, then so does the limiting distribution of the other, and they are equal.

$$\times I_{1-z^2/(2r)^2}\left(\frac{n+1}{2},\frac{1}{2}\right)dz\right\} \le k_3(n)\left\{r^{n(1+\delta)}+\right.$$

$$\left.+r^n\int_{(2r)^\delta}^{2r}z^{n-1}B^{m+1}(z)I_{1-z^2/(2r)^2}\left(\frac{n+1}{2},\frac{1}{2}\right)dz\right\}.$$

Divide the right-hand side of the above inequality by $\sigma_m^2(r)$. In view of the assumption VII,

$$\lim_{r\to\infty}r^{n(1+\delta)}/\sigma_m^2(r)=0,$$

$$\left\{r^n\int_{(2r)^\delta}^{2r}z^{n-1}B^{m+1}(z)I_{1-z^2/(2r)^2}\left(\frac{n+1}{2},\frac{1}{2}\right)dz\right\}\Big/\sigma_m^2(r)\le$$

$$\le k_4(n)\sup_{z\ge(2r)^\delta}\{B(z)\}\to 0,\ r\to\infty.$$

The above bounds imply that $\lim_{r\to\infty}\mathrm{var}\,S_2(r)=0$. Thus, the limiting distribution of $X_r(1)$ is the same as that of $X_{m,r}(1)=S_1(r)/[|C_m|\sigma_m^2(r)(m!)^{-1}]$, as required. ∎

VIII. There exist $\delta\in(0,1)$ and an integer $m\ge 1$ such that

$$\lim_{r\to\infty}\frac{1}{r^{(n-1)\delta}}\int_0^{2r}z^{n-2}B^m(z)\left(1-\frac{z^2}{(2r)^2}\right)^{(n-3)/2}dz=\infty,$$

where $B(z)\downarrow 0$ as $z\to\infty$.

Theorem 2.2.2. Let I, III, VIII hold for $n\ge 2$. Then the limit distributions of the r.v.'s

$$\tilde{X}_r(1)=\frac{\int_{s(r)}G(\xi(x))dm(x)-C_0r^{n-1}|s(1)|}{|C_m|\tilde{\sigma}_m(r)\frac{1}{m!}}$$

and the r.v.'s

$$\tilde{X}_{m,r}(1)=\frac{[\int_{s(r)}H_m(\xi(x))dm(x)]\mathrm{sgn}\{C_m\}}{\tilde{\sigma}_m(r)}$$

coincide as $r\to\infty$, where $\tilde{\sigma}_m(r)$ is given by formula (2.1.20).

The proof of Theorem 2.2.2 is analogous to that of Theorem 2.2.1.

Remark 2.2.1. For $m=1$ and as $r\to\infty$, the distributions of the r.v.'s $X_{m,r}(1)$ and $\tilde{X}_{m,r}(1)$ converge to the standard normal distribution. For $m\ge 2$ the limit distributions are not Gaussian. §2.10 contains examples of the limit distributions of the r.v.'s (2.2.1) and $\tilde{X}_{m,r}(1)$ for $m\ge 2$ (see Theorem 2.10.4).

Remark 2.2.2. Using Lemma 1.4.6, analogues to the theorems of this section are readily obtained for $n = 2$ and the integrals are computed over squares.

We now present some results on the rate of convergence to the normal law for integral functionals of homogeneous isotropic Gaussian random fields under strong dependence.

Let assumptions I, II, III hold with $C_1 \neq 0$, that is, rank $G = 1$. Consider

$$S_r = \left[\int_{v(r)} G(\xi(x))dx - C_0 r^n |v(1)| \right] / [|C_1|\sigma_1(r)],$$

where $\sigma_1^2(r)$ is defined by relation (2.1.6) for $m = 1$.

Introduce the uniform distance between distribution functions

$$\mu_r = \sup_t |P\{S_r < t\} - \Phi(t)|.$$

Theorem 2.2.3. Let assumptions I, II, III (for $m = 1$), IV with $\alpha \in (0, n/2)$ hold. Then as $r \to \infty$ there exists

$$\varlimsup_{r \to \infty} \sqrt[3]{\frac{1}{B(r)}} \mu_r = \varlimsup_{r \to \infty} \sqrt[3]{\frac{r^\alpha}{L(r)}} \mu_r$$

which does not exceed

$$\frac{3}{2} \sqrt[3]{f_1(n, \alpha) f_2(G)/\pi},$$

where

$$f_1(n, \alpha) = \frac{(n - \alpha)\Gamma(\frac{n - 2\alpha + 1}{2})\Gamma(\frac{2n - \alpha + 2}{2})}{(n - 2\alpha)2^\alpha \Gamma(n - \alpha + 1)\Gamma(\frac{n - \alpha + 1}{2})}$$

and

$$f_2(G) = \left[\sum_{q=2}^{\infty} \frac{C_q^2}{q!} \right] \frac{1}{C_1^2} = \left[\int_{-\infty}^{\infty} G^2(u)\phi(u)du - C_0^2 - C_1^2 \right] \frac{1}{C_1^2}.$$

Before proving the theorem, we formulate the following lemma due to V.V. Petrov [106, p.28].

Lemma 2.2.1. Let X, Y be random variables. Then for any $\epsilon > 0$,

$$\sup_t |P\{X + Y < t\} - \Phi(t)| \leq \sup_t |P\{X < t\} - \Phi(t)| + P\{|Y| \geq \epsilon\} + \frac{\epsilon}{\sqrt{2\pi}}.$$

Proof. We now prove Theorem 2.2.3. To apply Lemma 2.2.1, we represent S_r as

$$S_r = X_r + Y_r, \tag{2.2.2}$$

where the random variable

$$X_r = C_1 \int_{v(r)} H_1(\xi(x))dx/[\|C_1|\sigma_1(r)]$$

has the standard normal distribution for any $r > 0$ in view of assumption I and the fact that $H_1(u) = u$ and

$$Y_r = \left[\sum_{q=2}^{\infty}(C_q/q!) \int_{v(r)} H_q(\xi(x))dx\right] / [\|C_1|\sigma_1(r)].$$

Under assumptions I, II, III, IV and $\alpha \in (0, n/2)$, we obtain

$$\text{var } Y_r = \sum_{q=2}^{\infty}(C_q^2/q!) \int_{v(r)} \int_{v(r)} B^q(|x - y|)dx dy/[C_1^2\sigma_1^2(r)] \le$$

$$\le f_2(G) \int_0^{2r} z^{n-1}B^2(z)I_{1-z^2/(2r)^2}\left(\frac{n+1}{2}, \frac{1}{2}\right)dz/ \int_0^{2r} z^{n-1} \times$$

$$\times B(z)I_{1-z^2/(2r)^2}\left(\frac{n+1}{2}, \frac{1}{2}\right)dz.$$

Changing the variables $u = z/2r$ we arrive at

$$\text{var } Y_r \le f_2(G)\frac{L(r)}{r^{\alpha}}K_r(n, \alpha), \qquad (2.2.3)$$

where

$$K_r(n, \alpha) = \left[\int_0^2 \frac{L^2(2ru)}{L^2(r)} \frac{du}{u^{2\alpha}} \int_0^{1-u^2} t^{(n-1)/2}(1 - t)^{-1/2}dt\right] \times$$

$$\times \left[\int_0^2 \frac{L(2ru)}{L(r)} \frac{du}{u^{\alpha}} \int_0^{1-u^2} t^{(n-1)/2}(1 - t)^{-1/2}dt\right]^{-1} \to f_2(n, \alpha). \qquad (2.2.4)$$

We observe that Theorem 2.2.1 has been used above. Applying Lemma 2.2.1 and the Chebyshev inequality, we obtain from (2.2.2), (2.2.3)

$$\mu_r \le f_2(G)B(r)K_r(n, \alpha)\frac{1}{\epsilon^2} + \frac{\epsilon}{\sqrt{2\pi}}.$$

To minimize the right-hand side of the last inequality in $\epsilon > 0$, set

$$\epsilon = \{B(r)2\sqrt{2\pi}f_2(G)K_r(n, \alpha)\}^{1/3}.$$

Thus the following inequality is derived.

$$\mu_r \le \frac{3}{2}\{B(r)K_r(n,\alpha)f_2(G)/\pi\}^{1/3}.$$

The latter relation and (2.2.4) imply the assertion of Theorem 2.2.3. ■

Example 2.2.1. Let $G_1^{(A)}(u) = \chi(u \ge A)$, $A \ge 0$; then $C_0 = 1 - \Phi(A)$, $C_1 = \phi(A)$, $f_2(G_1^{(A)}) = \{\Phi(A)[1-\Phi(A)]-\phi^2(A)\}\phi^{-2}(A)$. We also have, for example, for $A = 0$,

$$C_0 = \frac{1}{2}, \ C_1 = \frac{1}{\sqrt{2\pi}}, \ f_2(G_1^{(0)}) = \frac{\pi - 2}{2}.$$

Consider a set of zero level

$$U_r = |\{x \in v(r) : \xi(x) > 0\}| = \int_{v(r)} \chi(\xi(x) > 0)dx.$$

Then under assumptions I, IV, $\alpha \in (0, n/2)$, there exists

$$\overline{\lim_{r\to\infty}} \sqrt[3]{\frac{1}{B(r)}} \sup_t |P(\{U_r - |v(1)|r^n/2\}/[\sigma_1(r)/\sqrt{2\pi}] < t) - \Phi(t)|,$$

not exceeding

$$\frac{3}{2} \sqrt[3]{f_1(n,\alpha)\frac{(\pi - 2)}{2\pi}}.$$

Example 2.2.2. Let $G_2^{(A)}(u) = \max\{0, u - A\}$, $A \ge 0$.

Integration by parts yields

$$C_0 = \phi(A) - A[1 - \Phi(A)], \ C_1 = 1 - \Phi(A), \ f_2(G_2^{(A)}) = \{(1 - \Phi(A))\times$$
$$\times[\Phi(A) - A^2\Phi(A) + 2A\phi(A)] - \phi(A)A - \phi^2(A)\}[1 - \Phi(A)]^{-2}.$$

Thus, with $A = 0$,

$$C_0 = \frac{1}{\sqrt{2\pi}}, \ C_1 = \frac{1}{2}, \ f_2(G_2^{(0)}) = \frac{\pi - 2}{\pi}.$$

Consider the functional

$$V_r = \int_{v(r)} \max\{0, \xi(x)\}dx.$$

Theorem 2.2.3. implies that under assumptions I, IV and $\alpha \in (0, n/2)$ there exists

$$\overline{\lim_{r\to\infty}} \sqrt[3]{\frac{1}{B(r)}} \sup_t |P\{[V_r - r^n|v(1)|/\sqrt{2\pi}]/[\sigma_1(r)/2] < t\} - \Phi(t)|$$

not exceeding

$$\frac{3}{2} \sqrt[3]{f_1(n,\alpha)\frac{(\pi - 2)}{\pi^2}}.$$

Now let assumption I hold but not necessarily assumption IV.

Theorem 2.2.4. Let assumptions I, II, III (with $m = 1$) hold and suppose that there exists $\delta \in (0,1)$ such that as $r \to \infty$

$$r^{n(1+\delta)}/\sigma_1^2(r) \to 0, \quad B(r) \downarrow 0. \tag{2.2.5}$$

Then for any $r > 0$

$$\mu_r \leq \sqrt[3]{\Psi(r)} \frac{3}{2} \sqrt[3]{\frac{f_2(G)}{\pi}},$$

where

$$\Psi(r) = B((2r)^\delta) + \frac{2^{n\delta} r^{n\delta}}{n} \left\{ \int_0^{2r} z^{n-1} B(z) I_{1-z^2/4r^2} \left(\frac{n+1}{2}, \frac{1}{2} \right) dz \right\}^{-1}.$$

Remark 2.2.3. Under condition (2.2.5), $\lim_{r \to \infty} \Psi(r) = 0$. Under assumption IV and $\alpha \in (0,n)$, condition (2.2.5) holds for $\delta \in (0, 1 - \frac{\alpha}{n})$. If $\alpha > n$, condition (2.2.5) does not hold. We now prove Theorem 2.2.4.

Proof. In the same manner as in the proof of Theorem 2.2.3 we obtain the inequality

$$\sum_{q=2}^\infty \frac{C_q^2}{q!} \int_{v(r)} \int_{v(r)} B^q(|x-y|) dx dy \leq$$

$$\leq c_1(n,1) \left(\sum_{q=2}^\infty \frac{C_q^2}{q!} \right) r^n \int_0^{2r} z^{n-1} B^2(z) I_{1-z^2/4r^2} \left(\frac{n+1}{2}, \frac{1}{2} \right) dz. \tag{2.2.6}$$

Choose $\delta \in (0,1)$ from condition (2.2.5). Then

$$\left[\int_0^{(2r)^\delta} + \int_{(2r)^\delta}^{2r} \right] z^{n-1} B^2(z) I_{1-z^2/4r^2} \left(\frac{n+1}{2}, \frac{1}{2} \right) dz \leq$$

$$\leq \frac{2^{n\delta} r^{n\delta}}{n} + \sup\{ B(z), \ z \geq (2r)^\delta \} \int_{(2r)^\delta}^{2r} z^{n-1} B(z) I_{1-z^2/4r^2} \left(\frac{n+1}{2}, \frac{1}{2} \right) dz. \tag{2.2.7}$$

Represent S_r in the form (2.2.2). In view of (2.2.6), (2.2.7) we obtain

$$\operatorname{var} Y_r = \sum_{q=2}^\infty \frac{C_q^2}{q!} \int_{v(r)} \int_{v(r)} B^q(|x-y|) dx dy / [C_1^2 \sigma_1^2(r)] \leq$$

$$\leq f_2(G) \left\{ \frac{(2r)^{n\delta}}{n} \left[\int_0^{2r} z^{n-1} B(z) I_{1-z^2/4r^2} \left(\frac{n+1}{2}, \frac{1}{2} \right) dz \right]^{-1} + \right.$$

$$+B((2r)^\delta)\tilde{K}_r\Big\} = f_2(G)\Psi(r), \qquad\qquad (2.2.8)$$

since for any $r > 0$

$$\tilde{K}_r = \int_{(2r)^\delta}^{2r} z^{n-1}B(z)I_{1-z^2/4r^2}(\tfrac{n+1}{2}, \tfrac{1}{2})dz \Big/ \int_0^{2r} z^{n-1}B(z)\times$$

$$\times I_{1-z^2/4r^2}(\tfrac{n+1}{2}, \tfrac{1}{2})dz \leq 1.$$

Applying Lemma 2.2.1, we have

$$\mu_r \leq f_2(G)\Psi(r)\frac{1}{\epsilon^2} + \frac{\epsilon}{\sqrt{2\pi}}.$$

Choosing $\epsilon = \{\Psi(r)2\sqrt{2\pi}f_2(G)\}^{1/3}$, we obtain the inequality

$$\mu_r \leq \sqrt[3]{\Psi(r)}\,\frac{3}{2}\sqrt[3]{\frac{f_2(G)}{\pi}}. \quad\blacksquare$$

We now state analogues to Theorems 2.2.3 and 2.2.4 for spherical averages. Let assumptions I, II, III hold with $C_1 \neq 0$, that is, rank $G = 1$. Consider

$$\tilde{S}_r = \Big[\iint_{s(r)} G(\xi(x))dm(x) - C_0 r^{n-1}|s(1)|\Big]\,[|C_1|\tilde{\sigma}_1(r)]^{-1}$$

where $\tilde{\sigma}_1^2(r)$ is given by formula (2.1.20).

Let

$$\tilde{\mu}_r = \sup_t |P\{\tilde{S}_r < t\} - \Phi(t)|.$$

Theorem 2.2.5. Let assumptions I, II, III (with $m = 1$), IV (with $\alpha \in (0, \frac{n-1}{2})$) hold for $n \geq 2$. Then as $r \to \infty$,

$$\varlimsup_{r\to\infty} \sqrt[3]{\frac{1}{B(r)}}\tilde{\mu}_r$$

exists and does not exceed

$$\frac{3}{2}\sqrt[3]{\frac{f_3(n,\alpha)f_2(G)}{\pi}},$$

where

$$f_3(n,\alpha) = \frac{\Gamma(\frac{n-1-2\alpha}{2})\Gamma(n-1-\frac{\alpha}{2})}{2^\alpha\Gamma(n-1-\alpha)\Gamma(\frac{n-1-\alpha}{2})}.$$

The proof of Theorem 2.2.5 is analogous to that of Theorem 2.2.3 using formula (2.1.20) instead of formula (2.1.6).

Theorem 2.2.6. Let assumptions I, II, III (with $m = 1$) hold for $n \geq 2$ and let $\delta \in (0,1)$ exist such that as $r \to \infty$

$$r^{(n-1)\delta} \Big/ \int_0^{2r} z^{n-2} B(z) \Big(1 - \Big(\frac{z}{2r} \Big)^2 \Big)^{(n-3)/2} dz \to 0, \quad B(r) \downarrow 0.$$

Then for any $r > 0$,

$$\tilde{\mu}_r \leq \sqrt[3]{\tilde{\psi}(r)} \, \frac{3}{2} \sqrt[3]{\frac{f_2(G)}{\pi}},$$

where

$$\tilde{\psi}(r) = B((2r)^\delta) + W(r),$$

$$W(r) = 2^\delta r^\delta \sqrt{1 - r^{2(\delta-1)}} \Big/ \int_0^{2r} B(z) \Big(1 - \Big(\frac{z}{2r} \Big)^2 \Big)^{-1/2} dz$$

for $n = 2$, and

$$W(r) = 2^{(n-1)\delta} \frac{1}{n-1} r^{(n-1)\delta} \Big/ \int_0^{2r} z^{n-2} B(z) \Big(1 - \Big(\frac{z}{2r} \Big)^2 \Big)^{(n-3)/2} dz$$

for $n \geq 3$.

The proof of Theorem 2.2.6 is similar to that of Theorem 2.2.4.

We now provide some generalizations. Under the conditions of Theorems 2.2.3–2.2.6, non-uniform bounds may be derived. As an example, consider Theorem 2.2.4.

Theorem 2.2.7. Let assumptions I, II, III (with $m = 1$), IV (with $\alpha \in (0, n/2)$) hold. Suppose that the inequality $0 < \Delta_r < \frac{1}{\sqrt{e}}$ holds for some r_0 with $r \geq r_0$. Then for all t and $r \geq r_0$,

$$|P\{S_r < t\} - \Phi(t)| \leq \frac{1}{1+t^2} \Big[2(5 + \sqrt{\frac{e}{\pi}} \Gamma(\tfrac{3}{2})) \Delta_r \log \frac{1}{\Delta_r} + \Psi(r) f_2(G) \Big],$$

where $\Psi(r)$ is defined in Theorem 2.2.4.

Proof. Apply Theorem 9 from [106,p.151] with the distribution function of S_r in place of $F(t)$. Then (2.2.8) implies

$$\int_{-\infty}^{\infty} t^2 dF(t) = 1 + \operatorname{var} Y_n \leq 1 + f_2(G)\Psi(r).$$

We have only to take $p = 2$ in the notation of the above mentioned theorem and note that

$$\lambda_p = \lambda_2 = \left| \int_{-\infty}^{\infty} |t|^2 dF(t) - \int_{-\infty}^{\infty} |t|^2 d\Phi(t) \right| \le f_2(G)\Psi(r). \quad \blacksquare$$

Theorems 2.2.3–2.2.6 may be extended to the case of sets other than balls. As an example consider Theorem 2.2.3.

Under assumptions I, II, III (with $m = 1$) consider

$$\zeta_r(\Delta) = \left[\int_{\Delta(\lambda)} G(\xi(x))dx - C_0 \lambda^n |\Delta| \right] \Big/ [|C_1|\sigma_1(\lambda)],$$

where in the notation of §2.1.1,

$$\sigma_1^2(\lambda) = |\Delta|^2 \lambda^{2n-\alpha} L(\lambda) \int_0^{d(\Delta)} \left[\frac{L(\lambda u)}{L(u)} \right] \frac{dF_\Delta(u)}{u^\alpha}, \quad \lambda > 0.$$

Theorem 2.2.8. Let assumptions I, II, III (with $m = 1$), IV (with $\alpha \in (0, n/2)$) hold. Then as $\lambda \to \infty$ there exists

$$\varlimsup_{\lambda \to \infty} \sqrt[3]{\frac{1}{B(\lambda)}} \sup_t |P\{\zeta_r(\Delta) < t\} - \Phi(t)|,$$

not exceeding $\frac{3}{2} \sqrt[3]{f_3(n, \alpha, \Delta) f_2(G)/\pi}$,

$$f_3(n, \alpha, \Delta) = \left[\int_0^{d(\Delta)} \frac{dF_\Delta(z)}{z^{2\alpha}} \right] \Big/ \left[\int_0^{d(\Delta)} \frac{dF_\Delta(z)}{z^\alpha} \right].$$

Remark 2.2.4. Clearly $f_3(n, \alpha, v(1)) = f_1(n, \alpha)$. In general, the constant $f_3(n, \alpha, \Delta)$ depends on the geometry of the set Δ.

The proof of Theorem 2.2.8 is similar to that of Theorem 2.2.3.

2.3. Central Limit Theorem for Non-Linear Transformations of Gaussian Fields

We shall investigate the conditions for asymptotic normality of spherical averages of local functionals of weakly dependent Gaussian fields.

Theorem 2.3.1. Let assumptions I, III, V hold with $c_6(n, m) > 0$. The finite-dimensional distributions of the random processes

$$Y_r(t) = \frac{1}{r^{n/2}} \left\{ \int_{v(rt^{1/n})} G(\xi(x))dx - C_0 r^n t |v(1)| \right\}, \quad t \in [0, 1]$$

converge as $r \to \infty$ to finite-dimensional distributions of the Brownian motion process $w_b(t)$, $t \in [0,1]$, where $b = c_6(n,m)$ is defined in (2.1.17).

Proof. We shall subdivide the proof into stages.

1. For any $q \geq 1$ choose points $0 = t_0 < t_1 < \ldots < t_q \leq 1$. Set $\zeta_r = \sum_{j=1}^q \lambda_j Y_r(t_j) = \sum_{j=1}^q a_j (Y_r(t_j) - Y_r(t_{j-1}))$, where $a_j = \lambda_j + \ldots + \lambda_q$, $\lambda_1, \ldots, \lambda_q \in \mathbf{R}^1$. Consider the r.v.

$$\zeta = \sum_{j=1}^q \lambda_j w_b(t_j) = \sum_{j=1}^q a_j (w_b(t_j) - w_b(t_{j-1})).$$

By the Markov moment method, it suffices to show that for any integer $p \geq 2$

$$\lim_{r \to \infty} E\zeta_r^p = E\zeta^p = \begin{cases} (p-1)!! [\sum_{j=1}^q b a_j^2 (t_j - t_{j-1})]^{p/2}, & p = 2\nu, \\ 0, & p = 2\nu + 1, \end{cases}$$
$$(2.3.1)$$

where $\nu = 1, 2, \ldots$.

2. Assume without loss of generality that $C_0 = 0$. Let

$$A_j = v(rt_j^{1/n}) \backslash v(rt_{j-1}^{1/n}), \quad \eta_j^l(r) = \int_{A_j} H_l(\xi(x)) dx, \quad \gamma_j^l(r) = \operatorname{var} \eta_j^l(r).$$

Note that in view of (2.1.18),

$$\lim_{r \to \infty} \{ r^{-n} \gamma_j^l(r) \} = c_6(n,l)(t_j - t_{j-1}). \tag{2.3.2}$$

Write $\zeta_r = \zeta_r' + \zeta_r''$ for $N \geq m$, where

$$\zeta_r' = \sum_{j=1}^q a_j \sum_{l=1}^N C_l r^{-n/2} \eta_j^l(r) / l!,$$

$$\zeta_r'' = \sum_{j=1}^q \lambda_j \sum_{l>N} C_l r^{-n/2} (l!)^{-1} \int_{v(rt_j^{1/n})} H_l(\xi(x)) dx. \tag{2.3.3}$$

According to (2.1.4)

$$\operatorname{var} \zeta_r'' \leq 2 \sum_{j=1}^q \lambda_j^2 \sum_{l>N} C_l^2 \sigma_l^2 (rt_j^{1/n}) r^{-n} / (l!)^2 \leq$$

$$\leq k_1 \left(\sum_{l>N} \frac{C_l^2}{l!} \right) B^{N-m}(0) \int_0^{2r} z^{n-1} |B(z)|^m dz. \tag{2.3.4}$$

In view of (2.1.2) and also V, we obtain from (2.3.4) that for any $\epsilon > 0$ there exists N_0, r_0 such that for $l > N_0$, $r > r_0$, $\operatorname{var} \zeta_r'' < \epsilon$. Henceforth we assume that $N = N_0$ in (2.3.3). It suffices then to show that (2.3.1) holds provided ζ_r is substituted by ζ_r' and b by $b' = \sum_{\nu=m}^{N} C_\nu^2 c_5(n,\nu)/(\nu!)^2$.

3. A graph Γ with $u_1 + \ldots + u_p$ vertices is called a diagram of order $(u_1, \ldots, u_p)$ if:

a) the set of vertices V of the graph Γ is of the form $V = \bigcup_{j=1}^{p} W_j$, where $W_j = \{(j,l) : 1 \leq l \leq u_j\}$ is the jth level of the graph Γ, $1 \leq j \leq p$ (if $u_j = 0$ assume $W_j = \emptyset$);

b) each vertex is of degree 1;

c) if $((j_1,l_1),(j_2,l_2)) \in \Gamma$ then $j_1 \neq j_2$, that is, the edges of the graph Γ may only connect different levels.

Let $T = T(u_1, \ldots, u_p)$ be a set of diagrams Γ of order $(u_1, \ldots, u_p)$. Denote by $R(V)$ the set of edges of a graph $\Gamma \in T$. For the edge $w = ((j_1,l_1),(j_2,l_2)) \in R(V)$, $j_1 < j_2$, we set $d_1(w) = j_1$, $d_2(w) = j_2$. We call a diagram Γ regular if its levels can be split into pairs in such a manner that no edge connects the levels belonging to different pairs. The set of regular diagrams $T^* \subseteq T(u_1, \ldots, u_p)$. If p is odd, then $T^* = \emptyset$. The following assertion will be used [156, 222].

Let $(\xi_1, \ldots, \xi_p)$, $p \geq 2$, be a Gaussian vector having $E\xi_j = 0$, $E\xi_j^2 = 1$, $E\xi_i\xi_j = r(i,j)$, $i,j = 1, \ldots, p$, and $H_{l_1}(u), \ldots, H_{l_p}(u)$ the Chebyshev-Hermite polynomials. Then $E\{\prod_{j=1}^{p} H_{l_j}(\xi_j)\} = \sum_{\Gamma \in T} I_\Gamma$, where $T = T(l_1, \ldots, l_p)$ and $I_\Gamma = \prod_{w \in R(V)} r(d_1(w), d_2(w))$.

The above diagram formula coincides for $p = 2$ with (2.1.4).

4. Let $D_p = \{(J,L) : J = (j_1, \ldots, j_p), 1 \leq j_i \leq q, i = 1, \ldots, p; L = (l_1, \ldots, l_p), 1 \leq l_\nu \leq N, \nu = 1, \ldots, p\}$, $K(J,L) = \prod_{i=1}^{p}\{a_{j_i} C_{l_i}/l_i!\}$, $\int^{(p)} \ldots = \prod_{j=1}^{p} \int_{A_j}, \ldots$. Then according to the diagram formula we obtain from (2.3.3) for $p \geq 2$:

$$E(\zeta_r')^p = \sum_{D_p} K(J,L) r^{-np/2} \int^{(p)} E\left\{\prod_{j=1}^{p} H_{l_j}(\xi(x^{(j)}))\right\} dx^{(1)} \ldots dx^{(p)} =$$

$$= \sum_{D_p} K(J,L) r^{-np/2} \int^{(p)} \left\{\sum_{\Gamma \in T} \prod_{w \in R(V)} B(x^{(d_1(w))} - x^{(d_2(w))})\right\} dx^{(1)} \ldots dx^{(p)} =$$

$$= \sum_{D_p} K(J,L) \sum_{\Gamma \in T} F_\Gamma(J,L,r), \qquad (2.3.5)$$

where

$$F_\Gamma(J,L,r) = r^{-np/2} \int^{(p)} \prod_{w \in R(V)} B(x^{(d_1(w))} - x^{(d_2(w))}) dx^{(1)} \ldots dx^{(p)}.$$

Let T^* be the set of regular diagrams. Split the sum $\sum_{\Gamma \in T} \ldots = \sum_{\Gamma \in T^*} \ldots + \sum_{\Gamma \in T \setminus T^*} \ldots$. Then the sum (2.3.5) is subdivided into two parts. The first one corresponding to $\sum_{\Gamma \in T^*} \ldots$ is denoted by $\sum_p^*(r)$ and the second one corresponding to $\sum_{\Gamma \in T \backslash T^*} \ldots$ is denoted by $\sum_p(r)$. We shall study their behaviour as $r \to \infty$ separately.

5. If $p = 2\nu+1$, $\nu = 1, 2, \ldots$, $T^* = \emptyset$ and $\lim_{r \to \infty} \sum_p^*(r) = 0$. For even $p = 2\nu$ and a regular diagram $\Gamma \in T^*$ consider the pairs $(i(1), i(2)), \ldots, (i(p-1), i(p))$, where $(i(1), \ldots, i(p))$ is a permutation of the set $(1, \ldots, p)$ such that the graph edges connect only neighbouring levels $i(2m - 1)$ and $i(2m)$, $m = 1, 2, \ldots, \nu$. Suppose that the levels $i(2m - 1)$ and $i(2m)$ have cardinality $r(m)$, $m = 1, \ldots, \nu$. Then the integral $\int^{(p)} \ldots$ is subdivided into a product of pairs of integrals. Similarly to (2.3.2),

$$\lim_{r \to \infty} r^{-n} \int_{A_{j_{i(2m-1)}}} \int_{A_{j_{i(2m)}}} B^{r(m)}(|u - v|) du dv =$$

$$= \delta_{j_{i(2m-1)}}^{j_{i(2m)}} c_5(n, r(m))(t_{j_{i(2m)}} - t_{j_{i(2m)}-1}). \qquad (2.3.6)$$

Whence

$$F_\Gamma(J, L, r) = \prod_{m=1}^{\nu} \{ r^{-n} \int_{A_{j_{i(2m-1)}}} \int_{A_{j_{i(2m)}}} B^{r(m)}(|u - v|) du dv \},$$

and from (2.3.6) follows:

$$\lim_{r \to \infty} F_\Gamma(J, L, r) = \begin{cases} \prod_{m=1}^{\nu} c_5(n, r(m))(t_{j_{i(2m)}} - t_{j_{i(2m)}-1}), & \text{if } j_{i(2m)} = j_{i(2m-1)}, \\ & m = 1, \ldots, \nu, \\ 0 & \text{otherwise.} \end{cases}$$

Hence

$$I(\Gamma) = \lim_{r \to \infty} \sum_J K(J, L) F_\Gamma(J, L, r) =$$

$$= \left(\sum_{j=1}^{q} a_j^2 (t_j - t_{j-1}) \right)^{\nu} \prod_{m=1}^{\nu} C_{r(m)}^2 c_5(n, r(m)) / \{ [r(m)]! \}^2$$

and $\lim_{r \to \infty} \sum_p^*(r) = \sum_{\Gamma \in T^*} I_\Gamma$, where the summation is carried out over all the regular diagrams having p levels.

The number of regular diagrams containing $2m_j$ levels of cardinality r_j, $j = 1, \ldots, s$, $\sum_{j=1}^{s} m_j = \nu$ (all the m_j are different) is equal to

$$\frac{(2\nu)!}{(2m_1)! \ldots (2m_s)!} \left[\prod_{j=1}^{s} (2m_j - 1)(2m_j - 3) \ldots 3 \cdot 1 \right] \prod_{i=1}^{s} (r_i!)^{m_i} =$$

$$= \frac{(2\nu - 1)!!\nu!}{m_1!\dots m_s!} \prod_{i=1}^{s}(r_i!)^{m_i}.$$

For such diagrams (with fixed s and r_i, $i = 1, \dots, s$) we have

$$I(\Gamma) = \left(\sum_{j=1}^{q} a_j^2(t_j - t_{j-1})\right)^{\nu} \prod_{i=1}^{s}\left\{\frac{C_{r(i)}^2 c_5(n, r(i))}{[r(i)!]^2}\right\}^{m_i}.$$

Recalling that $p = 2\nu$ we obtain

$$\lim_{r \to \infty}\sideset{}{^*}\sum_p(r) = \sum_{1 \le s \le \nu}\sum_{\substack{m_1 + \dots + m_s = \nu \\ 1 \le i \le s}}\sum_{1 \le r_i \le N}(2\nu - 1)!!\frac{\nu!}{m_1!\dots m_s!} \times$$

$$\times \left(\sum_{j=1}^{q} a_j^2(t_j - t_{j-1})\right)^{\nu} \prod_{i=1}^{s}\left\{\frac{C_{r(i)}^2 c_5(n, r(i))}{[r(i)!]^2}\right\}^{m_i} =$$

$$= (p - 1)!!\left(\sum_{j=1}^{q} a_j^2(t_j - t_{j-1})\right)^{p/2}\left(\sum_{i=m}^{N} C_i^2 c_5(n, i)/(i!)^2\right)^{p/2}.$$

6. We show that

$$\lim_{r \to \infty}\sum_p(r) = 0. \tag{2.3.8}$$

The assertion of the theorem will follow from (2.3.7), since due to (2.3.5), the relation $\lim_{r \to \infty}(\zeta_r')^p = \lim_{r \to \infty}\sideset{}{^*}\sum_p(r)$ follows from (2.3.8).

By introducing a permutation π of the set $(1, \dots, p)$, the diagram $\Gamma \in T(l_1, \dots, l_p)$ may be transformed to a diagram $\pi\Gamma$ which possesses the following properties: the $\pi(j)$th level of diagram $\pi\Gamma$ has the cardinality l_j (that is, the cardinality of level j of diagram Γ), $j = 1, \dots, p$, and $w = ((j_1, \nu_1), (j_2, \nu_2)) \in R(V)$ if and only if $\pi(w) = ((\pi(j_1), \nu_1), (\pi(j_2), \nu_2)) \in \pi R(V)$. Denote by $q_\Gamma(j)$ the number of edges $w \in R(V)$ such that $d_1(w) = j$. Note that for $\Gamma \in T(l_1, \dots, l_p)$, $J = (j_1, \dots, j_p)$, $L = (l_1, \dots, l_p)$,

$$F_\Gamma(J, L, r) = F_{\pi\Gamma}(\pi(J), \pi(L), r), \tag{2.3.9}$$

where $\pi(J) = (\pi(j_1), \dots, \pi(j_p))$, $\pi(L) = (\pi(l_1), \dots, \pi(l_p))$. For any diagram Γ a permutation π exists such that $\Gamma' = \pi\Gamma$ possesses the property

$$\Gamma' \in T(l_1', \dots, l_p'), \quad l_1' \le l_2' \le \dots \le l_p'. \tag{2.3.10}$$

In view of (2.3.9) it suffices to prove (2.3.8) for a diagram with property (2.3.10). We write

$$|F_\Gamma(J, L, r)| \le r^{-np/2}\int_{Aj_1}\dots\int_{Aj_p}\prod_{i=1}^{p} \times$$

$$\times \prod_{\substack{w \in R(V) \\ d_1(w)=i}} |B(|x^{(i)} - x^{(d_2(w))}|)| dx^{(1)} \dots dx^{(p)}. \qquad (2.3.11)$$

The inner product in (2.3.11) contains $q_\Gamma(i)$ terms with

$$\prod_{\substack{w \in R(V) \\ d_1(w)=i}} |B(|x^{(i)} - x^{(d_2(w))}|)| \le$$

$$\le \frac{1}{q_\Gamma(i)} \sum_{\substack{w \in R(V) \\ d_1(w)=i}} |B(|x^{(i)} - x^{(d_2(w))}|)|^{q_\Gamma(i)}.$$

Fix $x^{(2)}, \dots, x^{(p)}$ and integrate (2.3.11) with respect to $dx^{(1)}$. Proceed in this manner with respect to $dx^{(2)}, \dots, dx^{(p)}$ having chosen k_2 such that $v(k_2 r) \supset A_{j\nu}$, $\nu = 1, \dots, p$. Since $y - z \in v(2k_2 r)$ provided that $y \in A_{j\nu}$, $\nu = 1, \dots, p$, $z \in v(k_2 r)$, we have

$$|F_\Gamma(J, L, r)| \le r^{-np/2} \prod_{i=1}^{p} \int_{v(2k_2 r)} |B(|y|)|^{q_\Gamma(i)} dy \le$$

$$\le k_3 r^{-np/2} \prod_{i=1}^{p} \int_0^{2k_2 r} z^{n-1} |B(z)|^{q_\Gamma(i)} dz. \qquad (2.3.12)$$

If $l_i \ge m$, then

$$\int_0^{2k_2 r} z^{n-1} |B(z)|^{q_\Gamma(i)} dz \le k_4 r^{[1-g(i)]n}, \qquad (2.3.13)$$

with $q_\Gamma(i) = 0$ or $q_\Gamma(i) = l_i$, $g(i) = q_\Gamma(i)/l_i$. On the other hand,

$$\int_0^{2k_2 r} z^{n-1} |B(z)|^{q_\Gamma(i)} dz = o(r^{[1-g(i)]n}), \quad 0 < q_\Gamma(i) < l_i, \qquad (2.3.14)$$

since assumption V implies the existence of an r_ϵ such that $\int_{r_\epsilon}^{\infty} z^{n-1} |B(z)|^{l_i} dz < \epsilon$, $\epsilon > 0$. Hence it follows from Hölder's inequality that

$$\int_0^{2k_2 r} z^{n-1} |B(z)|^{q_\Gamma(i)} dz \le k_5(\epsilon) + \epsilon^{g(i)} (4k_2 r)^{(1-g(i))n}.$$

Thus denoting $\mu = n[p/2 - \sum_{1 \le i \le p} g(i)]$ we see from (2.3.13) and (2.3.12) that $|F_\Gamma(J, L, r)| \le k_6 r^\mu$. From (2.3.14) and (2.3.12) it follows that $|F_\Gamma(J, L, r)| = o(r^\mu)$ provided $0 < q_\Gamma(i) < l_i$ for some i. If $\Gamma \in T \backslash T^*$, then either $0 < q_\Gamma(i) < l_i$ for some i or the diagram contains an edge connecting levels of differing cardinality. Choose an edge $w \in R(V)$ and define the numbers $p_1(w)$ and

$p_2(w)$ as the cardinalities of levels $d_1(w)$ and $d_2(w)$ respectively. In view of (2.3.10), $p_1(w) \leq p_2(w)$ for any $w \in R(V)$. Taking into account the definition of $g(i)$, we obtain

$$2 \sum_{i=1}^{p} \frac{q_\Gamma(i)}{l_i} = 2 \sum_{w \in R(V)} \frac{1}{p_1(w)} \geq \sum_{w \in R(V)} \left\{ \frac{1}{p_1(w)} + \frac{1}{p_2(w)} \right\} = p, \qquad (2.3.15)$$

since the quantity $1/l_j$ appears exactly l_j times in the summation of $1/p_1(w)$ and $1/p_2(w)$. If the diagram contains edges connecting levels of different cardinality, then (2.3.15) is a strict inequality. Formula (2.3.15) now implies that $\lim_{r \to \infty} F_\Gamma(J, L, r) = 0$ with $\Gamma \in T \backslash T^*$ for any J, which yields (2.3.8). ∎

Theorem 2.3.2. Let assumptions I, III, VI hold with $c_8(n, m) \neq 0$. Then the r.v.

$$\widetilde{Y}_r(1) = \frac{1}{r^{n/2} L_1^{1/2}(r)} \left\{ \int_{v(r)} G(\xi(x)) dx - C_0 r^n |v(1)| \right\}$$

has an asymptotically normal distribution $N(0, c_8(n, m))$ as $r \to \infty$.

Proof. The outline of the proof is similar to that of Theorem 2.3.1. Slight differences are due to the fact that only one-dimensional distributions appear and $t = 1$. These differences merely simplify the proof. Also, assumption VI and Theorem 2.1.1 are used. For example, the right-hand side of (2.3.13) is multiplied by $[L_1(r)]^{g(i)}$. When proving (2.3.14), the integral should be subdivided into two parts:

$$\int_0^{2k_2 r} z^{n-1} |B(z)|^{q\Gamma(i)} dz = \left[\int_0^{\epsilon r} + \int_{\epsilon r}^{2k_2 r} \right] z^{n-1} |B(z)|^{q\Gamma(i)} dz = S_1 + S_2.$$

Next we apply Hölder's inequality to both summands and use the relation

$$\int_{\epsilon r}^{2k_2 r} z^{n-1} |B(z)|^{l_i} dz = o(L_1(r)) \text{ as } r \to \infty$$

to estimate S_2. ∎

Let assumption V hold and $\alpha \in (n/3, n/2)$. If the functionals $\eta_i(x) = G_i(\xi(x))$, $i = 1, 2, 3$, are such that rank $G_i = i$, then in view of Theorems 2.2.1 and 2.3.1, the central limit theorem holds for $G_1(\xi(x))$ and $G_3(\xi(x))$ but not for $G_2(\xi(x))$ (cf. the result of §1.7).

2.4. Approximation for Distribution of Geometric Functionals of Gaussian Fields

We shall examine "random volumes" constrained by a Gaussian surface as well as measures of the excess over a fixed level.

Consider the functionals

$$V_1(r) = \int_{v(r)} \max\{0, \xi(x)\}dx, \quad V_2(r) = \int_{v(r)} |\xi(x)|dx$$

which have a clear geometric meaning for continuous random fields when $n = 2$. For example, for $n = 2$, $V_1(r)$ is a "random volume" constrained by a realization of the field $\xi(x)$, $x \in \mathbf{R}^2$, a strip of a circle $v(r) \subset \mathbf{R}^2$ and a cylindrical surface passing through the boundary of this circle perpendicular to the plane of the latter.

Theorem 2.4.1. If assumptions I, III (with $m = 1$) hold, then the r.v.

$$\frac{V_1(r) - (2\pi)^{-1/2}r^n|v(1)|}{\sigma_1(r)/2} \xrightarrow{\mathcal{D}} N(0, 1)$$

as $r \to \infty$, where $\sigma_1(r)$ is defined in (2.1.6). In particular, if IV holds with $\alpha \in (0, n)$, then $\sigma_1(r)$ is defined by (2.1.10) with $m = 1$.

Proof. We calculate the first two coefficients of the expansion of the function $G(u) = \max\{0, u\}$ in a series in Chebyshev-Hermite polynomials. These are respectively $C_0 = \frac{1}{\sqrt{2\pi}}$ and $C_2 = \frac{1}{2}$. Thus, Theorem 2.4.1 is a corollary to Theorem 2.2.1 with $m = 1$. A more precise result is given by Theorem 2.2.4 and Example 2.2.2. ∎

Theorem 2.4.2. If assumptions I, V (with $m = 1$) hold, then

$$\frac{V_1(r) - (2\pi)^{-1/2}r^n|v(1)|}{r^{n/2}} \xrightarrow{\mathcal{D}} N(0, c_{12}(n))$$

as $r \to \infty$, where

$$c_{12}(n) = n|v(1)|^2 \{ \frac{1}{4} \int_0^\infty z^{n-1}B(z)dz + \sum_{j=2}^\infty \frac{[\Phi^{(j-1)}(0)]^2}{j!} \times$$
$$\times \int_0^\infty z^{n-1}B^j(z)dz \}, \quad \Phi(z) = \int_{-\infty}^z \phi(u)du.$$

Proof. In view of (2.1.3),

$$E[V_1(r)]^2 = \int_{v(r)} \int_{v(r)} dx dy \int_0^\infty \int_0^\infty us\phi(u, s, B(|x - y|))duds.$$

Using the well-known identity [76,§10.8] we have

$$\int_u^\infty \int_u^\infty (v - u)^q(w - u)^q \phi(u, w, \rho)dudw = \sum_{j=0}^q \frac{\rho^j}{j!} \times$$

$$\times \left[\frac{1}{q!(q-j)!} \int_u^\infty (v-u)^{q-j} \phi(v) dv \right]^2 + (q!)^2 \sum_{j=q+1}^\infty \frac{\rho^j}{j!} [\Phi^{(j-q)}(u)]^2; \quad (2.4.1)$$

setting $q = 1$, $u = 0$, we obtain analogously to Lemma 1.5.1

$$\text{var } V_1(r) = n|v(1)|^2 r^n \int_0^{2r} z^{n-1} I_{1-z^2/4r^2} \left(\tfrac{n+1}{2}, \tfrac{1}{2} \right) \times$$

$$\times \left\{ \frac{B(z)}{4} + \sum_{j=2}^\infty \frac{B^j(z)}{j!} [\Phi^{(j-1)}(0)]^2 \right\} dz =$$

$$= c_{12}(n) r^n (1 + o(1)). \quad (2.4.2)$$

Taking into account the values of the coefficients C_0, C_1, we conclude that Theorem 2.4.2 is a corollary to Theorem 2.2.1 with $m = 1$, $t = 1$. ∎

Theorem 2.4.3. If assumptions I, VI (with $m = 1$) hold, then the r.v.

$$\frac{V_1(r) - (2\pi)^{-1/2} r^n |v(1)|}{r^{n/2} L_1(r)} \xrightarrow{\mathcal{D}} N(0, c_{13}(n))$$

as $r \to \infty$. The value of $c_{13}(n)$ is defined by $c_8(n,1)$ (see Lemma 2.1.7) by setting $C_q = \int_0^\infty u\phi(u) H_q(u) du$, $q = 1, 2, \ldots$.

Theorem 2.4.3 is a corollary to Theorem 2.3.2 with $m = 1$ and Lemma 2.1.7.

If assumption IV holds, Theorems 2.4.1–2.4.3 describe the nature of the normal approximation of the functional $V_1(r)$. The form of normalization and the normalizing constants differ in the cases $\alpha \in (0,n)$, $\alpha = n$, $\alpha > n$ (cf. Lemma 2.1.3).

Theorem 2.4.4. If assumptions I, VII (with $m = 2$) hold, then the limiting distributions of the r.v. $\{V_2(r) - (\tfrac{2}{\pi})^{1/2} r^n |v(1)|\}/\{\sigma_2(r)\sqrt{2\pi}\}$ coincide with those of

$$\frac{1}{\sigma_2(r)} \int_{v(r)} H_2(\xi(x)) dx \quad (2.4.3)$$

as $r \to \infty$, where $\sigma_2(r)$ is defined by (2.1.6). In particular, if IV holds with $\alpha \in (0, n/2)$, then $\sigma_2(r)$ is defined by (2.1.10) with $m = 2$.

Proof. We calculate the first three coefficients of the expansion of the function $G(u) = |u|$ in $L_2(\mathbf{R}^1, \phi(u) du)$. These are respectively $C_0 = (\tfrac{2}{\pi})^{1/2}$, $C_1 =$

0, $C_2 = (\frac{2}{\pi})^{1/2}$. An application of Theorem 2.2.1 with $m = 2$ completes the proof. ∎

Theorem 2.4.5. If assumptions I, V (with $m = 2$) hold, then the r.v.

$$\frac{V_2(r) - \sqrt{\frac{2}{\pi}}r^n|v(1)|}{r^{n/2}} \xrightarrow{\mathcal{D}} N(0, c_{14}(n)),$$

as $r \to \infty$, where

$$c_{14}(n) = c_{15}(n) \int_0^\infty z^{n-1}\left\{\sqrt{1 - B^2(z)} - 1 + B(z)\sin^{-1}\{B(z)\}\right\}dz,$$

$$c_{15}(n) = 8\pi^{n-1}/[\Gamma^2(\tfrac{n}{2})n]. \tag{2.4.4}$$

Proof. Write

$$E[V_2(r)]^2 = \int_{v(r)} \int_{v(r)} E|\xi(x)\xi(y)|dxdy, \quad \rho = B(|x - y|).$$

Furthermore,

$$E|\xi(x)\xi(y)| = \int_{\mathbf{R}^2} |uz|\phi(u, z, \rho)dudz = \left[\int_0^\infty \int_0^\infty - \int_{-\infty}^0 \int_0^\infty\right]uz \times$$

$$\times \exp\{-(u^2 + z^2 - 2\rho uz)/[2(1 - \rho^2)]\}dudz\frac{1}{\pi\sqrt{1 - \rho^2}} =$$

$$= \frac{1}{\pi}(1 - \rho^2)^{3/2}\frac{d}{d\rho}\left(\left[\int_0^\infty \int_0^\infty - \int_{-\infty}^0 \int_0^\infty\right]\exp\{-(u^2 + z^2 - 2\rho uz)/2\}dudz\right) =$$

$$= \frac{(1 - \rho^2)^{3/2}}{\pi}\frac{d}{d\rho}\left[2\sin^{-1}\frac{\rho}{\sqrt{1 - \rho^2}}\right] = 2(\sqrt{1 - \rho^2} + \rho\sin^{-1}\rho)/\pi,$$

whence we obtain the expression for the variance

$$\text{var}\, V_2(r) = c_{15}(n)r^n \int_0^{2r} z^{n-1}I_{1-z^2/4r^2}(\tfrac{n+1}{2}, \tfrac{1}{2}) \times$$

$$\times(\sqrt{1 - B^2(z)} - 1 + B(z)\sin^{-1}\{B(z)\})dz = c_{14}(n)r^n(1 + o(1))$$

as $r \to \infty$. Recalling the values of the coefficients in the expansion of the function $G(u) = |u|$ in $L_2(\mathbf{R}^1, \phi(u)du)$ which were obtained when proving Theorem 2.4.4 we derive our theorem from Theorem 2.3.1 with $t = 1$, $m = 2$. ■

If IV holds, then the distribution of functional $V_2(r)$ cannot be approximated by a normal law for $\alpha \in (0, n/2)$ (Theorem 2.4.4) while it is possible to do so for $\alpha > n/2$ (Theorem 2.4.5). For $\alpha = n/2$, a normal approximation is also possible. The form of normalization and the constants may be obtained from Theorem 2.3.2.

Let $a \geq 0$. Consider the geometric functionals

$$W_1(r) = |\{x \in v(r) : \xi(x) > a\}| = \int_{v(r)} \chi(\xi(x) > a)dx,$$

$$W_2(r) = |\{x \in v(r) : |\xi(x)| > a\}| = \int_{v(r)} \chi(|\xi(x)| > a)dx.$$

The coefficients of the expansion of the function $G_1(u) = \chi(u > a)$ in $L_2(\mathbf{R}^1, \phi(u)du)$ are

$$C_q(a) = \begin{cases} 1 - \Phi(a), & q = 0, \\ \phi(a)H_{q-1}(a), & q \geq 1, \end{cases} \tag{2.4.5}$$

and the coefficients of the expansion of the function $G_2(u) = \chi(|u| > a)$ may be represented as:

$$C_q(a) = \begin{cases} 2(1 - \Phi(a)), & q = 0, \\ 2\phi(a)H_{q-1}(a), & q = 2r, \\ 0, & q = 2r + 1, \ r = 1, 2, \ldots. \end{cases} \tag{2.4.6}$$

Let $\zeta_1(r) = \{W_1(r) - (1 - \Phi(a))r^n|v(1)|\}$, $\zeta_2(r) = \{W_2(r) - 2r^n|v(1)|(1 - \Phi(a))\}$.

Theorem 2.4.6. Let I hold and $r \to \infty$. If VII holds (for $m = 1$), then

$$\zeta_1(r)/[\phi(a)\sigma_1(r)] \xrightarrow{\mathcal{D}} N(0, 1),$$

where $\sigma_1(r)$ is defined in (2.1.6) for $m = 1$. Under assumption V (for $m = 1$),

$$\frac{\zeta_1(r)}{r^{n/2}} \xrightarrow{\mathcal{D}} N(0, c_{16}(n, a)),$$

where

$$c_{16}(n, a) = \phi^2(a) \sum_{q=1}^{\infty} H_{q-1}^2(a) \frac{1}{(q!)^2} c_1(n, q) \int_0^{\infty} z^{n-1} B^q(z)dz.$$

A more precise result is given by Theorem 2.2.4 and Example 2.2.1.

Theorem 2.4.7. If I, VII (with $m = 2$) hold, the limit distribution of the r.v. $\zeta_2(r)/[a\phi(a)\sigma_2(r)]$ coincides as $r \to \infty$ with that given by (2.4.3).

If V holds (with $m = 2$), the r.v. $\zeta_2(r)r^{-n/2} \xrightarrow{D} N(0, c_{17}(n, a))$, where

$$c_{17}(n, a) = 4\phi^2(a) \sum_{q=1}^{\infty} H_{2q-1}^2(a) \frac{1}{[(2q)!]^2} c_1(n, 2q) \int_0^{\infty} z^{n-1} B^{2q}(z) dz.$$

Theorems 2.4.6, 2.4.7 are derived from Theorems 2.2.1, 2.3.1 and relations (2.4.4), (2.4.5). We note only that under assumption IV, the first assertion of each theorem involves the case $\alpha \in (0, n)$ (respectively $\alpha \in (0, n/2)$) and the second one the case $\alpha > n$ (or $\alpha > n/2$). However, Theorem 3.2.2 permits us readily to obtain a normal approximation of the functional $W_1(r)$ for $\alpha = n$ and $W_2(r)$ for $\alpha = n/2$.

In view of Lemma 1.4.6 the analogues of the results of this section may be obtained for the functionals $\int_{\Delta(\lambda)} G(\xi(x)) dx$ with $n = 2$, where $\Delta(1)$ is a square.

In Theorems 2.4.6 and 2.4.7 the Gaussian assumption may be relaxed. Then the c.l.t. of §1.7 may be used. For example, consider the functional $\Lambda(\Delta) = \int_{\Delta} \chi(\xi(x) > a) dx$. Then

$$E\Lambda(\Delta) = \int_{\Delta} P\{\xi(x) > a\} dx, \quad \sigma^2(\Delta) = \operatorname{var} \Lambda(\Delta) =$$

$$= \int_{\Delta} \int_{\Delta} P\{\xi(x) > a, \ \xi(y) > a\} dx dy - [E\Lambda(\Delta)]^2.$$

Let $\sigma^2(\Delta) \asymp |\Delta|$ as $\Delta \xrightarrow{V.H.} \infty$ and the mixing rate $\alpha_\eta(r)$ of the bounded field $\eta(x) = \chi(\xi(x) > a)$ satisfy the condition $\alpha_\eta(r) = O(r^{-p})$, $p > n$. Then $[\Lambda(\Delta) - E\Lambda(\Delta)]/\sigma(\Delta) \xrightarrow{D} N(0, 1)$ as $\Delta \xrightarrow{V.H.} \infty$. Note that in the case of a Gaussian process $\xi(x)$ $(n = 1)$, the mixing rate of $\eta(x)$ may be estimated by means of the cor.f. $B(x)$ [107].

For $\operatorname{Re}(\mu - \lambda + \frac{1}{2}) > 0$, $\operatorname{Re} z > 0$, we introduce the Whittaker functions [145]:

$$W_{\lambda,\mu}(z) = z^{\mu+1/2} e^{-z/2} \Gamma^{-1}(\mu - \lambda + \tfrac{1}{2}) \int_0^{\infty} e^{-zu} u^{\mu-\lambda-1/2} (1+u)^{\mu+\lambda-1/2} du.$$

Theorem 2.4.8. Let $\xi(t)$, $t \in \mathbf{R}^1$, be a stationary Gaussian process having continuous sample paths, $E\xi(t) = 0$, $E\xi^2(t) = 1$ and suppose that we are given a twice differentiable even cor.f. $B(t) = L(t)/t^\alpha$, $\alpha > 0$, $L \in \mathcal{L}$, such that the function $\tilde{B}(t) = -B''(t)$ satisfies the conditions: $\sigma^2 = \tilde{B}(0) \neq 0$,

$$\tilde{B}(t) = \alpha(\alpha+1)L(t)/t^{\alpha+2}. \tag{2.4.7}$$

Then as $T \to \infty$, the functional $L(T) = \int_{-T}^{T} \sqrt{1 + [\xi'(t)]^2}\,dt$ representing the "length of a realization" of the process $\xi(t)$, $-T \leq t \leq T$ admits a normal approximation:

$$\frac{L(T) - 2\sqrt{2}\sigma T W_{1/2,1/2}(\frac{1}{2\sigma^2})e^{1/4\sigma^2}}{\sqrt{c_{19}T}} \xrightarrow{\mathfrak{D}} N(0,1),$$

where

$$c_{19} = \frac{16}{\sigma^2} \sum_{\nu=1}^{\infty} \frac{1}{(2\nu)![B''(0)]^{2\nu}} \left[\int_0^\infty \sqrt{1+u^2}\Phi^{(2\nu+1)}\left(\frac{u}{\sigma}\right)du\right]^2 \times$$
$$\times \int_0^\infty [B''(\tau)]^{2\nu}\,d\tau.$$

Proof. The process $\dot\xi'(t)$ is Gaussian possessing the cor.f. $\tilde{B}(t)$. One-dimensional distributions of the process have the density function $\phi_{0,\sigma^2}(u)$ (see (1.1.3)). We shall examine the first three coefficients of the expansion of the function $G(u) = \sqrt{1+u^2}$ into a Fourier series in the space $L_2(\mathbf{R}^1, \phi_{0,\sigma^2}(u)du)$: $C_0 = \sqrt{2}\sigma W_{1/2,1/2}(\frac{1}{2\sigma^2})e^{1/4\sigma^2} \neq 0$, $C_1 = 0$, $C_2 \neq 0$.

Applying the relation (2.1.5) to the function (2.1.3), the variance can be expressed as

$$\text{var}\,L(T) = \frac{4}{\sigma^2} \sum_{\nu=1}^{\infty} \frac{1}{(2\nu)![B''(0)]^{2\nu}} \left[\int_0^\infty \sqrt{1+u^2}\Phi^{(2\nu+1)}\left(\frac{u}{\sigma}\right)du\right]^2 \times$$
$$\times \int_{-T}^{T}\int_{-T}^{T} [B''(|t-s|)]^{2\nu}\,dt\,ds.$$

From Lemma 1.4.2 with $n = 1$ (cf. Remark 1.4.1) we obtain

$$\int_{-T}^{T}\int_{-T}^{T} [B''(t-s)]^{2\nu}\,dt\,ds = 4T\int_0^{2T}\left(1 - \frac{u}{2T}\right)[B''(u)]^{2\nu}\,du,$$

whence $\text{var}\,L(T) = c_{19}T(1 + o(1))$ as $T \to \infty$. The assertion of the theorem now follows from Theorem 2.3.1. ∎

We provide sufficient conditions for the validity of (2.4.7). Let the cor.f. $B(t) = L(t)/t^\alpha$, $L \in \mathcal{L}$, $0 < \alpha < 1$ be twice differentiable. We are interested in the conditions on $B'(t)$ under which

$$B'(t) = -\alpha L(t)/t^{\alpha+1}. \tag{2.4.8}$$

Following W. Feller's arguments [169], we have for $0 < a < b$ and $t \to \infty$

$$\int_a^b \frac{B'(ty)t}{B(t)}dy = \frac{B(tb) - B(ta)}{B(t)} =$$

$$= \frac{L(tb)/(tb)^\alpha - L(ta)/(ta)^\alpha}{L(t)/t^\alpha} \to \frac{1}{b^\alpha} - \frac{1}{a^\alpha}. \qquad (2.4.9)$$

Let the quantity $\psi(t, y) = B'(ty)t/B(t)$ be bounded for fixed $y > 0$. By the Helly theorem [169], there exists a sequence $\{t_r\}$, $r = 1, 2, \ldots$, such that $\lim_{t_r \to \infty} \psi(t_r, y) = \psi(y)$ at the continuity points of $\psi(y)$. Formula (2.4.9) implies $\psi(y) = -\alpha y^{-\alpha-1}$. The last relation is independent of the choice of the sequence $\{t_r\}$, therefore $\lim_{t \to \infty} \psi(t, y) = -\alpha y^{-\alpha-1}$. For $y = 1$ we obtain (2.4.8). Repeating this argument, we conclude that (2.4.7) holds if the functions $\psi(t, y)$ and $\tilde{\psi}(t, y) = B''(ty)t/B'(t)$ are bounded for any $y > 0$. This holds, for example, when the functions $B(t)$, $-B'(t)$, $B''(t)$ are non-negative and monotonic for $t > t_0$, where $t_0 \geq 0$.

2.5. Reduction Conditions for Weighted Functionals

We shall study functionals of the form

$$\int_{\Delta(\lambda)} g(x)G(\xi(x))dx,$$

where $g(x)$ is a weight function, $\Delta(\lambda)$ is an expanding collection of sets and $\xi(x)$, $x \in \mathbf{R}^n$, is a homogeneous isotropic Gaussian field.

Consider the random processes

$$U_\lambda(t) = \int_{\Delta(\lambda t^{1/n})} g(x)\{G(\xi(x)) - C_0\}dx, \ t \in [0, 1],$$

under the assumptions I, III, IV, where $g(x)$ is a non-random function.

IX. Let $g(x) = g(|x|)$ be a radial continuous function positive for $|x| > 0$ and such that for some integer $m = \text{rank}G \geq 1$ there exists a positive limit

$$c_{20} = c_{20}(n, m, g, \Delta) = m! \lim_{\lambda \to \infty} \int_\Delta \int_\Delta \frac{g(\lambda|x|)g(\lambda|y|)}{g^2(\lambda)} \times$$

$$\times \frac{B^m(\lambda|x - y|)}{B^m(\lambda)}dxdy,$$

where $B(|x|) = \frac{L(|x|)}{|x|^\alpha}$ is from assumption IV with $\alpha \in (0, \frac{n}{m})$.

If assumption IX holds, the function $g(x)$ will be called admissible for the cor.f. $B(|x|)$. For example, the function $g_1(|x|) = |x|^\beta L_2(|x|)$, $\beta \geq 0$, $L_2 \in \mathcal{L}$, is admissible for the cor.f. $B(|x|) = L(|x|)/|x|^\alpha$, $\alpha \in (0, n/m)$, $L \in \mathcal{L}$, and

$$c_{20}(n, m, g_1, \Delta) = m! \int_\Delta \int_\Delta \frac{|x|^\beta |y|^\beta}{|x - y|^{\alpha m}} dx dy.$$

For $\beta = 0$, $c_{20}(n, m, g_1, \Delta) = c_2(n, m, \alpha, \Delta)$ (see (2.1.8)). In particular,

$$c_{20}(1, m, g_1, [0, 1]) = c_2(1, m, \alpha, [0, 1]) = 2/[(1 - m\alpha)(2 - m\alpha)].$$

Theorem 2.5.1. If assumptions I, III, IV, IX hold, we have

$$\operatorname{var} U_\lambda(1) = A_m^2(\lambda) c_{20}(n, m, g, \Delta) \, (1 + o(1)) \qquad (2.5.1)$$

as $\lambda \to \infty$, where $A_m^2(\lambda) = \lambda^{2n} B^m(\lambda) g^2(\lambda) C_m^2/(m!)^2$ and the limit distributions of the r.v. $Y_{\lambda,g}(1) = U_\lambda(1)/[A_m(\lambda)\sqrt{c_{20}}]$ and the r.v.

$$Y_{m,\lambda,g}(1) = \left[\int_{\Delta(\lambda)} g(x) H_m(\xi(x)) dx\right] \operatorname{sgn}\{C_m\} \lambda^{-n} B^{-m/2}(\lambda) g^{-1}(\lambda) \frac{1}{\sqrt{c_{20}}}$$

coincide.

Proof. Relation (2.5.1) with $G(u) = H_m(u)$, $m \geq 1$ follows from IV with $\alpha \in (0, n/m)$ and IX. In the general case,

$$\operatorname{var}\left[\int_{\Delta(\lambda)} g(x) G(\xi(x)) dx\right] = S_1(\lambda) + S_2(\lambda),$$

where in view of Lemma 2.1.1 and assumption IX,

$$S_1(\lambda) = \frac{C_m^2}{(m!)^2} E\left[\int_{\Delta(\lambda)} g(x) H_m(\xi(x)) dx\right]^2 = A_m^2(\lambda)(1 + o(1)) \qquad (2.5.2)$$

as $\lambda \to \infty$. Utilizing (2.1.2) we obtain

$$S_2(\lambda) = \sum_{q=m+1}^{\infty} \frac{C_q^2}{(q!)^2} E\left[\int_{\Delta(\lambda)} g(x) H_q(\xi(x)) dx\right]^2 \leq$$

$$\leq \left(\sum_{q=m+1}^{\infty} C_q^2/q!\right) \int_{\Delta(\lambda)} \int_{\Delta(\lambda)} g(x) g(y) B^{m+1}(|x - y|) dx dy \leq$$

$$\leq k_1(n, m, \alpha, g, \Delta) A_{m+1}^2(\lambda). \qquad (2.5.3)$$

Using the same method as in the proof of Lemma 2.1.4, one can show that

$$E\left[\int_{\Delta(\lambda)} g(x)H_{m+1}(\xi(x))dx\right]^2 = o\left(E\left[\int_{\Delta(\lambda)} g(x)H_m(\xi(x))dx\right]^2\right)$$

as $\lambda \to \infty$. Then (2.5.2), (2.5.3) imply (2.5.1), whence the relation

$$\lim_{\lambda\to\infty} E[Y_{\lambda,g}(1) - Y_{m,\lambda,g}(1)]^2 = 0$$

is readily obtained. ■

Theorem 2.5.2. If assumptions I, III, IV hold for $\alpha \in (0, \frac{n}{m})$ and $g(x) = g(|x|) \in \mathcal{L}$ is a radial continuous function positive for $|x| > 0$, then

$$\mathrm{var}\, U_\lambda(1) = A_m^2(\lambda)c_2(n, m, \alpha, \Delta)\,(1 + o(1)) \tag{2.5.4}$$

as $\lambda \to \infty$. The limits of finite-dimensional distributions of the random processes $Y_{\lambda,g}(t) = U_\lambda(t)/[A_m(\lambda)\sqrt{c_{20}}]$, $t \in [0,1]$ and the random processes $Y_{m,\lambda}(t) = \lambda^{-n}B^{-m/2}(\lambda)\{c_2\}^{-1/2}\mathrm{sgn}\{C_m\}\int_{\Delta(\lambda t^{1/n})} H_m(\xi(x))dx$, $t \in [0,1]$, coincide.

Proof. Using Theorem 2.2.2 and the same reasoning as in the proof of Lemma 2.1.4 we arrive at (2.5.4). Let $t_j \in [0,1]$, $a_j \in \mathbf{R}^1$, $\mathbf{R}_\lambda(t_j) = Y_{\lambda,g}(t_j) - Y_{m,\lambda}(t_j)$, $j = 1,\ldots,p$. Then $E(\sum_{j=1}^p a_j\mathbf{R}_\lambda(t_j))^2 \le \sum_{j=1}^p a_j^2 \sum_{j=1}^p E\mathbf{R}_\lambda^2(t_j)$. Proceeding along the same lines as in the proof of Lemma 2.1.4, and utilizing Theorem 2.1.1 one readily obtains

$$\lim_{\lambda\to\infty} E\mathbf{R}_\lambda^2(t_j) \le \lim_{\lambda\to\infty}\left\{k_2\left[\int_{\Delta(t_j^{1/n})}\int_{\Delta(t_j^{1/n})}\left|\frac{g(\lambda|x|)}{g(\lambda)} - 1\right|\times\right.\right.$$

$$\left.\left.\times\left|\frac{g(\lambda|y|)}{g(\lambda)} - 1\right|\frac{B^m(\lambda|x-y|)}{B^m(\lambda)}dxdy\right]^{1/2} + k_3\left[\frac{A_{m+1}^2(\lambda t_j^{1/n})}{A_m^2(\lambda)}\right]^{1/2}\right\}^2 = 0. ■$$

Examples of limiting distributions of the r.v. $Y_{m,\lambda,g}(1)$, $\lambda \to \infty$, are presented in §2.10.

2.6. Reduction Conditions for Functionals
Depending on a Parameter

We shall investigate functionals of the form

$$\int_{\Delta(\lambda)} G(\xi(x), \lambda)dx; \quad \int_{\Delta(\lambda)} G(\xi(x), x)dx,$$

where $\Delta(\lambda)$ is an expanding collection of sets and $\xi(x)$, $x \in \mathbf{R}^n$, is a homogeneous isotropic Gaussian field.

We shall consider two types of functionals depending on a parameter. The first type satisfies the condition

X. $G(u, \lambda)$, $u \in \mathbf{R}^1$, $\lambda > 0$, is a real-valued non-random function such that $EG^2(\xi(0), \lambda) < \infty$, $EG(\xi(0), \lambda) = C_0(\lambda) < \infty$.

Under assumption X, the function $G(u, \lambda)$ can be expanded into a series

$$G(u, \lambda) = \sum_{q=0}^{\infty} C_q(\lambda) H_q(u)/q!; \ C_q(\lambda) = \int_{\mathbf{R}^1} G(u, \lambda)\phi(u) H_q(u) du, \ q = 0, 1, \ldots$$

converging in the space $L_2(\mathbf{R}^1, \phi(u)du)$.

We suppose that asssumptions I, X hold and set

$$T_\lambda(t) = \int_{\Delta(\lambda t^{1/n})} G(\xi(x), \lambda) dx - C_0(\lambda) t \lambda^n |\Delta|, \ t \in [0, 1].$$

We are interested in conditions under which one-dimensional (or finite-dimensional) limit distributions of the processes

$$X_\lambda(t) = T_\lambda(t) \, / \, [|C_m(\lambda)| \sigma_m(\lambda)/m!], \ t \in [0, 1]$$

and the processes

$$X_{m,\lambda}(t) = \mathrm{sgn}\{C_m(\lambda)\} \sigma_m^{-1}(\lambda) \int_{\Delta(\lambda t^{1/n})} H_m(\xi(x)) dx, \ t \in [0, 1]$$

coincide for some integer $m \geq 1$.

XI. There exists an integer $m \geq 1$ such that

$$\varlimsup_{\lambda \to \infty} \operatorname{var} T_\lambda(1)/[C_m^2(\lambda)\sigma_m^2(\lambda)/(m!)^2] \leq 1.$$

Lemma 2.6.1. Under assumptions I, X, XI, the limiting distributions of the r.v. $X_\lambda(1)$ and the r.v. $X_{m,\lambda}(1)$ coincide as $\lambda \to \infty$.

Proof. Set $R_\lambda = X_\lambda(1) - X_{m,\lambda}(1)$. In view of Lemma 2.1.1, $\operatorname{var} X_\lambda(1) = 1 + \operatorname{var} R_\lambda$. The inequality $\operatorname{var} T_\lambda(1) \geq C_m^2(\lambda)\sigma_m^2(\lambda)/(m!)^2$ and assumption XI imply that $\lim_{\lambda \to \infty} \operatorname{var} X_\lambda(1) = 1$ therefore $\lim_{\lambda \to \infty} \operatorname{var} R_\lambda = 0$ Q.E.D. ∎

XII. There exist an integer $m \geq 1$ and $\alpha \in (0, \frac{n}{m})$ such that for each $t \in [0, 1]$

$$\varlimsup_{\lambda \to \infty} \operatorname{var} T_\lambda(1) \, / \, [C_m^2(\lambda)\sigma_m^2(\lambda)(m!)^{-2}] \leq t^{2-m\alpha/n}.$$

Lemma 2.6.2. Under assumptions I, IV (for $\alpha \in (0, n/m)$), X, XII, as $\lambda \to \infty$ the limits of collections of finite-dimensional distributions of the processes $\{X_\lambda(t),\ t \in [0,1]\}$ and $\{X_{m,\lambda}(t),\ t \in [0,1]\}$ coincide.

Proof. Let $c_p \in \mathbf{R}^1$, $t_p \in [0,1]$, $p = 1, \ldots, N$. Then

$$E\left[\sum_{p=1}^N c_p(X_\lambda(t_p) - X_{m,\lambda}(t_p))\right]^2 \leq 2 \sum_{p=1}^N c_p^2 E(X_\lambda(t_p) - X_{m,\lambda}(t_p))^2.$$

We shall show that $\lim_{\lambda \to \infty} E(X_\lambda(t_p) - X_{m,\lambda}(t_p))^2 = 0$. Note that $E(X_\lambda(t_p) - X_{m,\lambda}(t_p))^2 = EX_\lambda^2(t_p) - EX_{m,\lambda}^2(t_p)$. In view of (2.1.9), IV implies that

$$\sigma_m^2(\lambda t^{1/n}) = c_2 \lambda^{2n-m\alpha} L^m(\lambda t^{1/n}) t^{2-m\alpha/n}(1 + o(1)) \text{ as } \lambda \to \infty;$$

hence $EX_{m,\lambda}^2(t_p) \to t_p^{2-m\alpha/n}$ as $\lambda \to \infty$. Assumption XII implies that

$$EX_\lambda^2(t_p) \to t^{2-m\alpha/n} \text{ as } \lambda \to \infty. \quad \blacksquare$$

Functionals of the second type satisfy the condition:

XIII. Let $G(u, x)$, $u \in \mathbf{R}^1$, $x \in \mathbf{R}^n$, be a real non-random function such that $EG^2(\xi(0), x) < \infty$, $EG(\xi(0), x) = C_0(x) < \infty$. Then the function $G(u, x)$ can be expanded in the space $L_2(\mathbf{R}^1, \phi(u)du)$ in the series

$$G(u, x) = \sum_{q=0}^\infty C_q(x) H_q(u)/q!,$$

$$C_q(x) = \int_{\mathbf{R}^1} G(u, x) H_q(u) \phi(u) du, \quad q = 0, 1, 2, \ldots.$$

Using Lemma 2.1.1 we obtain for $m \geq 1$, $q \geq 1$

$$\tau_m^2(\lambda) = E \int_{\Delta(\lambda)} C_m(x) H_m(\xi(x)) dx \int_{\Delta(\lambda)} C_q(y) H_q(\xi(y)) dy =$$

$$= \delta_m^q m! \int_{\Delta(\lambda)} \int_{\Delta(\lambda)} C_m(x) C_m(y) B^m(|x - y|) dx dy.$$

Let

$$N_\lambda(t) = \int_{\Delta(\lambda t^{1/n})} \{G(\xi(x), x) - C_0(x)\} dx, \quad t \in [0, 1].$$

Consider under the assumptions I, XIII the random processes

$$Z_\lambda(t) = m! N_\lambda(t)/\tau_m(\lambda)$$

and the processes

$$Z_{m,\lambda}(t) = \frac{1}{\tau_m(\lambda)} \int_{\Delta(\lambda t^{1/n})} C_m(x) H_m(\xi(x)) dx, \; t \in [0,1].$$

XIV. There exists an integer $m \geq 1$ such that

$$\varlimsup_{\lambda \to \infty} \text{var } N_\lambda(1)/[\tau_m^2(\lambda)(m!)^{-2}] \leq 1.$$

Lemma 2.6.3. Under assumptions I, XIII, XIV and $\lambda \to \infty$ the limiting distributions of the r.v.'s $Z_\lambda(1)$ and $Z_{m,\lambda}(1)$ coincide.

The proof of the lemma is analogous to that of Lemma 2.6.1.

XV. There exists an integer $m \geq 1$ and $\alpha \in (0, n/m)$ such that for any $t \in [0,1]$

$$\varlimsup_{\lambda \to \infty} \text{var } N_\lambda(t)/[\tau_m^2(\lambda)(m!)^{-2}] \leq t^{2-m\alpha/n}.$$

Lemma 2.6.4. If assumptions I, IV are fulfilled for $\alpha \in (0, n/m)$ XIII, XV and $C_m(x) = C_m(|x|) \in \mathcal{L}$, then the limits of finite-dimensional distributions of the processes $\{Z_\lambda(t), \; t \in [0,1]\}$ and $\{Z_{m,\lambda}(t), \; t \in [0,1]\}$ coincide as $\lambda \to \infty$.

Proof. Choosing in Theorem 2.5.2 $G(u) = H_m(u)$, $g(x) = C_m(x)$, we obtain that $\tau_m^2(\lambda) = c_2(n, m, \alpha, \Delta) C_m^2(\lambda) B^m(\lambda)(1 + o(1))$ as $\lambda \to \infty$. The subsequent arguments are analogous to those in Lemma 2.6.2. ∎

Theorem 2.5.2 and Lemma 2.6.4 yield the following assertion.

Corollary 2.6.1. Under the conditions of Lemma 2.6.4, the finite-dimensional distributions of the processes $\{Z_{m,\lambda}(t), \; t \in [0,1]\}$ and the processes

$$\lambda^{-n} B^{-m/2}(\lambda) c_2^{-1/2} \int_{\Delta(\lambda t^{1/n})} H_m(\xi(x)) dx, \; t \in [0,1]$$

have the same limits.

We present expressions for variances of integrals of local Gaussian functionals depending on a parameter.

Let the function $G(u, \lambda)$ satisfy assumption X, $\Delta = v(1)$. Consider a function $G_r(u) = G(u, r)$, $r > 0$.

Lemma 2.6.5. Let X hold and suppose that the function $G_r(u)$ is differentiable in u and that $G_r'(u)$ is integrable over $\mathbf{R}^1$. Then under assumption I,

$$\text{var} \left[\int_{v(r)} G_r(\xi(x)) dx \right] = c_1(n,1) r^n \int_0^{2r} z^{n-1} I_{1-z^2/4r^2}\left(\tfrac{n+1}{2}, \tfrac{1}{2}\right) dz \times$$

$$\times \int_0^{B(z)} d\rho \iint_{\mathbf{R}^2} G_r'(u)G_r'(w)\phi(u,w,\rho)dudw,$$

where $\phi(u,w,\rho)$ is defined in accordance with (2.1.3) and $I_x(p,q)$ in accordance with (1.4.8).

Proof. It is well known that the function $\phi(u,w,\rho)$ (see (2.1.3)) satisfies equation $\frac{\partial}{\partial\rho}\phi(u,v,\rho) = \frac{\partial^2}{\partial u\partial v}\phi(u,v,\rho)$. Then

$$\mathrm{var}\left[\int_{v(r)} G_r(\xi(x))dx\right] = \int_{v(r)}\int_{v(r)} dxdy \iint_{\mathbf{R}^2} G_r(u)G_r(w)\times$$

$$\times \phi(u,w,B(|x-y|))dudw = \int_{v(r)}\int_{v(r)} dxdy \int_0^{B(|x-y|)} d\rho\times$$

$$\times \iint_{\mathbf{R}^2} G_r(u)G_r(w)\frac{\partial}{\partial\rho}\phi(u,w,\rho)dudw = c_1(n,1)r^n \int_0^{2r} z^{n-1}\times$$

$$\times I_{1-z^2/4r^2}(\tfrac{n+1}{2},\tfrac{1}{2})dz \int_0^{B(z)} d\rho \iint_{\mathbf{R}^2} G_r(u)G_r(w)\frac{\partial^2}{\partial u\partial w}\phi(u,w,\rho)dudw =$$

$$= c_1(n,1)r^n \int_0^{2r} z^{n-1}I_{1-z^2/4r^2}(\tfrac{n+1}{2},\tfrac{1}{2})dz\times$$

$$\times \int_0^{B(z)} d\rho \iint_{\mathbf{R}^2} G_r'(u)G_r'(w)\phi(u,w,\rho)dudw.$$

Let the function $G(u,x)$ satisfy the assumption XIII and suppose that $G_x(u) = G(u,x) - C_0(x)$.

Lemma 2.6.6. If assumptions I, XIII are valid and $G_x(u)$ is a function of a bounded variation as a function in u, then

$$\mathrm{var}\left[\int_{v(r)} G_x(\xi(x))dx\right] = \int_{v(r)}\int_{v(r)} dxdy \int_0^{B(|x-y|)} d\rho\times$$

$$\times \iint_{\mathbf{R}^2} \phi(u,w,\rho)dG_x(u)dG_y(w).$$

The proof is analogous to the preceding one.

2.7. Reduction Conditions for Measures of Excess
over a Moving Level

We shall examine measures of excess above "high" and "low" levels for realizations of homogeneous isotropic Gaussian random fields.

We consider specific functionals depending on parameters.

XVI. Let a, $b : \mathbf{R}^1 \to (0, \infty)$ be continuous monotonic functions such that $a(r) \to \infty$, $b(r) \to 0$ as $r \to \infty$.

We introduce the functional

$$T_r^{(1)} = |\{x \in v(r) : \xi(x) > a(r)\}| = \int_{v(r)} \chi(\xi(x) > a(r))dx.$$

Using formula (2.4.5), we obtain the following expansion for this functional in $L_2(\Omega)$:

$$T_r^{(1)} = (1 - \Phi(a(r)))|v(r)| + \phi(a(r)) \sum_{j=1}^{\infty} H_{j-1}(a(r)) \times$$

$$\times \frac{1}{j!} \int_{v(r)} H_j(\xi(x))dx.$$

Using the well-known formula [161]

$$\int_a^{\infty} \int_a^{\infty} \phi(u, w, \rho)dudw = \left(\int_0^{\infty} \phi(u)du \right)^2 +$$

$$+\frac{1}{2\pi} \int_0^{\rho} \exp\left\{ -\frac{a^2}{1+v} \right\} \frac{1}{\sqrt{1-v^2}} dv, \qquad (2.7.1)$$

we obtain in the same manner as when proving Lemma 2.6.5 that

$$\operatorname{var} T_r^{(1)} = c_1(n, 1) \frac{r^n}{2\pi} \int_0^{2r} z^{n-1} I_{1-z^2/4r^2}\left(\tfrac{n+1}{2}, \tfrac{1}{2} \right)dz \times$$

$$\times \int_0^{B(z)} \exp\left\{ -\frac{a^2(r)}{1+y} \right\} \frac{1}{\sqrt{1-y^2}} dy.$$

The validity of the following result can be verified by setting $m = 1$ in assumption XI and applying Lemma 2.6.1 to the functional under consideration.

Theorem 2.7.1. If assumptions I, XVI and the condition

XVII. $\varlimsup_{r \to \infty} \operatorname{var} T_r^{(1)} / [\sigma_1^2(r)\phi^2(a(r))] \leq 1$

hold, then the distribution of the r.v.

$$\frac{T_r^{(1)} - r^n |v(1)|(1 - \Phi(a(r)))}{\sigma_1(r)\phi(a(r))} \qquad (2.7.2)$$

is asymptotically $(0,1)$-normal as $r \to \infty$.

We formulate a more specific assertion.

Theorem 2.7.2. *If assumptions I, VII with $m = 1$ hold and there exists $\beta \in (0, \delta)$ (the δ is from assumption VII) such that*

$$a^2(r) = o(\ln r), \quad a^2(r) = o(B^{-1}(r^\beta)) \qquad (2.7.3)$$

as $r \to \infty$, then the r.v. (2.7.2) *has the normal distribution $N(0,1)$ as $r \to \infty$.*

Proof. We shall show that conditions (2.7.3) imply assumption XVII. Then Theorem 2.7.2 will be a corollary to Theorem 2.7.1. Since

$$(2\pi)^{-1} \exp\{-a^2(r)/(1+y)\} / \phi^2(a(r)) = \exp\{a^2(r)y/(1+y)\},$$

the relation in assumption XVII may be presented as:

$$\left[c_1(n,1)\frac{r^n}{2\pi} \int_0^{2r} z^{n-1} I_{1-z^2/4r^2}(\tfrac{n+1}{2}, \tfrac{1}{2})dz \times \right.$$
$$\times \int_0^{B(z)} \left[\exp\left\{-\frac{a^2(r)}{1+y}\right\}\frac{1}{\sqrt{1-y^2}}dy\right] \{c_1(n,1)\frac{r^n}{2\pi} \times$$
$$\left. \times \exp\{-a^2(r)\} \int_0^{2r} z^{n-1} B(z) I_{1-z^2/4r^2}(\tfrac{n+1}{2}, \tfrac{1}{2})dz\right\}^{-1} =$$
$$= \left\{ r^n \left[\int_0^{(2r)^\beta} + \int_{(2r)^\beta}^{2r}\right] z^{n-1} I_{1-z^2/4r^2}(\tfrac{n+1}{2}, \tfrac{1}{2})dz \times \right.$$
$$\times \int_0^{B(z)} \exp\left\{\frac{ya^2(r)}{1+y}\right\}\frac{dy}{\sqrt{1-y^2}}\right\} \times$$
$$\times \left[r^n \int_0^{2r} z^{n-1} B(z) I_{1-z^2/4r^2}(\tfrac{n+1}{2}, \tfrac{1}{2})dz\right]^{-1}. \qquad (2.7.4)$$

It follows from conditions (2.7.3) that for any $\epsilon > 0$, $a^2(r) = o(\ln r^{n\epsilon})$. Applying the inequalities $y/(1+y) \le 1, \sin^{-1}\{B(z)\} \le \frac{\pi}{2}$ for $0 \le z \le (2r)^\beta$ and using the fact that the integral $\int_0^{(2r)^\beta} z^{n-1} I_{1-z^2/4r^2}(\tfrac{n+1}{2}, \tfrac{1}{2})dz$ may be computed by changing the variables $u = z/2r$, we obtain the following bound for the first term in the numerator of (2.7.4):

$$r^n \int_0^{(2r)^\beta} z^{n-1} I_{1-z^2/4r^2}(\tfrac{n+1}{2}, \tfrac{1}{2})dz \int_0^{B(z)} \cdot \exp\left\{\frac{a^2(r)y}{1+y}\right\}\frac{dy}{\sqrt{1-y^2}} \le$$

$$\le r^n e^{a^2(r)} \int_0^{(2r)^\beta} z^{n-1} I_{1-z^2/4r^2}(\tfrac{n+1}{2}, \tfrac{1}{2}) \sin^{-1}\{B(z)\} dz \le$$

$$\le k_1(n, \epsilon) r^{n(1+\epsilon)} \int_0^{(2r)^\beta} z^{n-1} I_{1-z^2/4r^2}(\tfrac{n+1}{2}, \tfrac{1}{2}) dz \le$$

$$\le k_2(n, \epsilon, \beta) r^{n(1+\epsilon+\beta)}.$$

Dividing this expression by the denominator of (2.7.4) for $\epsilon = \delta - \beta$ yields

$$k_3 r^{n(1+\epsilon+\beta)} \Big/ \Big[r^n \int_0^{2r} z^{n-1} I_{1-z^2/4r^2}(\tfrac{n+1}{2}, \tfrac{1}{2}) B(z) dz \Big] =$$

$$= k_3 \Big[\frac{1}{r^{n\delta}} \int_0^{2r} z^{n-1} I_{1-z^2/4r^2}(\tfrac{n+1}{2}, \tfrac{1}{2}) B(z) dz \Big]^{-1}.$$

The latter expression tends to zero in view of assumption VII with $m = 1$.

Evaluate the second term in the numerator of (2.7.4) for $(2r)^\beta \le z \le 2r$ using relations $\frac{1}{1+y} \le 1$, $(1-y^2)^{-1/2} \sim 1$ (for $0 \le y \le B(z)$, $B(z) \searrow 0$). Then

$$r^n \int_{(2r)^\beta}^{2r} z^{n-1} I_{1-z^2/4r^2}(\tfrac{n+1}{2}, \tfrac{1}{2}) \int_0^{B(z)} \exp\Big\{\frac{a^2(r)y}{1+y}\Big\} \frac{dy}{\sqrt{1-y^2}} \le$$

$$\le r^n \int_{(2r)^\beta}^{2r} z^{n-1} I_{1-z^2/4r^2}(\tfrac{n+1}{2}, \tfrac{1}{2}) \int_0^{B(z)} \exp\{a^2(y)y\} dy \le$$

$$\le r^n \int_{(2r)^\beta}^{2r} z^{n-1} I_{1-z^2/4r^2}(\tfrac{n+1}{2}, \tfrac{1}{2}) e^{a^2(r)B(z)} \int_0^{B(y)} dy \le$$

$$\le r^n \exp\{a^2(r) \sup_{z \ge (2r)^\beta} \{B(z)\}\} \int_{(2r)^\beta}^{2r} z^{n-1} I_{1-z^2/4r^2}(\tfrac{n+1}{2}, \tfrac{1}{2}) B(z) dz. \quad (2.7.5)$$

However, in view of (2.7.3), $\lim_{r \to \infty} \exp\{a^2(r) B((2r)^\beta)\} = 1$. Therefore for r large, expression (2.7.5) does not exceed

$$r^n \int_{(2r)^\beta}^{2r} z^{n-1} I_{1-z^2/4r^2}(\tfrac{n+1}{2}, \tfrac{1}{2}) B(z) dz$$

which, clearly, does not exceed the denominator of (2.7.4). Thus the upper limit of relation (2.7.4) does not exceed 1 as $r \to \infty$. Hence XVII holds. Whence the assertion of the theorem follows. ■

Consider the functional

$$T_r^{(2)} = |\{x \in v(r) : |\xi(x)| > a(r)\}| = \int_{v(r)} \chi(|\xi(x)| > a(r)) dx,$$

which may be expanded in $L_2(\Omega)$ as follows:

$$T_r^{(2)} = 2(1 - \Phi(a(r)))|v(r)| + 2\phi(a(r)) \sum_{q=1}^{\infty} \frac{H_{2q-1}(a(r))}{(2q)!} \int_{v(r)} H_{2q}(\xi(x))dx$$

in view of (2.4.6), provided assumption XVI holds.

Direct computations or an application of Lemma 2.6.5 allow us to obtain the variance of this functional as follows:

$$\text{var } T_r^{(2)} = c_1(n,2)r^n \int_0^{2r} z^{n-1} I_{1-z^2/4r^2}\left(\tfrac{n+1}{2}, \tfrac{1}{2}\right)dz \times$$

$$\times \int_0^{B(z)} [\phi(a(r), a(r), y) - \phi(a(r), a(r), -y)]dy =$$

$$= c_1(n,2)\frac{r^n}{2\pi} \int_0^{2r} z^{n-1} I_{1-z^2/4r^2}\left(\tfrac{n+1}{2}, \tfrac{1}{2}\right)dz \times$$

$$\times \int_0^{B(z)} \left[\exp\left\{-\frac{a^2(r)}{1+y}\right\} - \exp\left\{-\frac{a^2(r)}{1-y}\right\}\right] \frac{dy}{\sqrt{1-y^2}}.$$

On choosing $m = 2$ in assumption XI, we obtain the following assertion from Lemma 2.6.1.

Theorem 2.7.3. If assumptions I, XVI and assumption

XVIII. $\varlimsup_{r \to \infty} \text{var } T_r^{(2)} / [a^2(r)\phi^2(a(r))\sigma_2^2(r)] \leq 1$

hold, then the r.v.

$$\frac{T_r^{(2)} - 2r^n|v(1)|(1 - \Phi(a(r)))}{a(r)\phi(a(r))\sigma_2(r)} \qquad (2.7.6)$$

has the same distribution as the r.v.

$$\overline{X}_{2,r}(1) = \frac{\int_{v(r)} H_2(\xi(x))dx}{\sigma_2(r)} \qquad (2.7.7)$$

as $r \to \infty$.

We formulate yet another assertion similar to the preceding one.

Theorem 2.7.4. If assumptions I, VIII (for $m = 2$) and XVI hold and there exists $\beta \in (0, \delta)$ (δ is from assumption VII) such that relations (2.7.3) hold, then the r.v. (2.7.6) has the same distribution as the r.v. (2.7.7) as $r \to \infty$.

Proof. It suffices to show that the conditions of the theorem imply assumption XVIII. Then Theorem 2.7.4 will follow from Theorem 2.7.3. We present the limiting relation in assumption XVIII in the form

$$\frac{\operatorname{var} T_r^{(2)}}{a^2(r)\phi^2(a(r))\sigma_2^2(r)} = \left\{ r^n \left[\int_0^{(2r)^\beta} + \int_{(2r)^\beta}^{2r} \right] z^{n-1} \times \right.$$

$$\times I_{1-z^2/4r^2}(\tfrac{n+1}{2}, \tfrac{1}{2})dz \int_0^{B(z)} \left[\exp\left\{\frac{a^2(r)y}{1+y}\right\} - \exp\left\{-\frac{a^2(r)y}{1-y}\right\}\right] \frac{dy}{\sqrt{1-y^2}} \right\} \times$$

$$\times \left[a^2(r)r^n \int_0^{2r} z^{n-1}I_{1-z^2/4r^2}(\tfrac{n+1}{2}, \tfrac{1}{2})B^2(z)dz\right]^{-1}. \qquad (2.7.8)$$

Using relations $a^2(r) \le \ln r^{n\epsilon}$, $\epsilon > 0$; $\frac{y}{1+y} \le 1$, $\frac{y}{y-1} < 0$; $\sin^{-1}\{B(z)\} \le \frac{\pi}{2}$, we estimate the first term in the numerator of (2.7.8) for $0 \le z \le (2r)^\beta$ as follows:

$$r^n \int_0^{(2r)^\beta} z^{n-1}I_{1-z^2/4r^2}(\tfrac{n+1}{2}, \tfrac{1}{2})dz \int_0^{B(z)} \left[\exp\left\{\frac{a^2(r)y}{1+y}\right\} - \right.$$

$$\left. - \exp\left\{-\frac{a^2(r)y}{1-y}\right\}\right] \frac{dy}{\sqrt{1-y^2}} \le r^n \int_0^{(2r)^\beta} z^{n-1}I_{1-z^2/4r^2}(\tfrac{n+1}{2}, \tfrac{1}{2})dz \times$$

$$\times \int_0^{B(z)} \exp\{a^2(r)\} \frac{dy}{\sqrt{1-y^2}} \le$$

$$\le r^n e^{a^2(r)} \int_0^{(2r)^\beta} z^{n-1}I_{1-z^2/4r^2}(\tfrac{n+1}{2}, \tfrac{1}{2})\sin^{-1}\{B(z)\}dz \le$$

$$\le k_4(n, \epsilon, \beta)r^{n(1+\epsilon+\beta)}$$

where $k_4 > 0$. For $\epsilon = \delta - \beta$ this expression divided by the denominator of (2.7.8) tends to zero in view of assumption VII for $m = 2$.

The second term in the numerator of (2.7.8) is estimated for $(2r)^\beta \le z \le 2r$ using the relation $\psi_y(r) = \exp\{a^2(r)y/(1+y)\} - \exp\{-a^2(r)y/(1-y)\} \sim 2a^2(r)y$ as $y \to 0$. In this case

$$\lim_{r\to\infty} \left\{ r^n \int_{(2r)^\beta}^{2r} z^{n-1}I_{1-z^2/4r^2}(\tfrac{n+1}{2}, \tfrac{1}{2})dz \times \right.$$

$$\times \int_0^{B(z)} \left[\psi_y(r) \frac{dy}{\sqrt{1-y^2}}\right]\Big\} \left[a^2(r)r^n \int_0^{2r} z^{n-1} B^2(z) I_{1-z^2/4r^2}\left(\tfrac{n+1}{2}, \tfrac{1}{2}\right) dz\right]^{-1} =$$

$$= \overline{\lim_{r\to\infty}} \Big\{ \int_{(2r)^\beta}^{2r} z^{n-1} I_{1-z^2/4r^2}\left(\tfrac{n+1}{2}, \tfrac{1}{2}\right) \times$$

$$\times \int_0^{B(y)} 2y\,dy \Big\} \left[\int_0^{2r} z^{n-1} I_{1-z^2/4r^2}\left(\tfrac{n+1}{2}, \tfrac{1}{2}\right) B^2(z) dz\right]^{-1} \leq 1.$$

Thus, assumption XVIII is verified. ∎

Consider the functional

$$T_r^{(3)} = |\{x \in v(r) : |\xi(x)| < b(r)\}| = \int_{v(r)} \chi(|\xi(x)| < b(r)) dx,$$

which is expanded in the space $L_2(\Omega)$ in the form

$$T_r^{(3)} = 2|v(r)|(\Phi(b(r)) - \tfrac{1}{2}) - 2\phi(b(r)) \sum_{q=1}^\infty \frac{H_{2q-1}(b(r))}{(2q)!} \int_{v(r)} H_{2q}(\xi(x)) dx.$$

From Lemma 2.6.5 or by a direct computation of the variance of this functional one can derive the expression

$$\operatorname{var} T_r^{(3)} = c_1(n,2) \frac{r^n}{2\pi} \int_0^{2r} z^{n-1} I_{1-z^2/4r^2}\left(\tfrac{n+1}{2}, \tfrac{1}{2}\right) dz \times$$

$$\times \int_0^{B(z)} \left[\exp\left\{-\frac{b^2(r)}{1+y}\right\} - \exp\left\{-\frac{b^2(r)}{1-y}\right\}\right] \frac{dy}{\sqrt{1-y^2}}.$$

XIX. Let XVI be valid, let assumption VIII hold for $m = 2$, $b^2(r) \geq k_5 r^{-\gamma}$ for large r, where $k_5 > 0$, $\gamma \in (0, \delta n)$ (δ is from assumption VII).

Theorem 2.7.5. If assumptions I, XIX are valid, then the limiting distribution of the r.v. $[T_r^{(3)} - 2r^n |v(1)|(\phi(b(r)) - \tfrac{1}{2})]/[b(r)\phi(b(r))\sigma_2(r)]$ coincides with that of $-\overline{X}_{2,r}(1)$ (see (2.7.7)).

Proof. Using the relation

$$\psi_r(y) = \left[\exp\left\{-\frac{b^2(r)}{1+y}\right\} - \exp\left\{-\frac{b^2(r)}{1-y}\right\}\right] e^{-b^2(r)} =$$

$$= \exp\left\{\frac{b^2(r)y}{1+y}\right\} - \exp\left\{-\frac{b^2(r)y}{1-y}\right\}$$

we write

$$\frac{\operatorname{var} T_r^{(3)}}{b^2(r)\phi^2(b(r))\sigma_2^2(r)} = \left\{ r^n \left[\int_0^{(2r)^\beta} + \int_{(2r)^\beta}^{2r} \right] z^{n-1} \times \right.$$

$$\left. \times I_{1-z^2/4r^2}\left(\tfrac{n+1}{2}, \tfrac{1}{2}\right) dz \int_0^{B(z)} \frac{\psi_r(y)}{\sqrt{1-y^2}} dy \right\} \times$$

$$\times \left[2b^2(r)r^n \int_0^{2r} z^{n-1} B^2(z) I_{1-z^2/4r^2}\left(\tfrac{n+1}{2}, \tfrac{1}{2}\right) dz \right]^{-1}. \qquad (2.7.9)$$

Utilizing the inequalities $\psi_r(y) < \exp\{b^2(r)\}$, $\sin^{-1}\{B(z)\} \le \tfrac{\pi}{2}$, we estimate the first term in the numerator of (2.7.9) for $0 \le z \le (2r)^\beta$:

$$r^n \int_0^{(2r)^\beta} z^{n-1} I_{1-z^2/4r^2}\left(\tfrac{n+1}{2}, \tfrac{1}{2}\right) dz \int_0^{B(z)} \frac{\psi_r(y)dy}{\sqrt{1-y^2}} \le$$

$$\le r^n e^{b^2(r)} \int_0^{(2r)^\beta} z^{n-1} I_{1-z^2/4r^2}\left(\tfrac{n+1}{2}, \tfrac{1}{2}\right) \times$$

$$\times \sin^{-1}\{B(z)\} dz \le k_6(n, \beta) r^{n(1+\beta)} e^{b^2(r)}. \qquad (2.7.10)$$

Divide (2.7.10) by the denominator of (2.7.9). The expression obtained does not exceed

$$k_7(n, \beta) r^{n\beta+\gamma} \left[\int_0^{2r} z^{n-1} B^2(z) I_{1-z^2/4r^2}\left(\tfrac{n+1}{2}, \tfrac{1}{2}\right) dz \right]^{-1},$$

where $k_7 > 0$. When $\beta = \delta - \gamma/n$, the latter expression tends to zero in view of assumption VII for $m = 2$ provided $r \to \infty$. In the same manner as in the proof of Theorem 2.7.4, one may conclude that the upper limit of the second term of (2.7.9) does not exceed 1, that is, $\varlimsup\limits_{r\to\infty} \operatorname{var} T_r^{(3)} / [b^2(r)\phi^2(b(r))\sigma_2^2(r)] \le 1$. The assertion of the theorem now follows from Lemma 2.6.1 with $m = 2$. ∎

Consider the functional $T_r^{(4)} = T_r^{(2)} + T_r^{(3)}$. It is easy to verify that $ET_r^{(4)} = 2r^n |v(1)| (\tfrac{1}{2} - \Phi(a(r)) + \Phi(b(r)))$ and the expansion of $T_r^{(4)}$ in $L_2(\Omega)$ is of the form

$$T_r^{(4)} = ET_r^{(4)} + 2 \sum_{q=1}^\infty \left[\frac{\phi(a(r))H_{2q-1}(a(r))}{(2q)!} - \right.$$

$$\left. - \frac{\phi(b(r))H_{2q-1}(b(r))}{(2q)!} \right] \int_{v(r)} H_{2q}(\xi(x)) dx.$$

Using Lemma 2.6.5 one can show that

$$\operatorname{var} T_r^{(4)} = c_1(n, 2) r^n \int_0^{2r} z^{n-1} I_{1-z^2/4r^2}\left(\tfrac{n+1}{2}, \tfrac{1}{2}\right) dz \times$$

$$\times \int_0^{B(z)} L(a(r), b(r), y) dy,$$

where $L(a, b, y) = [\phi(a, a, y) - \phi(a, a, -y)] - 2[\phi(a, b, y) - \phi(a, b, -y)] + [\phi(b, b, y) - \phi(b, b, -y)] = L_1 + L_2 + L_3$.

XX. Let XIX be valid. Then for any $r > 0$, $0 < b(r) < a(r) < \infty$,

$$a^2(r) = o(B^{-1}(r^\beta)), \quad a^2(r) = o(\ln r), \quad r \to \infty, \qquad (2.7.11)$$

where $\beta = \delta - \gamma/n$ and for r sufficiently large there exists a constant $k_8 > 0$ such that

$$\left| \frac{a(r)\phi(a(r))}{b(r)\phi(b(r))} - 1 \right| > k_8. \qquad (2.7.12)$$

Theorem 2.7.6. Under assumptions I, XX, the limiting distribution of the r.v.

$$\frac{T_r^{(4)} - ET_r^{(4)}}{A_1(r)},$$
$$A_1^2(r) = [a(r)\phi(a(r)) - b(r)\phi(b(r))]^2 \sigma_2^2(r),$$

coincides with that of the r.v. $\overline{X}_{2,r}(1)[\mathrm{sgn}(a(r)\phi(a(r)) - b(r)\phi(b(r)))]$.

Proof. We write $\mathrm{var}\, T_r^{(4)} = S_1(r) + S_2(r)$,

$$S_1(r) = c_1(n, 2) r^n \int_0^{(2r)^\beta} z^{n-1} I_{1-z^2/4r^2}\left(\tfrac{n+1}{2}, \tfrac{1}{2}\right) \times$$

$$\times \int_0^{B(z)} [L_1 + L_2 + L_3] dy.$$

Denote the part of $S_1(r)$ corresponding to L_i by $S_1^{(i)}(r)$, $i = 1, 2, 3$, and instead of $a(r), b(r)$ we shall simply write a, b. Then as in the proof of Theorem 2.7.2 we obtain

$$S_1^{(1)}(r) = c_1(n, 2) r^n \phi^2(a) \int_0^{(2r)^\beta} z^{n-1} I_{1-z^2/4r^2}\left(\tfrac{n+1}{2}, \tfrac{1}{2}\right) dz \times$$

$$\times \int_0^{B(z)} \frac{\phi(a, a, y) - \phi(a, a, -y)}{\phi^2(a)} dy \le k_9(n, \beta) r^{n(1+\delta)} \phi^2(a).$$

Since by condition (2.7.12),

$$\phi^2(a) / [a\phi(a) - b\phi(b)]^2 = \{a^2[1 - b\phi(b)/a\phi(a)]\}^{-1} \to 0 \text{ as } r \to \infty,$$

it follows that

$$S_1^{(1)}(r)/A_1^2(r) \le k_{10}(n,\beta)[r^{n(1+\delta)}/\sigma_2^2(r)].$$

The latter quantity tends to zero in view of VII for $m = 2$ as $r \to \infty$.

For $|y| \le 1$,

$$\phi(a,b,y)/[\phi(a)\phi(b)] \le \exp\{aby/(1+y)\}(1-y^2)^{-1/2} \le \exp\{ab\}/\sqrt{1-y^2},$$

therefore

$$S_1^{(2)}(r) = c_1(n,2)r^n \int_0^{(2r)^\beta} z^{n-1} I_{1-z^2/4r^2}\left(\tfrac{n+1}{2},\ \tfrac{1}{2}\right)dz \times$$

$$\times \int_0^{B(z)} [\phi(a,b,y) - \phi(a,b,-y)]dy \le k_{11}(n,\beta)\phi(a)\phi(b)e^{ab}r^{n(1+\beta)}.$$

Thus

$$S_1^{(2)}(r)/A_1^2(r) \le k_{11}(n,\beta)[\{1 - (b\phi(b)/a\phi(a))\} \times$$

$$\times \{[a\phi(a)/b\phi(b)] - 1\}]^{-1} \exp\{ab\}r^{n(1+\beta)}/[ab\sigma_2^2(r)].$$

Assumption XX implies that the latter expression tends to zero. Since $e^{b^2(r)} \to 1$,

$$S_1^{(3)}(r)/A_1^2(r) \le k_{12}(n,\beta)[\phi^2(b)r^{n(1+\beta)}/A_1^2(r)] \le$$

$$\le k_{12}(n,\beta)[(a\phi(a)/b\phi(b)) - 1]^{-2}r^{n(1+\beta)}/[b^2\sigma_2^2(r)] \le$$

$$\le k_{13}(n,\beta)r^{n(1+\beta)}/\sigma_2^2(r) \to 0,\ r \to \infty.$$

We have thus proved that $\lim_{r\to\infty} S_1(r)/A_1^2(r) \to 0$. Consider now

$$S_2(r) = c_1(n,2)r^n \int_{(2r)^\beta}^{2r} z^{n-1} I_{1-z^2/4r^2}\left(\tfrac{n+1}{2},\ \tfrac{1}{2}\right)dz \times$$

$$\times \int_0^{B(z)} L(a,b,y)dy = c_1(n,2)r^n \int_{(2r)^\beta}^{2r} z^{n-1} I_{1-z^2/4r^2}\left(\tfrac{n+1}{2},\ \tfrac{1}{2}\right)dz \times$$

$$\times \int_0^{B(z)} \left[\left\{\phi^2(a)\psi_y^{(1)}(a) + 2\phi(a)\phi(b)\exp\{-(a^2+b^2)y^2/2(1-y)^2\} \times\right.$$

$$\times \psi_y^{(2)}(a,b) + \phi^2(b)\psi_y^{(1)}(b)\Big\}/\sqrt{1-y^2}\Big]dy,$$

where

$$\psi_y^{(1)}(a) = \exp\left\{\frac{a^2 y}{1+y}\right\} - \exp\left\{-\frac{a^2 y}{1-y}\right\} \sim$$

$$\sim 2ya^2 - a^2 y^3 \frac{a^4 - 6a^2 + 6}{3}; \tag{2.7.13}$$

$$\psi_y^{(2)}(a,b) = \exp\left\{\frac{aby}{1-y^2}\right\} - \exp\left\{-\frac{aby}{1-y^2}\right\} \sim$$

$$\sim 2aby + aby^3 \frac{(ab)^2 + 6}{3} \tag{2.7.14}$$

as $a^2 y \to 0, aby \to 0, b^2 y \to 0$ (the latter follows from assumption XX). Then

$$S_2(r)/A_1^2(r) \le \left\{c_1(n,2)r^n \int_{(2r)^\beta}^{2r} z^{n-1} I_{1-z^2/4r^2}(\tfrac{n+1}{2}, \tfrac{1}{2}) dz \times \right.$$

$$\left. \times \int_0^{B(z)} 2(a\phi(a) - b\phi(b))^2 y dy\right\} / A_1^2(r).$$

Hence $\varlimsup_{r \to \infty} S_2(r)/A_1^2(r) \le 1$.

Thus the theorem follows from Lemma 2.6.1 for $m = 2$. ∎

Theorem 2.7.7. Let assumptions I, VIII for $m = 4$ and XX be valid, where δ is given in assumption VII and relation (2.7.11) is replaced by the relation $0 < b(r) < 1 < a(r) < \infty$ and (2.7.12) by the relation $a(r)\phi(a(r)) \sim b(r)\phi(b(r))$ as $r \to \infty$. Then the limiting distribution of the r.v. $[T_r^{(4)} - ET_r^{(4)}]/A_2(r)$, where $A_2(r) = \{a(r)\phi(a(r))[a^2(r) - b^2(r)]\sigma_4(r)/12\}$, coincides with that of

$$\frac{1}{\sigma_4(r)} \int_{v(r)} H_4(\xi(x)) dx. \tag{2.7.15}$$

Proof. Note that under the conditions of the theorem,

$$C_4(r) \sim 2(a^2 - b^2)a^2 \phi^2(a).$$

Thus

$$\operatorname{var} T_r^{(4)} \Big/ [C_4^2(r)\sigma_4^2(r)(4!)^{-2}] = c_1(n,2)r^n \times$$

$$\times \left[\int_0^{(2r)^\beta} + \int_{(2r)^\beta}^{2r}\right] z^{n-1} I_{1-z^2/4r^2}(\tfrac{n+1}{2}, \tfrac{1}{2}) dz \int_0^{B(z)} (L_1 + L_2 + L_3) dy,$$

as $r \to \infty$ and, using the notation introduced in the left-hand sides of (2.7.13) and (2.7.14), we obtain

$$\left\{(12)^2 c_1(n,2)r^n \int_{(2r)^\beta}^{2r} z^{n-1} I_{1-z^2/4r^2}(\tfrac{n+1}{2}, \tfrac{1}{2}) dz \times\right.$$

$$\times \int_0^{B(z)} \frac{1}{\sqrt{1-y^2}} \Big[\frac{\psi_y^{(1)}(a)}{a^2} - \frac{2}{ab} \exp\Big\{ -\frac{(a^2+b^2)y^2}{(1-y^2)2} \Big\} \times$$

$$\times \psi_y^{(2)}(a,b) + \frac{\psi_y^{(1)}(b)}{b^2} \Big] dy \Big\} \Big/ [(a^2-b^2)\sigma_4(r)]^2 \le$$

$$\le \Big\{ (12)^2 c_1(n,2)r^n \int_{(2r)^\beta}^{2r} z^{n-1} I_{1-z^2/4r^2}(\tfrac{n+1}{2}, \tfrac{1}{2}) dz \times$$

$$\times \int_0^{B(z)} [y^3 \{ (a^4 - 6a^2 + 6) - 2(a^2 b^2 + 6) + 6(a^2 + b^2) +$$

$$+ (b^4 - 6b^2 + 6) \}/3] dy \} [(a^2-b^2)\sigma_4(r)]^{-2} \le 1.$$

As above, one proves that the contribution to the variance var $T_r^{(4)}$ due to the integration over $0 \le z \le (2r)^\beta$ divided by $A_2^2(r)$ tends to zero; thus $\varlimsup_{r\to\infty} \text{var} T_r^{(4)}/A_2^2(r) \le 1$ and the theorem now follows from Lemma 2.6.1 with $m = 4$. ∎

Under assumptions I, XVI consider the functional

$$V_r^{(1)} = \int_{v(r)} \max\{0, \xi(x) - a(r)\} dx$$

which has a clear geometrical meaning for random fields with continuous realizations. For example, for $n = 2$ it is a part of the volume formed by a realization of the field $\xi(x)$, $x \in v(r)$, the plane $z = a(r)$ and a cylindrical surface passing through the boundary of the circle $v(r)$.

Below we shall show that

$$EV_r^{(1)} = A(r) = |v(1)| r^n \{ \phi(a(r)) - a(r)[1 - \Phi(a(r))] \}.$$

Theorem 2.7.8. Let assumptions I, XVI be valid and $B(|x|) \searrow 0$ as $|x| \to \infty$. Furthermore, suppose that there exists $\delta \in (0,1)$, $\beta \in (0,\delta)$ such that as $r \to \infty$

$$\sigma_1^2(r)/[r^{n(1+\delta)} a^2(r)] \to \infty, \tag{2.7.16}$$

$$a^2(r)/\log r \to 0; \quad a^2(r) B(r^\beta) \to 0. \tag{2.7.17}$$

Then the random variables

$$X_r^{(1)}(1) = \frac{V_r^{(1)} - A(r)}{\sigma_1(r)(1 - \Phi(a(r)))}$$

have asymptotically $(0, 1)$-normal distribution as $r \to \infty$, where $\sigma_1^2(r)$ is defined by the formula (2.1.6).

Proof. The function $G_r^{(1)}(u) = G(u, r) = \max\{0, u - a(r)\}$ satisfies assumption X with coefficients $C_k(r)$, $k = 0, 1$ computed by integration by parts:

$$C_0(r) = \int_{a(r)}^{\infty} (u - a(r))\phi(u)du = \phi(a(r)) - a(r)[1 - \Phi(a(r))], \qquad (2.7.18)$$

$$C_1(r) = \int_{a(r)}^{\infty} u(u - a(r))\phi(u)du = 1 - \Phi(a(r)). \qquad (2.7.19)$$

Thus

$$EV_r^{(1)} = \int_{v(r)} C_0(r)dx = |v(1)|r^n C_0(r) = A(r).$$

Let $\eta_r(x) = G_r^{(1)}(\xi(x)) = \max\{0, \xi(x) - a(r)\}$. Using (2.1.5) we integrate by parts and obtain

$$E\eta_r(x)\eta_r(y) = \int_{a(r)}^{\infty} \int_{a(r)}^{\infty} (u - a(r))(w - a(r))\phi(u, w; B(|x - y|))dudw =$$

$$= \sum_{j=0}^{\infty} \frac{[B(|x - y|)]^j}{j!} \left[\int_{a(r)}^{\infty} (u - a(r))\Phi^{(j+1)}(u)du \right]^2 =$$

$$= \left[\int_{a(r)}^{\infty} (u - a(r))\phi(u)du \right]^2 + B(|x - y|) \left[\int_{a(r)}^{\infty} \phi(u)du \right]^2 +$$

$$+ \sum_{j=2}^{\infty} \frac{[B(|x - y|)]^j}{j!} [\Phi^{(j-1)}(a(r))]^2. \qquad (2.7.20)$$

From (2.1.18), (2.1.20) and Lemma 1.4.2 we have

$$\text{var } V_r^{(1)} = \int_{v(r)} \int_{v(r)} E\eta_r(x)\eta_r(y)dxdy - (EV_r)^2 =$$

$$= \int_{v(r)} \int_{v(r)} \{B(|x - y|)[1 - \Phi(a(r))]^2 + \sum_{j=2}^{\infty} \frac{[B(|x - y|)]^j}{j!} [\Phi^{(j-1)}(a(r))]^2\}dxdy =$$

$$= c_2(n, 1)r^n \left\{ \int_0^{2r} z^{n-1} B(z) I_{1-z^2/4r^2}(\tfrac{n+1}{2}, \tfrac{1}{2})dz(1 - \Phi(a(r)))^2 + \right.$$

$$\left. + \sum_{j=2}^{\infty} \int_0^{2r} z^{n-1} B^j(z) I_{1-z^2/4r^2}(\tfrac{n+1}{2}, \tfrac{1}{2})dz[\Phi^{(j-1)}(a(r))]^2 \frac{1}{j!} \right\}. \qquad (2.7.21)$$

Using the relation (2.7.1) (see relation (10.8.3) from reference [161]) we obtain

$$\sum_{j=2}^{\infty} \frac{\rho^j}{j!} [\Phi^{(j-1)}(a)]^2 \le \frac{\rho}{2} \sum_{k=1}^{\infty} \frac{\rho^k}{k!} [\Phi^{(k)}(a)]^2 =$$

$$= \frac{\rho}{4\pi} \int_0^\rho \exp\left\{-\frac{a^2}{1+v}\right\} \frac{dv}{\sqrt{1-v^2}}. \qquad (2.7.22)$$

Formulas (2.7.21) and (2.7.22) yield:

$$\text{var } V_r^{(1)} \le c_2(n,1) r^n \left\{ \int_0^{2r} z^{n-1} B(z) I_{1-z^2/4r^2}\left(\tfrac{n+1}{2}, \tfrac{1}{2}\right) dz \times \right.$$

$$\times (1 - \Phi(a(r)))^2 + \int_0^{2r} z^{n-1} B(z) I_{1-z^2/4r^2}\left(\tfrac{n+1}{2}, \tfrac{1}{2}\right) dz \times$$

$$\times \frac{1}{4\pi} \int_0^{B(z)} \exp\left\{-\frac{a^2(r)}{1+v}\right\} \frac{dv}{\sqrt{1-v^2}}. \qquad (2.7.23)$$

Now apply Lemma 2.6.1 to the functional $V_r^{(1)} = \int_{v(r)} G_r^{(1)}(\xi(x)) dx$ with $m = 1$. Assumption X implies that if

$$\varlimsup_{r\to\infty} \text{var } V_r^{(1)} / [(1 - \Phi(a(r)))^2 \sigma_1^2(r)] \le 1, \qquad (2.7.24)$$

then the limiting distribution of the variables $[V_r^{(1)} - A(r)]/[(1 - \Phi(a(r)))\sigma_1(r)]$ as $r \to \infty$ is the same (in view of Lemma 2.6.1 with $m = 1$) as that of

$$\overline{X}_{1,r}(1) = \frac{1}{\sigma_1(r)} \int_{v(r)} \xi(x) dx,$$

that is, it is $(0, 1)$-normal.

Moreover, to prove Theorem 2.7.8, it suffices to verify relation (2.7.24) utilizing the assumption of the theorem.

Formula (2.7.23) yields

$$\varlimsup_{r\to\infty} \text{var } V_r^{(1)} / [(1 - \Phi(a(r)))^2 \sigma_1^2(r)] = 1 + \varlimsup_{r\to\infty} R_r \qquad (2.7.25)$$

where, taking into account the relation

$$1 - \Phi(a) \sim e^{-a^2/2} \frac{1}{a\sqrt{2\pi}}, \qquad a \to \infty$$

or its refinements as $a \to \infty$, we have

$$R_r = \left\{ \frac{c_2(n,1)}{4\pi} r^n \int_0^{2r} z^{n-1} B(z) I_{1-z^2/4r^2}\left(\tfrac{n+1}{2}, \tfrac{1}{2}\right) dz \times \right.$$

$$\times \int_0^{B(z)} \exp\left\{-\frac{a^2(r)}{1+v}\right\}\frac{dv}{\sqrt{1-v^2}}\right\}[(1-\Phi(a(r)))^2\sigma_1^2(r)]^{-1} =$$

$$= \{S_1(r) + S_2(r)\}\left[\int_0^{2r} z^{n-1}B(z)I_{1-z^2/4r^2}(\tfrac{n+1}{2},\ \tfrac{1}{2})dz\right]^{-1}(1+o(1)) \quad (2.7.26)$$

as $r \to \infty$, where

$$S_1(r) = \int_0^{(2r)^\beta} Q_r(z)dz, \quad S_2(r) = \int_{(2r)^\beta}^{2r} Q_r(z)dz,$$

$$Q_r(z) = a^2(r)z^{n-1}B(z)I_{1-z^2/4r^2}(\tfrac{n+1}{2},\ \tfrac{1}{2})\times$$

$$\times \frac{1}{2}\int_0^{B(z)} \exp\left\{\frac{v}{1+v}a^2(r)\right\}\frac{dv}{\sqrt{1-v^2}}$$

and β is given in the conditions of the theorem.

For $(2r)^\beta \leq z \leq 2r$ and $r \to \infty$, we have (for $\frac{1}{1+v} \leq 1$, $\frac{1}{\sqrt{1-v^2}} \leq 1$, $0 \leq v \leq B(z)$):

$$S_2(r) \leq a^2(r)\int_{(2r)^\beta}^{2r} z^{n-1}B(z)I_{1-z^2/4r^2}(\tfrac{n+1}{2},\ \tfrac{1}{2})dz\times$$

$$\times \frac{1}{2}\int_0^{B(z)} \exp\{a^2(r)v\}dv \leq \tfrac{1}{2}T(r)e^{T(r)}\int_{(2r)^\beta}^{2r} z^{n-1}B(z)I_{1-z^2/4r^2}(\tfrac{n+1}{2},\ \tfrac{1}{2})dz,$$

$$(2.7.27)$$

where in view of (2.7.17), as $r \to \infty$,

$$T(r) = a^2(r)\sup\{B(z),\ z \geq (2r)^\beta\} \to 0. \quad (2.7.28)$$

Formulas (2.7.27), (2.7.28) imply that

$$\varlimsup_{r\to\infty} S_2(r) \Big/ \left[\int_0^{2r} z^{n-1}B(z)I_{1-z^2/4r^2}(\tfrac{n+1}{2},\ \tfrac{1}{2})dz\right] \leq$$

$$\leq \varlimsup_{r\to\infty} \tfrac{1}{2}T(r)e^{T(r)}N(r) = 0, \quad (2.7.29)$$

since in view of assumption I,

$$N(r) = \left[\int_{(2r)^\beta}^{2r} z^{n-1}B(z)I_{1-z^2/4r^2}(\tfrac{n+1}{2},\ \tfrac{1}{2})dz\right]\times$$

$$\times \left[\int_0^{2r} z^{n-1} B(z) I_{1-z^2/4r^2}\left(\tfrac{n+1}{2}, \tfrac{1}{2}\right) dz \right]^{-1} \le 1.$$

For $0 \le z \le (2r)^\beta$ we use the relation $|B(z)| \le 1$, $\sin^{-1}\{B(z)\} \le \tfrac{\pi}{2}$, $\frac{v}{1+v} \le 1$. The first of the relations (2.7.17) implies that for any $\epsilon > 0$ as $r \to \infty$,

$$\overline{\lim}\, S_1(r) \le \tfrac{1}{2} \overline{\lim}\, a^2(r) e^{a^2(r)} \int_0^{(2r)^\beta} z^{n-1} B(z) \times$$

$$\times I_{1-z^2/4r^2}\left(\tfrac{n+1}{2}, \tfrac{1}{2}\right) \sin^{-1}\{B(z)\} dz \le \overline{\lim}\, c_{14}(n,\beta) a^2(r) r^{n(\epsilon+\beta)}. \qquad (2.7.30)$$

Choosing $\epsilon = \delta - \beta$ we obtain in view of (2.7.30) and (2.7.16) that

$$\overline{\lim_{r\to\infty}}\, S_1(r) \Big/ \left[\int_0^{2r} z^{n-1} B(z) I_{1-z^2/4r^2}\left(\tfrac{n+1}{2}, \tfrac{1}{2}\right) dz \right] \le$$

$$\le \overline{\lim_{r\to\infty}} \left[\frac{k_{15}(n,\beta)}{a^2(r) r^{n(\epsilon+\beta)}} \int_0^{2r} z^{n-1} B(z) I_{1-z^2/4r^2}\left(\tfrac{n+1}{2}, \tfrac{1}{2}\right) dz \right]^{-1} = 0. \qquad (2.7.31)$$

Substituting (2.7.29), (2.7.31) into (2.7.26), we obtain: $\overline{\lim}_{r\to\infty} R_r = 0$ and this together with relation (2.7.25) proves the relation (2.7.24). ∎

Under assumptions I, XVI consider the functional

$$V_r^{(2)} = \int_{v(r)} \max\{0, |\xi(x)| - a(r)\} dx.$$

Theorem 2.7.9. Let assumptions I, XVI be valid and $B(|x|) \searrow 0$ as $|x| \to \infty$. Furthermore, let there exist $\delta \in (0,1)$, $\beta \in (0,\delta)$ such that for $r \to \infty$,

$$\sigma_2^2(r)/r^{n(1+\delta)} \to \infty; \qquad (2.7.32)$$

$$a^2(r)/\log r \to 0; \quad a^2(r) B(r^\beta) \to 0. \qquad (2.7.33)$$

Then as $r \to \infty$ the limiting distributions of the random variables

$$X_r^{(2)}(1) = \frac{V_r^{(2)} - 2A(r)}{\phi(a(r))\sigma_2(r)}$$

and

$$\overline{X}_{2,r}(1) = \frac{1}{\sigma_2(r)} \int_{v(r)} (\xi^2(x) - 1) dx$$

coincide (provided one of them exists).

Proof. The function $G_r^{(2)}(u) = G^{(2)}(u,r) = \max\{0, |u| - a(r)\}$ satisfies assumption X with coefficients $C_k(r)$, $k = 0,1,2$ computed by integration by parts:

$$C_0(r) = 2\{\phi(a(r)) - a(r)(1 - \Phi(a(r)))\}, \tag{2.7.34}$$

$$C_1(r) = 0, \quad C_2(r) = 2\phi(a(r)). \tag{2.7.35}$$

In view of (2.7.35), $EV_r^{(2)} = \int_{v(r)} C_0(r)dx = 2A(r)$. Let $\zeta_r(x) = \max\{0, |\xi(x)| - a(r)\}$. Then

$$E\zeta_r(x)\zeta_r(y) = \int_{|u|>a(r)} \int_{|w|>a(r)} (|u| - a(r))(|w| - a(r)) \times$$

$$\times \phi(u, w, B(|x - y|))dudw = 2 \int_{a(r)}^{\infty} \int_{a(r)}^{\infty} (u - a(r))(w - a(r)) \times$$

$$\times [\phi(u, w, B(|x - y|)) + \phi(u, w, -B(|x - y|))]dudw. \tag{2.7.36}$$

From (2.1.18), (2.7.34), (2.7.36) we obtain

$$\text{var } V_r^{(2)} = \int_{v(r)} \int_{v(r)} E\zeta_r(x)\zeta_r(y)dxdy - (EW_r)^2 =$$

$$= 2c_2(n,1)r^n \int_0^{2r} z^{n-1} I_{1-z^2/4r^2}\left(\tfrac{n+1}{2}, \tfrac{1}{2}\right)\left\{\sum_{j=2}^{\infty} \frac{[B(z)]^j}{j!} \times\right.$$

$$\left.\times [\Phi^{(j-1)}(a(r))]^2 + \sum_{j=2}^{\infty} \frac{[-B(z)]^j}{j!}[\Phi^{(j-1)}(a(r))]^2\right\}dz. \tag{2.7.37}$$

Using inequalities of the type (2.7.22) we obtain from (2.7.37) that

$$\text{var } V_r^{(2)} \leq \frac{c_2(n,2)}{4\pi} r^n \int_0^{2r} z^{n-1} I_{1-z^2/4r^2}\left(\tfrac{n+1}{2}, \tfrac{1}{2}\right) B(z) \times$$

$$\times \left\{\int_0^{B(z)} \exp\left\{-\frac{a^2(r)}{1+v}\right\} + \exp\left\{-\frac{a^2(r)}{1-v}\right\} \frac{dv}{\sqrt{1-v^2}}\right\}dz. \tag{2.7.38}$$

Now apply Lemma 2.6.1 to the functional $V_r^{(2)} = \int_{v(r)} G_r^{(2)}(\xi(x))dx$ with $m = 2$. From (2.7.35) it follows that if

$$\varlimsup_{r \to \infty} \text{var } V_r^{(2)} / [\phi^2(a(r))\sigma_2^2(r)] \leq 1, \tag{2.7.39}$$

then Theorem 2.7.9 is a corollary to Lemma 2.6.1 with $m = 2$. Thus, it suffices to verify (2.7.39) using the conditions of the theorem.

Utilizing (2.7.38), we have

$$\frac{\text{var } V_r^{(2)}}{\phi^2(a(r))\sigma_2^2(r)} \leq \frac{S_1(r) + S_2(r)}{\int_0^{2r} z^{n-1} B^2(z) I_{1-z^2/4r^2}(\frac{n+1}{2}, \frac{1}{2}) dz}, \tag{2.7.40}$$

where for β given in the conditions of the theorem

$$S_1(r) = \int_0^{(2r)^\beta} Q_r(z) dz, \quad S_2(r) = \int_{(2r)^\beta}^{2r} Q_r(z) dz,$$

$$Q_r(z) = z^{n-1} I_{1-z^2/4r^2}(\tfrac{n+1}{2}, \tfrac{1}{2}) B(z) \times$$

$$\times \tfrac{1}{2} \int_0^{B(z)} \left[\exp\left\{ a^2(r) \frac{y}{1+y} \right\} + \exp\left\{ -a^2(r) \frac{y}{1-y} \right\} \right] \frac{dy}{\sqrt{1-y^2}}.$$

For $(2r)^\beta \leq z \leq 2r$ we use the relations

$$\exp\left\{ a^2 \frac{y}{1+y} \right\} + \exp\left\{ -a^2 \frac{y}{1-y} \right\} \sim 2, \quad \frac{1}{\sqrt{1-y^2}} \sim 1, \ y \to 0.$$

Then as $r \to \infty$,

$$\overline{\lim} \, S_2(r) \leq \overline{\lim} \int_{(2r)^\beta}^{2r} z^{n-1} I_{1-z^2/4r^2}(\tfrac{n+1}{2}, \tfrac{1}{2}) B(z) \times$$

$$\times \left[\int_0^{B(z)} dy \right] dz \leq \overline{\lim_{r \to \infty}} \int_{(2r)^\beta}^{2r} z^{n-1} B^2(z) I_{1-z^2/4r^2}(\tfrac{n+1}{2}, \tfrac{1}{2}) dz. \tag{2.7.41}$$

From (2.7.41) and assumption A we obtain that as $r \to \infty$,

$$\overline{\lim} \, S_2(r) \Big/ \left[\int_0^{2r} z^{n-1} B^2(z) I_{1-z^2/4r^2}(\tfrac{n+1}{2}, \tfrac{1}{2}) dz \right] \leq$$

$$\leq \overline{\lim} \left[\int_{(2r)^\beta}^{2r} z^{n-1} B^2(z) I_{1-z^2/4r^2}(\tfrac{n+1}{2}, \tfrac{1}{2}) dz \right] \Big/ \left[\int_0^{2r} z^{n-1} \times \right.$$

$$\left. \times B^2(z) I_{1-z^2/4r^2}(\tfrac{n+1}{2}, \tfrac{1}{2}) dz \right] \leq 1.$$

For $0 \leq z \leq (2r)^\beta$ we obtain in the same manner as in the derivation of (2.7.30) that as $r \to \infty$

$$\overline{\lim} \, S_1(r) \leq \overline{\lim} \tfrac{1}{2} e^{a^2(r)} \int_0^{(2r)^\beta} z^{n-1} B^2(z) \times$$

$$\times \sin^{-1}\{B(z)\} I_{1-z^2/4r^2}(\tfrac{n+1}{2}, \tfrac{1}{2}) dz. \qquad (2.7.42)$$

Formulas (2.7.32), (2.7.42) imply that for any $\epsilon > 0$ and $r \to \infty$,

$$\overline{\lim} \, S_1(r) \leq \overline{\lim} \, c_{16}(n, \beta) r^{n(\epsilon+\beta)}.$$

Choosing $\epsilon = \delta - \beta$ and using (2.7.32), we obtain as $r \to \infty$ that

$$\overline{\lim} \, S_1(r) \Big/ \Big[\int_0^{2r} z^{n-1} B^2(z) I_{1-z^2/4r^2}(\tfrac{n+1}{2}, \tfrac{1}{2}) dz \Big] \leq$$

$$\leq c_{17}(n, \beta) \overline{\lim} \Big[\frac{1}{r^{n(\epsilon+\beta)}} \int_0^{2r} z^{n-1} B^2(z) I_{1-z^2/4r^2}(\tfrac{n+1}{2}, \tfrac{1}{2}) dz \Big]^{-1} = 0. \quad (2.7.43)$$

Substituting (2.7.43), (2.7.41) into (2.7.40) yields relation (2.7.39).   ■

Remark 2.7.1. Limiting distributions of the variables (2.7.7) (see also Theorems 2..7.3, 2.7.4, 2.7.5, 2.7.6, 2.7.9) and also of the variables (2.7.15) (see Theorem 2.7.7) may be defined in terms of multiple stochastic integrals (see Theorems 2.10.1, 2.10.3 in this connection).

2.8. Reduction Conditions for Characteristics of the Excess over a Radial Surface

We shall investigate measures of excess over a radial surface for homogeneous isotropic strongly dependent Gaussian fields. Consider the functional

$$N_r^{(1)} = |\{x \in v(r) : \xi(x) > a(x)\}| = \int_{v(r)} \chi(\xi(x) > a(x)) dx,$$

where $a(x) = a(|x|)$ is a positive radial function satisfying assumption XVI. Clearly,

$$E N_r^{(1)} = |s(1)| \int_0^r z^{n-1}(1 - \Phi(a(z))) dz.$$

Theorem 2.8.1. Under the conditions of Theorem 2.7.2, the r.v. $[N_r^{(1)} - E N_r^{(1)}]/A_1(r)$, where

$$A_1^2(r) = \int_{v(r)} \int_{v(r)} \phi(a(x)) \phi(a(y)) B(|x - y|) dx dy,$$

has an asymptotically normal distribution $N(0,1)$ as $r \to \infty$.

Proof. The function $G^{(1)}(u,x) = \chi(u > a(x))$ satisfies assumption XIII with $C_0(x) = 1 - \Phi(a(x))$, $C_1(x) = \phi(a(x))$, where B is from (2.7.3). We subdivide the set $V = v(r) \times v(r)$ into two subsets: $\Delta_1 = \{(x,y) \in V : |x - y| \le (2r)^\beta\}$, $\Delta_2 = \{(x,y) \in V : |x - y| > (2r)^\beta\}$. Using Lemma 2.6.6, we obtain

$$\text{var } N_r^{(1)} = \int_{v(r)} \int_{v(r)} dx\,dy \int_0^{B(|x-y|)} \phi(a(x), a(y), \rho) d\rho = S_1(r) + S_2(r).$$

Since for $|B| \le 1$

$$\int_0^B \phi(u, w, \rho) d\rho \le \int_0^{|B|} \phi(u, w, \rho) d\rho \le \int_0^1 \phi(u, w, \rho) d\rho \le 1,$$

it follows that

$$S_1(r) = \iint_{\Delta_1} dx\,dy \int_0^{B(|x-y|)} \phi(a(x), a(y), \rho) d\rho \le \iint_{\Delta_1} dx\,dy \le$$

$$\le k_1(n, \beta) r^{n(1+\beta)}. \tag{2.8.1}$$

Under the conditions of the theorem,

$$A_1^2(r) \ge \frac{1}{2\pi} e^{-a^2(r)} \int_{v(r)} \int_{v(r)} B(|x - y|) dx\,dy \ge$$

$$\ge k_2(n) r^{-n(\delta-\beta)+n} \int_0^{2r} z^{n-1} I_{1-z^2/4r^2}\left(\tfrac{n+1}{2}, \tfrac{1}{2}\right) B(z) dz, \tag{2.8.2}$$

where δ and β are from (2.7.3).

Therefore it follows from (2.8.1) and (2.8.2) that

$$S_1(r)/A_1^2(r) \le k_1(n, \beta) r^{n(1+\beta)} \Big\{ k_2(n) r^{n(1+\beta-\delta)} \times$$

$$\times \int_0^{2r} z^{n-1} I_{1-z^2/4r^2}\left(\tfrac{n+1}{2}, \tfrac{1}{2}\right) B(z) dz \Big\}^{-1}.$$

The latter expression tends to zero as $r \to \infty$ in view of assumption VII with $m = 1$. Using the identity

$$\phi(a, b, \rho)/\phi(a)\phi(b) = \exp\left\{ \frac{(a^2 + b^2)\rho}{2(1+\rho)} \right\} \exp\left\{ -\frac{(a-b)^2 \rho}{2(1-\rho^2)} \right\} \frac{1}{\sqrt{1-\rho^2}},$$

for any $\epsilon > 0$ and large r, we obtain

$$S_2(r) = \iint_{\Delta_2} dx\,dy \int_0^{B(|x-y|)} \phi(a(x), a(y), \rho)d\rho \leq$$

$$\leq \iint_{\Delta_2} \phi(a(x))\phi(a(y))dx\,dy \int_0^{B(|x-y|)} (1+\epsilon)\exp\left\{\frac{[a^2(x)+b^2(y)]\rho}{2(1+\rho)}\right\}d\rho \leq$$

$$\leq \exp\{k_3(\beta)a^2(r)\sup_{z>r^\beta}\{B(z)\}\} \iint_{\Delta_2} \phi(a(x))\phi(a(y))B(|x-y|)dx\,dy. \quad (2.8.3)$$

Since $\lim_{r\to\infty} \exp\{k_3(\beta)a^2(r)\sup\{B(z), z > r^\beta\} = 1$, (2.8.3) does not exceed $A_1^2(r)$ for large r, which implies that $\overline{\lim}_{r\to\infty} \operatorname{var} N_r^{(1)}/A_1^2(r) \leq 1$ and the theorem follows from Lemma 2.6.3 with $m = 1$. ∎

Let $N_r^{(2)} = |\{x \in v(r) : |\xi(x)| > a(x)\}| = \int_{v(r)} \chi(|\xi(x)| > a(x))dx$. Then under assumption XVI, $EN_r^{(2)} = 2EN_r^{(1)}$.

Theorem 2.8.2. Under the conditions of Theorem 2.7.4, the r.v. $[N_r^{(2)} - EN_r^{(2)}]/A_2(r)$, where

$$A_2^2(r) = 2 \int_{v(r)} \int_{v(r)} \phi(a(x))\phi(a(y))a(x)a(y)B^2(|x-y|)dx\,dy \quad (2.8.4)$$

and the r.v.

$$\frac{1}{A^2(r)} \int_{v(r)} a(x)\phi(a(x))H_2(\xi(x))dx \quad (2.8.5)$$

have the same limiting distributions as $r \to \infty$.

Proof. The function $G^{(2)}(u, x) = \chi(|u| > a(x))$ satisfies XIII with $C_0(x) = 2(1 - \Phi(a(x)))$, $C_1(x) = 0$, $C_2(x) = 2a(x)\phi(a(x))$. It follows from Lemma 2.6.6 that

$$\operatorname{var} N_r^{(2)} = 2 \int_{v(r)} \int_{v(r)} dx\,dy \int_0^{B(|x-y|)} [\phi(a(x), a(y), \rho) -$$

$$- \phi(a(x), a(y), -\rho)]d\rho = S_1(r) + S_2(r),$$

where $S_1(r)$ and $S_2(r)$ correspond to partitioning of the first integral into two integrals over the sets Δ_1 and Δ_2 (cf. the proof of Theorem 2.8.1). Since for any $\epsilon > 0$,

$$\inf\{[a(x)a(y)\phi(a(x))\phi(a(y))], 1 \leq |x| \leq r, 1 \leq |y| \leq r\} \geq k_4(n)r^{-\epsilon n},$$

we have

$$A_2^2(r) \geq k_5(n) + k_6(n) r^{n(1-\epsilon)} \int_0^{2r} z^{n-1} B^2(z) \times$$

$$\times I_{1-z^2/4r^2}(\tfrac{n+1}{2}, \tfrac{1}{2}) dz.$$

If δ is from assumption VII for $m = 2$, setting $\epsilon = \delta - \beta$ we obtain

$$S_1(r)/A_2^2(r) \leq k_7(n, \beta) r^{n\delta} \Big[\int_0^{2r} z^{n-1} B^2(z) \times$$

$$\times I_{1-z^2/4r^2}(\tfrac{n+1}{2}, \tfrac{1}{2}) dz \Big]^{-1}.$$

The last expression tends to zero as $r \to \infty$ in view of assumption VII with $m = 2$.

Using the identity

$$\frac{[\phi(a, b, \rho) - \phi(a, b, -\rho)]}{\phi(a)\phi(b)} = \exp\Big\{ -\frac{(a^2 + b^2)\rho^2}{2(1 - \rho^2)} \Big\} \times$$

$$\times \Big[\exp\Big\{ \frac{ab\rho}{1 - \rho^2} \Big\} - \exp\Big\{ -\frac{ab\rho}{1 - \rho^2} \Big\} \Big] \frac{1}{\sqrt{1 - \rho^2}}, \qquad (2.8.6.)$$

the asymptotic equivalence (2.7.14) and the relation

$$\sup_{(x,y)\in v(r)\times v(r)} \{a(x)a(y)\} \sup_{s \geq r^\beta} \{B(s)\} \to 0,$$

which follows from (2.7.3), we conclude that the expression

$$\frac{S_2(r)}{A_2^2(r)} = \frac{1}{A_2^2(r)} \int\!\!\int_{\Delta_2} dx dy \int_0^{B(|x-y|)} \phi(a(x), a(y), \rho) d\rho$$

does not exceed

$$\frac{1}{A_2^2(r)} \int\!\!\int_{\Delta_2} \phi(a(x))\phi(a(y)) dx dy \int_0^{B(|x-y|)} 2\rho d\rho \leq 1$$

for r sufficiently large . Thus $\varlimsup_{r\to\infty} \text{var } N_r^{(2)}/A_2^2(r) \leq 1$ and the theorem follows from Lemma 2.6.3 with $m = 2$. $\blacksquare$

Let $N_r^{(3)} = |\{x \in v(r) : |\xi(x)| < b(x)\}| = \int_{v(r)} \chi(|\xi(x)| < b(x)) dx$, where $b(x) = b(|x|)$ is a positive radial function satisfying assumption XVI. Then

$$EN_r^{(3)} = 2|s(1)| \int_0^r z^{n-1} [\Phi(b(z)) - \tfrac{1}{2}] dz.$$

Theorem 2.8.3.   Under the conditions of Theorem 2.7.5 the r.v. $[N_r^{(3)} - EN_r^{(3)}]/A_3(r)$, where

$$A_3^2(r) = 2 \int_{v(r)} \int_{v(r)} b(x)b(y)\phi(b(x))\phi(b(y))B^2(|x-y|)dxdy,$$

and the r.v.

$$-\frac{1}{A_3(r)} \int_{v(r)} b(x)\phi(b(x))H_2(\xi(x))dx \qquad (2.8.7)$$

have the same limiting distributions as $r \to \infty$.

Proof.   For the function $G^{(3)}(u,x) = \chi(|u| < b(x))$, $C_0(x) = 2(\Phi(b(x)) - 1)$, $C_1(x) = 0$, $C_2(x) = -2\phi(b(x))b(x)$. In view of Lemma 2.6.3 with $m = 2$ and Lemma 2.6.6, it suffices to show that

$$\varlimsup_{r \to \infty} \frac{1}{A_3^2(r)} \int_{v(r)} \int_{v(r)} dxdy \int_0^{B(|x-y|)} [y(b(x),b(y),\rho) -$$

$$- \phi(b(x),b(y),-\rho)]d\rho \le 1.$$

Under the conditions of the theorem,

$$\inf_{(x,y)\in v(r)\times v(r)} \{b(x)b(y)\phi(b(x))\phi(b(y))\} \ge k_8 r^{-\gamma},$$

therefore

$$A_3^2(r) \ge k_9(n)r^{n-\gamma} \int_0^{2r} z^{n-1}B^2(z)I_{1-z^2/4r^2}(\tfrac{n+1}{2}, \tfrac{1}{2})dz. \qquad (2.8.8)$$

Let $\beta \in (0,1)$ and suppose that the sets Δ_1 and Δ_2 are as in the proof of Theorem 2.8.1. Using (2.8.8) in the same manner as in the proof of the preceding theorem, we obtain

$$\frac{1}{A_3^2(r)} \int\int_{\Delta_1} dxdy \int_0^{B(|x-y|)} [\phi(b(x),b(y),\rho) - \phi(b(x),b(y),-\rho)]d\rho \le$$

$$\le k_{10}(n,\beta)r^{n(1+\beta)}\Big\{k_9(n)r^{n-\gamma} \int_0^{2r} z^{n-1}B^2(z)I_{1-z^2/4r^2}(\tfrac{n+1}{2}, \tfrac{1}{2})dz\Big\}^{-1}.$$

Setting $\beta = \delta - \gamma/n$ (δ is from assumption VII with $m = 2$), we obtain that the last expression tends to zero. Using (2.8.8) and (2.7.14) one can show that

$$\varlimsup_{r \to \infty} \frac{1}{A_3^2(r)} \int\int_{\Delta_2} dxdy \int_0^{B(|x-y|)} [\phi(b(x),b(y),\rho) - \phi(b(x),b(y),-\rho)]d\rho \le 1,$$

which completes the proof. ■

Let $N_r^{(4)} = N_r^{(2)} + N_r^{(3)}$; then

$$EN_r^{(4)} = 2|s(1)| \int_0^r z^{n-1}\{\Phi(b(z)) - \Phi(a(z)) + \tfrac{1}{2}\}dz.$$

Theorem 2.8.4. Under the conditions of Theorem 2.7.7, the r.v. $[N_r^{(4)} - EN_r^{(4)}]/A_4(r)$, where

$$A_4^2(r) = \int_{v(r)} \int_{v(r)} a^3(x)b^3(y)\phi(a(x))\phi(b(y))B^4(|x-y|)dxdy/6$$

and the r.v.

$$\frac{1}{12A_4(r)} \int_{v(r)} a^3(x)\phi(a(x))H_4(\xi(x))dx \qquad (2.8.9)$$

have the same distributions as $r \to \infty$.

Proof. The function $G^{(4)}(u,x) = \chi(|u|) < b(x)) + \chi(|u| > a(x))$ satisfies XIII with $C_0(x) = 1 - 2\Phi(a(x)) + 2\Phi(b(x))$, $C_4(x) = 2[\phi(a(x))H_3(a(x)) - \phi(b(x))H_3(b(x))] \sim 2a^3(x)\phi(a(x))$ under the conditions of Theorem 2.7.7. By Lemma 2.6.6 we have

$$\text{var } T_r^{(4)} = 2 \int_{v(r)} \int_{v(r)} dxdy \int_0^{B(|x-y|)} \{L_1 + L_2 + L_3 + L_4\}d\rho, \qquad (2.8.10)$$

where

$$L_1 = \phi(a(x), a(y), \rho) - \phi(a(x), a(y), -\rho),$$
$$L_2 = -[\phi(a(x), b(y), \rho) - \phi(a(x), b(y), -\rho)],$$
$$L_3 = -[\phi(b(x), a(y), \rho) - \phi(b(x), a(y), -\rho)],$$
$$L_4 = [\phi(b(x), b(y), \rho) - \phi(b(x), b(y), -\rho].$$

Subdivide the first integral in (2.8.10) into two, corresponding to the sets Δ_1 and Δ_2 (see the proof of Theorem 2.8.1). The integral over Δ_1 is handled in the same manner as the one from Theorem 2.8.2. Using the relations

$$[\exp\left\{\frac{z\rho}{1-\rho^2}\right\} - \exp\left\{-\frac{z\rho}{1-\rho^2}\right\}]\frac{1}{z} = 2\rho + 2\rho^3 + \frac{z^2\rho^3}{3} + O(\rho^5 + z^4\rho^2),$$

$$\exp\left\{-\frac{\tilde{z}\rho^2}{2(1-\rho^2)}\right\} = 1 - \frac{\tilde{z}\rho^2}{2} + O(\tilde{z}\rho^4 + \tilde{z}^2\rho^4), \quad z\rho \to 0, \ \tilde{z}\rho \to 0, \ \rho \to 0,$$

we obtain

$$\frac{2}{A_4^2(r)} \int\int_{\Delta_2} dxdy \int_0^{B(|x-y|)} \left\{\sum_{i=1}^4 L_i\right\}d\rho \sim \frac{2}{A_4(r)} \int\int_{\Delta_2} a(x)a(y)\times$$

$$\times \phi(a(x))\phi(a(y))dxdy \int_0^{B(|x-y|)} \frac{1}{\sqrt{1-\rho^2}} \left[\exp\left\{-\frac{[a^2(x)+a^2(y)]\rho^2}{2(1-\rho^2)}\right\}\times\right.$$

$$\times \left[\exp\left\{\frac{a(x)a(y)\rho}{1-\rho^2}\right\} - \exp\left\{-\frac{a(x)a(y)\rho}{1-\rho^2}\right\}\right]\frac{1}{a(x)a(y)} -$$

$$-\frac{1}{a(x)b(y)}\exp\left\{-\frac{[a^2(x)+b^2(y)]\rho^2}{2(1-\rho^2)}\right\}\left[\exp\left\{\frac{a(x)b(y)\rho)}{1-\rho^2}\right\}-\right.$$

$$-\exp\left\{-\frac{a(x)b(y)\rho}{(1-\rho^2)}\right\}\right] - \frac{1}{b(x)a(y)}\exp\left\{-\frac{[b^2(x)+a^2(y)]\rho^2}{2(1-\rho^2)}\right\}\times$$

$$\times \left[\exp\left\{\frac{b(x)a(y)\rho}{1-\rho^2}\right\} - \exp\left\{-\frac{b(x)a(y)\rho}{1-\rho^2}\right\}\right] +$$

$$+\frac{1}{b(x)b(y)}\exp\left\{-\frac{(b^2(x)+b^2(y))\rho^2}{2(1-\rho^2)}\right\}\left[\exp\left\{\frac{b(x)b(y)\rho}{1-\rho^2}\right\}-\right.$$

$$\left.\left.-\exp\left\{-\frac{b(x)b(y)\rho}{1-\rho^2}\right\}\right]\right]d\rho \sim$$

$$\sim \frac{2}{A_4(r)}\iint_{\Delta_2} a(x)a(y)\phi(a(x))\phi(a(y))dxdy \int_0^{B(|x-y|)} \{(a^2(x)-b^2(x))\times$$

$$\times (a^2(y)-b^2(y))\rho^3/3 + R\}d\rho, \tag{2.8.11}$$

where for large r and any $\epsilon > 0$, $R \le \epsilon a^2(x)a^2(y)\rho^3$. Consequently, (2.8.11) does not exceed

$$\frac{2(1+\epsilon)}{A_4^2(r)}\iint_{\Delta_2} a^3(x)a^3(y)\phi(a(x))\phi(a(y))dxdy \int_0^{B(|x-y|)} \frac{\rho^3}{3}d\rho.$$

Since $\epsilon > 0$ is arbitrary, the last expression does not exceed 1. Thus

$$\varlimsup_{r\to\infty} \text{var } T_r^{(4)}/A_4^2(r) \le 1$$

and the theorem follows from Lemma 2.6.3 with $m = 4$ ∎

Remark 2.8.1. In view of Theorem 2.5.2, an investigation of the limiting distributions of the r.v.'s (2.8.5), (2.8.7), (2.8.9) is reduced to that of (2.7.7), (2.7.15) in the case when the functions $a(|x|)$, $a(|x|)\phi(a(|x|))$, $a^3(|x|)\phi(a(|x|))$ and $b(|x|)\phi(b(|x|))$ vary slowly as $|x| \to \infty$. This is satisfied if , for example, $a(|x|) = \sqrt{\log\log|x|}$, $|x| > e^e$ and $b(|x|) \downarrow 0$ where $b(|x|)$ is a positive slowly varying function as $|x| \to \infty$. Therefore examples of the limiting distributions of the r.v.'s (2.8.5), (2.8.7), (2.8.9) as well as the limiting distributions of the r.v.'s (2.7.7), (2.7.15) can be constructed based on Theorems 2.10.1 and 2.10.4.

2.9. Multiple Stochastic Integrals

This section presents a brief account of multiple stochastic integrals (m.s.i.) with respect to Gaussian random measures.

Let $F(\cdot)$ be the spectral measure of a homogeneous (possibly generalized) random field, that is, a σ-finite measure on $(\mathbf{R}^n, \mathcal{B}^n)$ such that $F(\Delta) = F(-\Delta)$, $\Delta \in \mathcal{B}^n$, and for some $p > 0$, $\int_{\mathbf{R}^n} (1 + |\lambda|)^{-p} F(d\lambda) < \infty$.

Suppose that $F(\cdot)$ is a non-atomic measure: $F(\{x\}) = 0$ for any $x \in \mathbf{R}^n$. Consider a complex Gaussian orthogonal measure $Z_F(\cdot)$ having the structure function $F(\cdot)$, that is, a collection of jointly Gaussian r.v.'s such that for any $\Delta_j \in \mathcal{B}^n$, $j = 1, \ldots, r$, having a finite $F(\cdot)$-measure, the properties 1)–4) of §1.2 are valid. Note that these properties imply: a) the r.v.'s $\operatorname{Re} Z_F(\Delta)$ and $\operatorname{Im} Z_F(\Delta)$ are independent and have the distribution $N(0, F(\Delta)/2)$; b) if $\Delta_1 \cup (-\Delta_1), \ldots, \Delta_r \cup (-\Delta_r)$ are disjoint sets, then the r.v.'s $Z_F(\Delta_1), \ldots, Z_F(\Delta_r)$ are independent; c) if $\Delta \cap (-\Delta) = \emptyset$, then the r.v.'s $Z_F(\Delta), Z_F(-\Delta)$ are independent and have the distribution $N(0, F(\Delta)/2)$.

If $F(\Delta) = |\Delta|$, $Z_F(\cdot) = W(\cdot)$ is called a Gaussian white noise in $\mathbf{R}^n$.

Denote by $\overline{L}_2(\mathbf{R}^{nm}, F)$ the real Hilbert space of the equivalence classes of complex-valued functions $f_m = f_m(x^{(1)}, \ldots, x^{(m)})$, $x^{(j)} \in \mathbf{R}^n$, $j = 1, \ldots, m$, satisfying the conditions:

$$f_m(-x^{(1)}, \ldots, -x^{(m)}) = \overline{f_m(x^{(1)}, \ldots, x^{(m)})};$$

$$\|f_m\|^2 = \int_{\mathbf{R}^{nm}} |f_m(x^{(1)}, \ldots, x^{(m)})|^2 \prod_{j=1}^m F(dx^{(j)}) < \infty.$$

Clearly, $\|f_m\|$ is the norm in $\overline{L}_2(\mathbf{R}^{nm}, F)$. Denote by $L_2(\mathbf{R}^{nm}, F) \subset \overline{L}_2(\mathbf{R}^{nm}, F)$ the space of functions $f_m \in \overline{L}_2(\mathbf{R}^{nm}, F)$ such that $f(x^{(1)}, \ldots, x^{(m)}) = f_m(x^{i_1}, \ldots, x^{(i_m)})$ for any permutation $\pi_m = (i_1, \ldots, i_m)$ from the permutation group Π_m. The norm in $L_2(\mathbf{R}^{nm}, F)$ is the same as in $\overline{L}_2(\mathbf{R}^{nm}, F)$.

In general, if $f_m \in \overline{L}_2(\mathbf{R}^{nm}, F)$, the symmetrization

$$\operatorname{sym}\{f_m\} = \sum_{\pi_m \in \Pi_m} f_m(x^{(i_1)}, \ldots, x^{(i_m)}) \frac{1}{m!}$$

belongs to $L_2(\mathbf{R}^{nm}, F)$ and $\|\operatorname{sym} f_m\| \leq \|f_m\|$.

Let $\{\Delta\}_r = \{\Delta_j, \ j = \pm 1, \pm 2, \ldots\}$ be measurable sets ($\Delta_j = -\Delta_{-j}$, $j = 1, 2, \ldots$), disjoint for each r and forming a monotonic sequence in r of countable partitions of $\mathbf{R}^n$ such that $\lim\limits_{r \to \infty} \sup\limits_{\Delta \in \{\Delta\}_r} \operatorname{diam}\{\Delta\} = 0$ and $\{\Delta\}_r^m = \{\Delta^m = \Delta_{j_1} \times \ldots \times \Delta_{j_m}, \ \Delta_j \in \{\Delta\}_r, \ j_1, \ldots, j_m = \pm 1, \pm 2, \ldots\}, \ r = 1, 2, \ldots$ is the induced sequence of partitions of the set $\mathbf{R}^{nm}$. The function $\bar{f}_m = \bar{f}_m(x^{(1)}, \ldots, x^{(m)})$ is said to be simple if for some $r \geq 1$, it assumes constant values on sets of the form $\Delta^m = \Delta_{j_1} \times \ldots \times \Delta_{j_m}, \ j_r = \pm 1, \pm 2, \ldots$; $\bar{f}_m$ is non-vanishing only on a finite number of such sets; $\bar{f}_m(x^{(1)}, \ldots, x^{(m)}) = 0$ if $(x^{(1)}, \ldots, x^{(m)}) \in N_m = \{\Delta^m = \Delta_{j_1} \times \ldots \times \Delta_{j_m}, \ j_r = \pm j_{r'} \text{ for some } r \neq r'\}$. The set of simple functions is denoted by $\hat{\bar{L}}_2(\mathbf{R}^{nm}, F) \subset L_2(\mathbf{R}^{nm}, F)$. If $\bar{f}_m \in \hat{\bar{L}}_2(\mathbf{R}^{nm}, F)$ then the m.s.i. is defined by the formula

$$I_m(\bar{f}) = \frac{1}{m!} \int'_{\mathbf{R}^{nm}} \bar{f}_m(x^{(1)}, \ldots, x^{(m)}) \prod_{j=1}^{m} Z_F(dx^{(j)}) =$$

$$= \frac{1}{m!} \sum_{j_1, \ldots, j_m} \bar{f}_m(x^{(j_1)}, \ldots, x^{(j_m)}) \prod_{r=1}^{m} Z_F(\Delta_{j_r}).$$

If $\bar{f}_m \in \hat{\bar{L}}_2(\mathbf{R}^{nm}, F) = \hat{\bar{L}}_2(\mathbf{R}^{nm}, F) \cap L_2(\mathbf{R}^{nm}, F)$, then

$$I_m(\bar{f}) = I_m(\operatorname{sym} \bar{f}), \quad E|I_m(f)|^2 = \|\bar{f}_m\|^2 / m!.$$

The set of simple functions $\hat{\bar{L}}(\mathbf{R}^{nm}, F)$ is dense in $\bar{L}_2(\mathbf{R}^{nm}, F)$. Therefore the constructed isometric (up to a factor) mapping $\hat{\bar{L}}(\mathbf{R}^{nm}, F) \to L_2(P)(\bar{f}_m \to I_m(\bar{f}))$ can be extended on the closure of all functions $\bar{f}_m \in \hat{\bar{L}}(\mathbf{R}^{nm}, F)$, which coincides with $\bar{L}_2(\mathbf{R}^{nm}, F)$. Thus, for an arbitrary function $f_m \in \bar{L}_2(\mathbf{R}^{nm}, F)$, the m.s.i.

$$I_m(f) = \frac{1}{m!} \int'_{\mathbf{R}^{nm}} f(x^{(1)}, \ldots, x^{(m)}) \prod_{j=1}^{m} Z_F(dx^{(j)})$$

is an isometric (up to a factor) mapping of $\bar{L}_2(\mathbf{R}^{nm}, F)$ into $L_2(P)$. The integral is defined as the limit $I_m(f) = \operatorname{l.i.m.} I_m(\bar{f}^{(r)})$ where $\bar{f}^{(r)} \in \hat{\bar{L}}_2(\mathbf{R}^{nm}, F)$ is a sequence of functions associated with the partition system $\{\Delta\}^r, \ r = 1, 2, \ldots$.

One can show that such a limit exists and is independent of both the partitioning mode and the choice of the sequence of simple functions. In a similar manner a m.s.i. may be defined for a subset of $\mathbf{R}^m$.

We note the following well-known properties of m.s.i.'s:

1) $E I_m(f) = 0, \ f \in \bar{L}_2(\mathbf{R}^{nm}, F)$;

2) $I_m(f) = I_m(\operatorname{sym} f)$ is a real-valued r.v. belonging to $L_2(P)$, where $\bar{f} \in \bar{L}_2(\mathbf{R}^{nm}, F)$;

3) $EI_m(f)I_r(\phi) = 0,\ r \neq m,\ f, \phi \in \bar{L}_2(\mathbf{R}^{nm}, F)$;

4) $E|I_m(f)|^2 = \|f_m\|^2/m!,\ f_m \in L_2(\mathbf{R}^{nm}, F)$;

5) $E|I_m(f)|^2 \leq \|f_m\|^2/m!,\ f_m \in \bar{L}_2(\mathbf{R}^{nm}, F)$.

We now state the most important property of the m.s.i., called the Itô formula: if $\phi_1, \ldots, \phi_r \in L_2(\mathbf{R}^n, F)$ is an orthogonal system of functions and $j_1, \ldots, j_r$ are positive integers such that $j_1 + \ldots + j_r = N$, define for all $i = 1, \ldots, N$, functions $g_i = \phi_s$ with $j_1 + \ldots + j_{s-1} < i \leq j_1 + \ldots + j_s$. Then

$$\prod_{r=1}^{m} H_{j_r}\left(\int \phi_r(x) Z_F(dx)\right) = \int' g_1(x^{(1)}) \ldots g_N(x^{(N)}) \prod_{r=1}^{N} Z_F(dx^{(r)})$$

$$= \int' \mathrm{sym}\Big[\prod_{r=1}^{N} g_r(x^{(r)})\Big] \prod_{r=1}^{N} Z_F(dx^{(r)}),$$

where $H_j(u)$ is the jth Chebyshev-Hermite polynomial.

A measure F on $(\mathbf{R}^n, \mathbf{B}^n)$ is called locally finite if $F(\Delta) < \infty$ for any bounded $\Delta \in \mathbf{B}^n$. A collection of locally finite measures $\{F_\mu\}$, $\mu > 0$, is said to converge locally weakly to the locally finite measure F_0 if

$$\lim_{\mu \to \infty} \int_{\mathbf{R}^n} f(u) F_\mu(du) = \int_{\mathbf{R}^n} f(u) F_0(du)$$

for any continuous function $f(\cdot)$ with a bounded support.

A collection $\{F_\mu\}$ of locally finite measures converges locally weakly to the measure F_0 if for any bounded $\Delta : F_0(\partial\Delta) = 0,\ \lim_{\mu \to \infty} F_\mu(\Delta) = F_0(\Delta)$.

We note an important property of the m.s.i. expressed in the following assertion.

Lemma 2.9.1. Let $\{F_\mu\}$, $\mu > 0$, be a collection of non-atomic spectral measures on $\mathbf{R}^n$ which converge locally weakly to a non-atomic spectral measure F_0 on $\mathbf{R}^n$ and $K_\mu = K_\mu(x^{(1)}, \ldots, x^{(m)}, \Delta), \mu > 0, |\Delta| < \infty$, a set of measurable functions on $\mathbf{R}^{nm}$ belonging to $L_2(\mathbf{R}^{nm}, F_\mu)$ and converging to a continuous function $K_0(x^{(1)}, \ldots, x^{(m)}, \Delta)$ uniformly in each set $\Delta^m(A) = \{x \in \mathbf{R}^{nm}, x^{(j)} \in \Delta(A), j = 1, \ldots, m\}$ and

$$\lim_{A \to \infty} \int_{\mathbf{R}^{nm} \backslash \Delta^m(A)} |K_\mu(x^{(1)}, \ldots, x^{(m)}, \Delta)|^2 \prod_{j=1}^{m} F_\mu(dx^{(j)}) = 0 \qquad (2.9.1)$$

uniformly in $\mu \in (0, \infty)$ and $\mu = 0$.

Then there exists a m.s.i.

$$\int_{\mathbf{R}^{nm}}' K_0(x^{(1)}, \ldots, x^{(m)}, \Delta) \prod_{j=1}^{m} Z_{F_0}(dx^{(j)}), \qquad (2.9.2)$$

that is, $K_0 \in L_2(\mathbf{R}^{nm}, F_0)$, and the m.s.i.

$$\int_{\mathbf{R}^{nm}}^{'} K_\mu(x^{(1)}, \ldots, x^{(m)}, \Delta) \prod_{j=1}^{m} Z_{F_\mu}(dx^{(j)})$$

converges weakly to the m.s.i. (2.9.2) as $\mu \to \infty$.

2.10. Conditions for Attraction of Functionals of Homogeneous Isotropic Gaussian Fields to Semi-Stable Processes

The examination above of limiting distributions for a wide class of functionals of strongly dependent Gaussian fields has been reduced to the analysis of the limiting distributions of functionals $\int_{\Delta(\mu)} g(x) H_m(\xi(x)) dx$ where $g(x)$ is a weight function, H_m is the mth Chebyshev-Hermite polynomial, $\Delta(\mu)$ is a collection of expanding sets.

We now describe limiting distributions of such functionals in correlation and spectrum terms. To do this, we introduce the following definition.

A random process $X : \Omega \times \mathbf{R}_+^1 \to \mathbf{R}^1$ is said to be semi-stable with a parameter $\kappa > 0$: a) in the strict sense, if for any $a > 0$ $X(at) \overset{d}{=} a^\kappa X(t)$; b) in the wide sense, if $EX(t) = 0$, $\rho(t, s) = EX(t)X(s) < \infty$ and $\rho(at, as) = a^{2\kappa} \rho(t, s)$.

Clearly, a) and b) are equivalent for Gaussian processes. The definition implies that $\rho(t, s) = (st)^\kappa \rho\left(\sqrt{\frac{t}{s}}, \sqrt{\frac{s}{t}}\right) = (st)^\kappa \tilde{\rho}\left(\frac{s}{t}\right)$

If $\zeta(t)$, $t \in \mathbf{R}^1$, is a strictly stationary process, the process $X(t) = t^\kappa \zeta(\ln t)$, $t > 0$, $X(0) = 0$ a.s., is a strictly semi-stable process on $\mathbf{R}_+^1$ having parameter $\kappa > 0$ since $X(at) \overset{d}{=} (at)^\kappa \zeta(\ln t + \ln a) = a^\kappa t^\kappa \zeta(\ln t) \overset{d}{=} a^\kappa X(t)$. Conversely, if $X(t)$, $t \in \mathbf{R}_+^1$, is a strictly semi-stable random process with parameter $\kappa > 0$, then the process $\zeta(t) = \exp\{-\kappa t\} X(\exp\{t\})$ is strictly stationary, since $\zeta(t + a) \overset{d}{=} e^{-\kappa t} e^{-\kappa a} X(e^t e^a) \overset{d}{=} e^{-\kappa t} X(e^t) \overset{d}{=} \zeta(t)$.

Semi-stable and stationary processes in the wide sense are related in a similar manner. Thus the mapping $X \to \zeta : \zeta(\ln t) = t^{-\kappa} X(t)$ defines a correspondence between sets of semi-stable processes on $\mathbf{R}_+^1$ and stationary processes on $\mathbf{R}^1$ in the strict or wide senses. If not specified, strictly semi-stable processes will be considered in what follows.

A simple example of a semi-stable process with a parameter $\kappa = \frac{1}{2}$ is provided by the Wiener process $w(t)$, $t \in \mathbf{R}_+^1$, that is a Brownian motion process with $b = 1$. Stable processes having parameter $\gamma \in (0, 2]$ are semi-stable processes with parameter $\kappa = \frac{1}{\gamma}$. These processes have independent increments on disjoint intervals. However, there are semi-stable processes which have only covariance of incrememts tending to zero. Thus, let $X(t)$, $t \in \mathbf{R}_+^1$, be a semi-stable process with parameter $\kappa \in (1/2, 1)$ having stationary increments and let $X(0) = 0$, $EX(t) = 0$, $EX^2(t) = 1$. Then $EX^2(t) = t^{2\kappa}$, $EX(t)X(s) = \frac{1}{2}\{|t|^{2\kappa} + |s|^{2\kappa} - |t - s|^{2\kappa}\}$, and $E(X(t+1) - X(t))(X(t+s+1) - X(t+s)) \sim \kappa(2\kappa - 1)s^{2\kappa-2}$ as $s \to \infty$. Semi-stable processes may be defined by means of m.s.i.'s.

In §§2.2–2.7 the examination of limiting distributions of local functionals of a Gaussian field was reduced to the analysis of limiting distributions of the r.v. (2.2.1) and the r.v. $Y_{m,\mu,g}(1)$ (cf. §2.5). We now provide examples of distributions of the above r.v.'s in spectral terms.

We introduce the assumption:

XXI. Let $\xi : \Omega \times \mathbf{R}^n \to \mathbf{R}^1$ be a m.s. continuous homogeneous isotropic Gaussian field, $E\xi(x) = 0$, $E\xi^2(x) = 1$, having a spectral density $f(|\lambda|) = h(|\lambda|)/|\lambda|^{n-\alpha}$, $\lambda \in \mathbf{R}^n$, such that $f(|\lambda|) = f(\rho) = |s(1)|\rho^{n-1}\tilde{g}(\rho)$, $\rho \in \mathbf{R}_+^1$, $\rho^{n-1}\tilde{g}(\rho) \in L_1(\mathbf{R}_+^1)$, $\tilde{g}(\rho) = h(\rho)/\rho^{n-\alpha}$, $\alpha \in (0, (n+1)/2)$, where $h(\rho)$, $\rho \in \mathbf{R}_+^1$, is a function continuous in a neighbourhood of zero, $h(0) \neq 0$ and $h(\rho)$ is bounded on $\mathbf{R}_+^1$.

We set $c_{21}(n, \alpha) = 2^\alpha \pi^{n/2} \Gamma(\frac{\alpha}{2})/\Gamma(\frac{n-\alpha}{2})$.

Lemma 2.10.1. Under assumption XXI (possibly without assuming that the field is Gaussian) and as $r \to \infty$, the cor.f. $B(r) = h(0)c_{21}(n, \alpha)\frac{1}{r^\alpha}(1 + o(1))$, $\alpha \in (0, \frac{n+1}{2})$.

Proof. The relation

$$\int_0^\infty J_\mu(\alpha t)t^{\rho-1}dt = \frac{2^{\rho-1}\Gamma(\frac{\mu+\rho}{2})}{\alpha^\rho \Gamma(1 + \frac{\mu-\rho}{2})},$$

$$-\mathrm{Re}\mu < \mathrm{Re}\rho < \tfrac{3}{2}, \quad \alpha > 0,$$

is well-known [145]. By (1.2.8) and Lemma 1.5.4,

$$B(r) = |s(1)| \int_0^\infty Y_n(\rho r)h(\rho)\rho^{\alpha-1}d\rho =$$

$$= h(0)c_{21}(n, \alpha)\frac{1}{r^\alpha}(1 + o(1))$$

as $r \to \infty$. ∎

Under assumption XXI in a manner analogous to the proof of Lemma 2.1.3, we obtain for $\alpha \in (0, \min\{\frac{n}{m}, \frac{n+1}{2}\})$, $m \geq 1$,

$$\sigma_m^2(r) = \text{var}\left[\int_{v(r)} H_m(\xi(x))dx\right] =$$

$$= c_{22}(n, m, \alpha)h^m(0)r^{2n-m\alpha}(1+o(1))$$

as $r \to \infty$ and $c_{22}(n, m, \alpha) = c_{21}^m(n, \alpha)c_2(n, m, \alpha, v(1))$.

Let $t \in [0, 1]$. Consider the random processes

$$X_m(t) = c_{23}(n, m, \alpha)\sqrt{t} \int_{\mathbf{R}^{nm}}' \frac{J_{n/2}(|\lambda^{(1)} + \ldots + \lambda^{(m)}|t^{1/n})}{|\lambda^{(1)} + \ldots + \lambda^{(m)}|^{n/2}} \times$$

$$\times \frac{W(d\lambda^{(1)})\ldots W(d\lambda^{(m)})}{|\lambda^{(1)}|^{(n-\alpha)/2}\ldots|\lambda^{(m)}|^{(n-\alpha)/2}} \tag{2.10.1}$$

where a m.s.i. with respect to a Gaussian white noise $W(\cdot)$ in $(\mathbf{R}^n, \mathbf{\mathscr{B}}^n)$ appears on the right-hand side of (2.10.1), $\alpha \in (0, \min\{\frac{n}{m}, \frac{n+1}{2}\})$, $c_{23}(n, m, \alpha) = (2\pi)^{n/2}/\sqrt{c_{22}(n, m, \alpha)}$.

Using the semi-stability of the Gaussian white noise of order $\kappa = \frac{n}{2}$ (formally $W(d(ax)) \stackrel{d}{=} a^{n/2}W(dx)$), we verify the strict semi-stability of processes $X_m(t)$ of order $\kappa = 1 - \frac{\alpha m}{2n}$ with $\kappa \in (\frac{1}{2}, 1)$. Applying the relations $J_\nu(z) = O(z^{-1/2})$, $z \to \infty$, $J_\nu(z) \sim k_1 z^\nu$, $z \to 0$, one can verify that

$$EX_m^2(t) = c_{23}^2(n, m, \alpha)t \int_{\mathbf{R}^{nm}} \frac{J_{n/2}^2(|\sum_{j=1}^m \lambda^{(j)}|t^{1/n})\prod_{j=1}^m d\lambda^{(j)}}{|\sum_{j=1}^m \lambda^{(j)}|^n \prod_{j=1}^m |\lambda^{(j)}|^{n-\alpha}} < \infty \tag{2.10.2}$$

for each $t \in [0, 1]$. We note that the processes $X_m(t)$, $t \in [0, 1]$ are not Gaussian for $m \geq 2$.

Theorem 2.10.1. Let assumption XXI hold and $\alpha \in (0, \min\{\frac{n}{m}, \frac{n+1}{2}\})$, $m \geq 1$. Then as $r \to \infty$, the finite-dimensional distributions of the random processes

$$X_{m,r}(t) = \frac{1}{r^{n-m\alpha/2}\sqrt{c_{22}(n, m, \alpha)h^m(0)}} \int_{v(rt^{1/n})} H_m(\xi(x))dx, \ t \in [0, 1]$$

converge weakly to finite-dimensional distributions of the process $X_m(t)$, $t \in [0, 1]$.

Proof. In view of (2.2.16) and the Itô formula (see §1.9), we obtain

$$H_m(\xi(x)) = \int_{\mathbf{R}^{nm}}' e^{i(x, \lambda^{(1)}+\ldots+\lambda^{(m)})}\left\{\prod_{j=1}^m \sqrt{f(|\lambda^{(j)}|)}\right\}\prod_{j=1}^m W(d\lambda^{(j)}).$$

Using (1.4.5), we have

$$X_{m,r}(t) \overset{d}{=} \left\{ \int_{\mathbf{R}^{nm}}' \frac{J_{n/2}(|\lambda^{(1)} + \ldots + \lambda^{(m)}| |t^{1/n} r)}{|\lambda^{(1)} + \ldots + \lambda^{(m)}|^{n/2}} \times \right.$$

$$\left. \times \sqrt{\prod_{j=1}^{m} \frac{h(|\lambda^{(j)}|)}{h(0)}} \frac{\prod_{j=1}^{m} W(d\lambda^{(j)})}{\prod_{j=1}^{m} |\lambda^{(j)}|^{(n-\alpha)/2}} \right\} \frac{\sqrt{t} c_{23}(n, m, \alpha)}{r^{n/2} r^{-m\alpha/2}}. \qquad (2.10.3)$$

Changing the variables $\lambda^{(j)} r = \tilde{\lambda}^{(j)}$, $j = 1, \ldots, m$, and using semi-stability of the Gaussian white noise, we derive from (2.10.3)

$$X_{m,r}(t) \overset{d}{=} \sqrt{t} c_{23}(n, m, \alpha) \int_{\mathbf{R}^{nm}}' J_{n/2}(|\lambda^{(1)} + \ldots + \lambda^{(m)}| t^{1/n}) \times$$

$$\times |\sum_{j=1}^{m} \lambda^{(j)}|^{-n/2} \prod_{j=1}^{m} |\lambda^{(j)}|^{(\alpha-n)/2} \left\{ \prod_{j=1}^{m} h(|\lambda^{(j)}|/r) h^{-1}(0) \right\}^{1/2} \prod_{j=1}^{m} W(d\lambda^{(j)}). \qquad (2.10.4)$$

It follows from (2.10.1), (2.10.4) that

$$R_{m,r}(t) = E(X_{m,r}(t) - X_m(t))^2 = c_{23}^2(n, m, \alpha) t \times$$

$$\times \int_{\mathbf{R}^{nm}} J_{n/2}^2 \left(|\sum_{j=1}^{m} \lambda^{(j)}| t^{1/n} \right) |\sum_{j=1}^{m} \lambda^{(j)}|^{-n} \times$$

$$\times \left\{ \prod_{j=1}^{m} |\lambda^{(j)}|^{\alpha-n} \right\} Q_r(\lambda^{(1)}, \ldots, \lambda^{(m)}) \prod_{j=1}^{m} d\lambda^{(j)}, \qquad (2.10.5)$$

where

$$Q_r(\lambda^{(1)}, \ldots, \lambda^{(m)}) = \left[\sqrt{\prod_{j=1}^{m} h(|\lambda^{(j)}| r^{-1})/h(0)} - 1 \right]^2.$$

Let $\psi(r) \to \infty$, but $\psi(r)/r \to 0$ as $r \to \infty$. Subdivide the integral in (2.10.5) into two parts; the integration in the first one of them (denoted by I_1) is carried out over the set $B_1 = \{\lambda^{(j)} \in \mathbf{R}^n : |\lambda^{(j)}| \leq \psi(r), \ j = 1, \ldots, n\}$ and over the set $B_2 = \mathbf{R}^{nm} \backslash B_1$ (denote it by I_2) in the second one. In view of assumption XXI, for any $\epsilon > 0$ there exists r_0 such that for $r > r_0$,

$$Q_r(\lambda^{(1)}, \ldots, \lambda^{(m)}) < \epsilon, \quad (\lambda^{(1)}, \ldots, \lambda^{(m)}) \in B_1.$$

We then obtain from (2.10.2) that $I_1 < \epsilon k_2$ and hence, this integral can be made arbitrarily small by choosing an appropriate $\epsilon > 0$. If $(\lambda^{(1)}, \ldots, \lambda^{(m)}) \in B_2$, assumption XXI implies that $Q_r(\lambda^{(1)}, \ldots, \lambda^{(m)}) < k_3$. Using the structure of the set B_2 and relation (2.10.2), one can show that $\lim_{r \to \infty} I_2 = 0$. Thus, $\lim_{r \to \infty} R_{m,r}(t) = 0$, whence it follows that for any $a_j \in \mathbf{R}^1$, $j = 1, \ldots, p$,

$$\lim_{r \to \infty} E\left(\sum_{j=1}^{p} a_j(X_{m,r}(t_j) - X_m(t_j))\right)^2 = 0. \quad \blacksquare$$

Other representations of the limiting process $X_m(t)$, $t \in [0,1]$ can be derived by using representations (1.2.19) (or (1.2.12) with $n = 2$ and (1.2.13) with $n = 3$) in place of formula (2.1.16). Decompositions of the processes $X_m(t)$, $t \in [0,1]$, will consist of series of m.s.i.'s with respect to jointly uncorrelated Gaussian white noises in $(\mathbf{R}_+^1, \mathcal{B}(\mathbf{R}_+^1))$.

Repeating the proof of Theorem 2.10.1 utilizing (1.2.12) and (6.561) from [42] for $m = 1$, $n = 2$, we obtain the Gaussian process

$$X_1(t) = c_{24}(\alpha)\sqrt{t} \int_0^\infty \rho^{(\alpha-3)/2} J_1(\rho\sqrt{t}) W(d\rho), \quad t \in [0,1],$$

where $W(\cdot)$ is the Gaussian white noise in $(\mathbf{R}_+^1, \mathcal{B}(\mathbf{R}_+^1))$ and

$$c_{24}(\alpha) = \frac{\{(2-\alpha)\Gamma(1-\frac{\alpha}{2})\Gamma(3-\frac{\alpha}{2})\}^{1/2}}{\{\Gamma(\frac{\alpha}{2})\Gamma(\frac{3-\alpha}{2})\}^{1/2}\pi^{1/4}}$$

for $0 < \alpha < 3/2$.

We now examine the limiting distributions of the r.v. $Y_{m,\mu,g}(1)$ (cf. §2.5). Let assumptions IX, XXI be valid. From Lemma 2.10.1 and Theorem 2.5.1 we obtain

$$\mathrm{var}\left[\int_{\Delta(\mu)} g(|x|)H_m(\xi(x))dx\right] = c_{25}h^m(0)\mu^{2n-m\alpha}g^2(\mu)(1+o(1))$$

as $\mu \to \infty$, where $c_{25} = c_{25}(n, m, \alpha, g, \Delta) = c_{21}^m(n, \alpha)c_{20}(n, m, g, \Delta)$.

XXII. A function $\bar{g}(|x|)$ exists such that

$$\int_{\mathbf{R}^{nm}} \prod_{j=1}^{m} |\lambda^{(j)}|^{\alpha-n} \left|\int_{\Delta(t^{1/n})} e^{i(\lambda^{(1)}+\ldots+\lambda^{(m)}, x)}\bar{g}(x)dx\right|^2 \prod_{j=1}^{m} d\lambda^{(j)} < \infty,$$

with

$$\lim_{\mu \to \infty} \int_{\mathbf{R}^{nm}} \prod_{j=1}^{m} |\lambda^{(j)}|^{\alpha-n} \left|\int_{\Delta(t^{1/n})} e^{i(\lambda^{(1)}+\ldots+\lambda^{(m)}, x)} \times\right.$$

$$\times\left\{\frac{g(\mu|x|)}{g(\mu)}\prod_{j=1}^{m}\sqrt{\frac{h(|\lambda^{(j)}|\mu^{-1})}{h(0)}}-\bar{g}(x)\right\}dx\Big|^2\prod_{j=1}^{m}d\lambda^{(j)}=0$$

for all $t \in [0,1]$.

Theorem 2.10.2. Let assumptions I, IX, XXII hold and $\alpha \in (0,\min\{\frac{n}{m},\frac{n+1}{2}\})$. Then as $\mu \to \infty$, the finite-dimensional distributions of the random processes

$$Y_{m,\mu,g}(t) = \left[\int_{\Delta(\mu t^{1/n})} g(|x|)H_m(\xi(x))dx\right]\times$$

$$\times\left[\mu^{n-m\alpha/2}g(\mu)\sqrt{c_{25}h^m(0)}\right]^{-1} \tag{2.10.6}$$

converge weakly to finite-dimensional distributions of the process

$$Y_{m,g}(t) = \frac{1}{\sqrt{c_{25}}}\int_{\mathbf{R}^{nm}}'\prod_{j=1}^{m}|\lambda^{(j)}|^{(\alpha-n)/2}\times$$

$$\times\left[\int_{\Delta(t^{1/n})} \bar{g}(x)e^{i(\lambda^{(1)}+\dots+\lambda^{(m)},x)}dx\right]\prod_{j=1}^{m}W(d\lambda^{(j)}). \tag{2.10.7}$$

The proof of Theorem 2.10.2 is analogous to that of Theorem 2.10.1.

In view of assumption XXII, $EY_{m,g}^2(t) < \infty$ for all $t \in [0,1]$.

If the function $g_1(|x|) = |x|^\beta$, $\beta \geq 0$, and the sets $\Delta(\mu)$ form a collection of balls $v(r)$, assumptions IX, XXII are valid. Moreover $\bar{g}_1(|x|) = |x|^\beta$, $\beta \geq 0$, $c_{25} = c_{21}^m(n,\alpha)m!\int_{v(1)}\int_{v(1)}|x|^\beta|y|^\beta|x-y|^{-\alpha m}dxdy$, $\alpha \in (0,\min\{\frac{n}{m},\frac{n+1}{2}\})$.

Changing to spherical coordinates (1.1.1), one obtains the relation

$$\int_{s(\rho)}|x|^\nu e^{i(\lambda,x)}dm(x) = \rho^\nu\left(\frac{2\pi\rho}{|\lambda|}\right)^{n/2}J_{(n-2)/2}(\rho|\lambda|), \quad \nu \geq 0.$$

Using the formula $\frac{d}{d\rho}\int_{v(\rho)}f(x)dx = \int_{s(\rho)}f(x)dm(x)$ we obtain from the last relation that

$$\int_{v(\rho)}|x|^\nu e^{i(\lambda,x)}dx = \int_0^\rho r^\nu\left(\frac{2\pi r}{|\lambda|}\right)^{n/2}J_{(n-2)/2}(r|\lambda|)dr.$$

The limiting process (2.10.7) may now be written as

$$Y_{m,g_1}(t) = \frac{1}{\sqrt{c_{25}}}\int_{\mathbf{R}^{nm}}'\prod_{j=1}^{m}|\lambda^{(j)}|^{(\alpha-n)/2}\times$$

$$\times \left[\int_{v(t^{1/n})} |x|^\beta e^{i(\lambda^{(1)}+\dots+\lambda^{(m)},x)} dx \right] \prod_{j=1}^{m} W(d\lambda^{(j)}) =$$

$$= (2\pi)^{n/2} \frac{1}{\sqrt{c_{25}}} \int_{\mathbf{R}^{nm}}' \prod_{j=1}^{m} |\lambda^{(j)}|^{(\alpha-n)/2} \left\{ \int_{0}^{t^{1/n}} r^{n/2+\beta} \times \right.$$

$$\times J_{(n-2)/2}\left(r|\sum_{j=1}^{m}\lambda^{(j)}|\right) dr \Big\} |\sum_{j=1}^{m}\lambda^{(j)}|^{-n/2} \prod_{j=1}^{m} W(d\lambda^{(j)}). \qquad (2.10.8)$$

The process $Y_{m,g_1}(t)$, $t \in [0,1]$, is strictly semi-stable with parameter $\kappa \in (\frac{1}{2} + \beta/n, \, 1 + \beta/n)$. For $\beta = 0$ the process $Y_{m,g_1}(t)$, $t \in [0,1]$, coincides with the process (2.10.1).

Corollary 2.10.1. Let assumption XXI be valid, $g_1(|x|) = |x|^\beta$, $\beta \geq 0$, $\alpha \in (0, \min\{\frac{n}{m}, \frac{n+1}{2}\})$.

Then finite-dimensional distributions of the random processes

$$Y_{m,r,g_1}(t) = \frac{1}{r^{n-m\alpha/2} r^\beta \sqrt{c_{25} h^m(0)}} \int_{v(rt^{1/n})} |x|^\beta \times$$

$$\times H_m(\xi(x)) dx, \quad t \in [0,1] \qquad (2.10.9)$$

converge weakly to finite-dimensional distributions of the process $Y_{m,g_1}(t)$, $t \in [0,1]$, as $r \to \infty$ (cf. (2.10.8)).

In the general case, the processes $Y_{m,g}(t)$ are semi-stable with parameter $\kappa \in (\frac{1}{2} + \beta/n, \, 1 + \beta/n)$ only when the function $\bar{g}(x)$ in assumption XXII is homogeneous: $\bar{g}(a|x|) = a^\beta \bar{g}(|x|)$, $\beta \geq 0$.

Theorems 2.10.1 and 2.10.2 have been stated in spectral terms. Imposing conditions on the cor.f. $B(|x|)$ of the field $\xi(x)$, $x \in \mathbf{R}^n$ one may provide more general assertions.

Let assumptions I and IV be valid for $\alpha \in (0, n)$. Consider the spectral representations (1.2.1) and (1.2.3) where $F(\cdot)$ is the spectral measure of the field $\xi(x)$, $x \in \mathbf{R}^n$, and $Z_F(\cdot)$ is an orthogonal Gaussian measure having the structure function $F(\cdot)$. Assume henceforth that $F(\cdot)$ is a non-atomic measure. We introduce the collection of locally finite measures

$$F_r(\Delta) = r^\alpha L^{-1}(r) F(r^{-1}\Delta), \quad \Delta \in \mathbf{B}^n. \qquad (2.10.10)$$

Consider the collection of random processes

$$\overline{X}_{m,r}(t) = \frac{1}{r^{n-m\alpha/2}L^{m/2}(r)\sqrt{c_2}} \int_{v(rt^{1/n})} H_m(\xi(x))dx, \qquad (2.10.11)$$

where $c_2 = c_2(n, m, \alpha, v(1))$ is given by formula (2.1.8). Using the Itô formula (cf. §2.9), we obtain

$$H_m(\xi(x)) = \int_{\mathbf{R}^{nm}}' \exp\left\{i\sum_{j=1}^m \langle \lambda^{(j)}, x\rangle\right\} \prod_{j=1}^m Z_F(d\lambda^{(j)}). \qquad (2.10.12)$$

Substituting (2.10.12) into (2.10.11) and using (1.4.5), we have

$$\overline{X}_{m,r}(t) \stackrel{d}{=} c_{25}\sqrt{t} \int_{\mathbf{R}^{nm}}' J_{n/2}\left(\Big|\sum_{j=1}^m \lambda^{(j)}\Big|t^{1/n}\right) \times$$

$$\times \Big|\sum_{j=1}^m \lambda^{(j)}\Big|^{-n/2} \prod_{j=1}^m Z_{F_r}(d\lambda^{(j)}), \qquad (2.10.13)$$

where $c_{25} = c_{25}(n, m, \alpha) = (2\pi)^n/\sqrt{c_2(n, m, \alpha, v(1))}$ and $Z_{F_r}(\cdot)$ is an orthogonal Gaussian random measure with the structure function $F_r(\cdot)$.

Below we are going to proceed to the limit in formula (2.10.13) as $r \to \infty$. Let

$$f_r(z^{(1)}, \ldots, z^{(m)}; t) = \int_{\mathbf{R}^{nm}} \exp\left\{i\sum_{j=1}^m \langle \lambda^{(j)}, z^{(j)}\rangle\right\} \times$$

$$\times K(\lambda^{(1)}, \ldots, \lambda^{(m)}; t) \prod_{j=1}^m F_r(d\lambda^{(j)}),$$

where $K(\lambda^{(1)}, \ldots, \lambda^{(m)}; t) = \dfrac{tJ_{n/2}^2(|\lambda^{(1)}+\ldots+\lambda^{(m)}|t^{1/n})}{|\lambda^{(1)}+\ldots+\lambda^{(m)}|^n}$.

Lemma 2.10.2. Under assumptions I, IV and $\alpha \in (0, n)$, for each $t \in [0, 1]$

$$\lim_{r\to\infty} f_r(z^{(1)}, \ldots, z^{(m)}; t) = f(z^{(1)}, \ldots, z^{(m)}; t) =$$

$$= \int_{v(t^{1/n})} \int_{v(t^{1/n})} \frac{dxdy}{\prod_{j=1}^m |x - y + z^{(j)}|^\alpha} \qquad (2.10.14)$$

uniformly in every finite interval, where $f(z^{(1)}, \ldots, z^{(m)})$ is a continuous function.

Proof. Performing the change of variables we obtain, in view of (1.4.5),

$$
f_r(z^{(1)}, \dots, z^{(m)}; t) = \frac{1}{r^{2n-m\alpha} L^m(r)} \int_{v(rt^{1/n})} \int_{v(rt^{1/n})} dx\,dy \times
$$

$$
\times \int_{\mathbf{R}^{nm}} \exp\Big\{ i \sum_{j=1}^{m} \langle \lambda^{(j)}, x - y + rz^{(j)} \rangle \Big\} \prod_{j=1}^{m} F(d\lambda^{(j)}) =
$$

$$
= \frac{1}{L^m(r) r^{-m\alpha}} \int_{v(t^{1/n})} \int_{v(t^{1/n})} \prod_{j=1}^{m} B(r|x - y + z^{(j)}|)\,dx\,dy =
$$

$$
= \int_{v(t^{1/n})} \int_{v(t^{1/n})} \Big\{ \prod_{j=1}^{m} |x - y + z^{(j)}|^{-\alpha} \Big\} \Big\{ \prod_{j=1}^{m} \frac{L(r|x - y + z^{(j)}|)}{L(r)} \Big\} dx\,dy.
$$

Using Theorem 2.1.1, the last relation implies the assertion of Lemma 2.10.2. ∎

Lemma 2.10.3. If assumptions I, IV are valid with $\alpha \in (0, n)$, then the collection of locally finite measures (2.10.10) converges locally weakly to a locally finite measure $F_0(\Delta)$, $\Delta \in \mathcal{B}^n$, satisfying the homogeneity condition $F_0(a\Delta) = a^\alpha F_0(\Delta)$, $\Delta \in \mathcal{B}^n$, $a \in (0, \infty)$, and the relation

$$
\int_{\mathbf{R}^n} \exp\{i\langle z, \lambda \rangle\} J_{n/2}^2(|\lambda|) |\lambda|^{-n} F_0(d\lambda) =
$$

$$
= \int_{v(1)} \int_{v(1)} \frac{1}{|x - y + z|^\alpha} dx\,dy. \qquad (2.10.15)
$$

Proof. Let $K(z) = K(z; 1)$. Using the theorem on the equivalence between weak convergence of measures and convergence of the corresponding Fourier transforms [30], we conclude from Lemma 2.10.1 for $m = 1$, $t = 1$, that the collection of finite measures $\mu_r(\Delta) = \int_\Delta K(z) F_r(dz)$, $\Delta \in \mathcal{B}^n$, converges weakly to a finite measure $\mu_0(\Delta)$, $\Delta \in \mathcal{B}^n$, such that the corresponding Fourier transform $\int_{\mathbf{R}^n} \exp\{i\langle \lambda, z \rangle\} \mu_0(dz) = f(z; 1)$ (cf. (2.10.14)).

Let γ be the first zero of the function $J_{n/2}(u) u^{-n}$. Then for any $\Delta \subset v(\gamma)$ such that $\mu_0(\partial\Delta) = 0$,

$$
\lim_{r \to \infty} F_r(\Delta) = \lim_{r \to \infty} \int_\Delta [K(\lambda)]^{-1} \mu_r(d\lambda) = \int_\Delta [K(\lambda)]^{-1} \mu_0(d\lambda) = F_0(\Delta),
$$

where $F_0(\cdot)$ is a measure on $\mathcal{B}(v(\gamma))$. For any $a > 0$, $\Delta \in \mathcal{B}^n$, $\Delta \subset v(\gamma)$ such that $F_0(\partial\Delta) = 0$,

$$
\lim_{r \to \infty} F_r(a\Delta) = a^\alpha F_0(\Delta). \qquad (2.10.16)
$$

Indeed, let $\nu = r/a$. As $r \to \infty$,

$$F_r(a\Delta) = (r/\nu)^\alpha (L(\nu)/L(r)) F_r(a\nu\Delta/r) \to a^\alpha F_0(\Delta),$$

whence (2.10.16) follows. Equality (2.10.16) implies that for any Δ_1, $\Delta_2 \subset v(\gamma)$, $\Delta_1 = a\Delta_2$ and for some $a > 0$, the relation

$$F_0(\Delta_1) = a^\alpha F_0(\Delta_2) \tag{2.10.17}$$

is valid.

Define a set function $F_0(\Delta)$ for any bounded $\Delta \in \mathcal{B}^n$ in the following manner: if $\Delta \subset v(s\gamma)$, then $F_0(\Delta) = s^\alpha F_0(\Delta/s)$. Having extended $F_0(\cdot)$ onto $\mathcal{B}^n$ we obtain a locally finite measure $F_0(\cdot)$. Relation (2.10.16) holds without the constraint $\Delta \subset v(\gamma)$. The measures F_r converge locally weakly to the locally finite measure F_0 and (2.10.17) is valid. Hence $\mu_0(\Delta) = \int_\Delta K(\lambda) F_0(d\lambda)$. By the uniqueness theorem for Fourier transforms of finite measures the Fourier transform of the measure μ_0 is $f(z; 1)$ (see (2.10.14)); this implies (2.10.15). ∎

It follows from Lemma 2.10.3 that the number $\alpha \in (0, n)$ defines the Fourier transform of the measure $K(z) F_0(dz)$ and hence, the measure F_0 itself.

Let $Z_{F_0}(\cdot)$ be an orthogonal Gaussian random measure with the structure function $F_0(\cdot)$. Consider the random processes

$$\overline{X}_m(t) = c_{25} \sqrt{t} \int_{\mathbf{R}^{nm}}' \frac{J_{n/2}(|\lambda^{(1)} + \ldots + \lambda^{(m)}| t^{1/n})}{|\lambda^{(1)} + \ldots + \lambda^{(m)}|^{n/2}} \times$$

$$\times \prod_{j=1}^m Z_{F_0}(d\lambda^{(j)}), \quad t \in [0, 1], \ m \geq 1,$$

which are semi-stable in the wide sense with parameter $\kappa = 1 - \frac{\alpha m}{2n} \in (\frac{1}{2}, 1)$.

Theorem 2.10.3. If assumptions I, IV with $\alpha \in (0, \frac{n}{m})$, $m \geq 1$ are valid, then the finite-dimensional distributions of the random processes $\overline{X}_{m,r}(t)$ converge to finite-dimensional distributions of the random processes $\overline{X}_m(t)$, $t \in [0, 1]$, as $r \to \infty$.

Proof. Let $t_j \in [0, 1]$, $b_j \in \mathbf{R}^1$, $j = 1, \ldots, p$. It suffices to show that

$$\zeta_r = \sum_{j=1}^p b_j \overline{X}_{m,r}(t_j) \xrightarrow{\mathcal{D}} \sum_{j=1}^p b_j \overline{X}_m(t_j) \tag{2.10.18}$$

as $r \to \infty$.

In view of (2.10.13),

$$\zeta_r = \int_{R^{nm}}' \overline{K}(\lambda^{(1)}, \ldots, \lambda^{(m)}) \prod_{j=1}^{m} Z_{F_r}(d\lambda^{(j)}),$$

where

$$\overline{K}(\lambda^{(1)}, \ldots, \lambda^{(m)}) = c_{25} \sum_{j=1}^{p} b_j \sqrt{t} J_{n/2}(|\lambda^{(1)} + \ldots + \lambda^{(m)}| t_j^{1/n}) \times$$

$$\times |\lambda^{(1)} + \ldots + \lambda^{(m)}|^{-n/2}.$$

Lemma 2.10.3 implies locally weak convergence of the measures F_r to the measure F_0 as $r \to \infty$; it follows from Lemma 2.10.2 that the collection of finite measures

$$\bar{\mu}_r(\Delta) = \int_\Delta [\overline{K}(\lambda^{(1)}, \ldots, \lambda^{(m)})]^2 \prod_{j=1}^{m} F_r(d\lambda^{(j)}), \quad \Delta \in \mathcal{B}^n,$$

converges to the finite measure

$$\bar{\mu}_0(\Delta) = \int_\Delta [\overline{K}(\lambda^{(1)}, \ldots, \lambda^{(m)})]^2 \prod_{j=1}^{m} F_0(d\lambda^{(j)}), \quad \Delta \in \mathcal{B}^n,$$

since their characteristic functions converge. Therefore condition (2.9.1) holds and (2.10.18) follows from Theorem 2.9.1. ∎

We now provide an analogue of Theorem 2.10.1 for spherical averages (see Theorem 2.2.2).

Let assumption XXI be valid with $\alpha \in (0, \frac{n+1}{2})$. Then proceeding in the same manner as in the proof of Lemma 2.10.1, we obtain for $\tilde{\sigma}_m^2(r)$ (see (2.1.10)) the following asymptotic expression as $r \to \infty$, $\alpha \in (0, \min\{\frac{n-1}{m}, \frac{n+1}{2}\})$, $m \geq 1$, $n \geq 2$:

$$\tilde{\sigma}_m^2(r) = \mathrm{var}\left[\int_{s(r)} H_m(\xi(x)) dm(x)\right] = r^{2n-2-m\alpha} h^m(0) \tilde{f}(n, m, \alpha)(1 + o(1)),$$

where

$$\tilde{f}(n, m, \alpha) = c_9(n, m) B\left(\frac{n - m\alpha - 1}{2}, \frac{n-1}{2}\right) \times$$

$$\times 2^{n-2} \left[\pi^{n/2} \Gamma\left(\frac{\alpha}{2}\right) / \Gamma\left(\frac{n-\alpha}{2}\right)\right]^m.$$

Consider for $n \geq 2$, $t \in [0, 1]$, the random processes

$$\tilde{X}_{m,r}(t) = \frac{1}{\sqrt{\tilde{f}(n, m, \alpha)} h^m(0) r^{n-1-m\alpha/2}} \int_{s(rt^{1/(n-1)})} H_m(\xi(x)) dm(x)$$

and the random processes

$$\tilde{X}_m(t) = \frac{(2\pi)^{n/2}}{\sqrt{\tilde{f}(n, m, \alpha)}} t^{n/2(n-1)} \times$$

$$\times \int'_{\mathbf{R}^{nm}} \frac{J_{(n-2)/2}(|\lambda^{(1)} + \ldots + \lambda^{(m)}| t^{1/(n-1)}) \prod_{j=1}^m W(d\lambda^{(j)})}{|\lambda^{(1)} + \ldots + \lambda^{(m)}|^{(n-2)/2} \prod_{j=1}^m |\lambda^{(j)}|^{(n-\alpha)/2}},$$

where the m.s.i. with respect to a Gaussian white noise $W(\cdot)$ in $(\mathbf{R}^n, \mathbf{B}^n)$ appears on the right-hand side and $\alpha \in (0, \min\{\frac{n-1}{m}, \frac{n+1}{2}\})$, $m \geq 1$, $n \geq 2$.

Note that $E\tilde{X}_m^2(t) < \infty$, $t \in [0, 1]$, for $n \geq 2$. The process $\tilde{X}_m(t)$, $t \in [0, 1]$, is semi-stable of order $\kappa = 1 - \frac{\alpha m}{2(n-1)} \in (\frac{1}{2}, 1)$.

Theorem 2.10.4. If assumption XXII is valid for $\alpha \in (0, \min\{\frac{n-1}{m}, \frac{n+1}{2}\})$, $m \geq 1$, $n \geq 2$, then as $r \to \infty$, the finite-dimensional distributions of the random processes $\tilde{X}_{m,r}(t)$, $t \in [0, 1]$, converge weakly to finite-dimensional distributions of $\tilde{X}_m(t)$, $t \in [0, 1]$.

The proof of Theorem 2.10.4 is analogous to that of Theorem 2.10.1 using formula (1.2.8) in place of (1.4.5) (see the derivation of relation (2.10.3)).

CHAPTER 3

Estimation of Mathematical Expectation

3.1. Asymptotic Properties of the Least Squares Estimators for Linear Regression Coefficients

Consider the random field

$$\xi(x) = \xi_\theta(x) = g(x, \theta) + \epsilon(x), \qquad (3.1.1)$$

where $g(x, \theta) : \mathbf{R}^n \times \Theta^c \to \mathbf{R}^1$ is a function depending on an unknown parameter $\theta \in \Theta \in \mathcal{B}^q$ and $\epsilon(x)$, $x \in \mathbf{R}^n$, is a random field with zero mean. Assume that in $\mathbf{R}^n$ a system of measurable bounded sets $\mathfrak{M} = \{\Delta\}$ is selected and $\Delta \to \infty$. The problem is to estimate the parameter θ from observations of the random field $\xi(x)$ on the sets $\Delta \to \infty$.

The regression model (3.1.1) will be called linear if the function $g(x, \theta)$ is a linear form in the variables $\theta = (\theta_1, \ldots, \theta_q)$: $g(x, \theta) = \sum_{j=1}^q \theta_j g_j(x)$. The vector-function $g'(x) = (g_1(x), \ldots, g_q(x))$ whose coordinates are linearly independent on $\Delta \in \mathfrak{M}$ is assumed to be given. In particular, with $q = 1$ the linear model becomes

$$\xi(x) = \theta g(x) + \epsilon(x) \qquad (3.1.2)$$

and includes estimation of the unknown mean θ of the field $\xi(x)$ $(g(x) \equiv 1)$.

In §§3.2–3.6 a non-linear model (3.1.1) will be discussed and asymptotic properties of the least squares and least moduli estimators of the parameter θ will be obtained as $\Delta \to \infty$. Most of the results to be derived are valid for a linear regression model as well. This section provides properties of the least squares estimators of the parameter θ for a linear function $g(x, \theta) = \sum_{j=1}^q \theta_j g_j(x)$.

Assume that $\epsilon(x), x \in \mathbf{R}^n$ is a measurable m.s. continuous homogeneous random field with cor. f. $B(x)$; let $g(x, \theta)$ appearing in (3.1.1) be measurable with respect to x for any $\theta \in \Theta^c$ and continuous with respect to θ for any $x \in \mathbf{R}^n$. Thus $g(x, \theta)$ is measurable with respect to the totality of the variables (x, θ). Set

$$G(x, \tau) = (\xi_\theta(x) - g(x, \tau))^2, \; L_\Delta(\tau) = \int_\Delta G(x, \tau) dx.$$

The least squares estimator of the parameter $\theta \in \Theta$ is defined as the r.v. $\hat{\theta}_\Delta = (\hat{\theta}_{1\Delta}, \ldots, \hat{\theta}_{q\Delta})$ with the property

$$L_\Delta(\hat{\theta}_\Delta) = \inf_{\tau \in \Theta^c} L_\Delta(\tau).$$

If $g(x, \theta)$ is a non-linear function or $g(x, \theta)$ is linear but $\Theta \neq R^q$ and is not defined by simple constraints, the estimator $\hat{\theta}_\Delta$ cannot in general be written down explicitly. If $g(x, \theta)$ is a linear function and $\Theta = R^q$, the estimator $\hat{\theta}_\Delta$ may then be obtained in an explicit form.

Suppose that the functions $g_j(x)$ of the linear model are square integrable, that is, $d_{j\Delta}^2 = \int_\Delta g_j^2(x) dx < \infty$, $j = 1, \ldots, q$, $\Delta \in \mathfrak{M}$. Then the matrices $V_\Delta = (\int_\Delta g_j(x) g_r(x) dx)_{j,r=1,\ldots,q}$ will be non-singular and if $\zeta_\Delta = \int_\Delta g(x) \xi(x) dx$, then $\hat{\theta}_\Delta = V_\Delta^{-1} \zeta_\Delta$. Clearly, $\hat{\theta}_\Delta$ is an unbiased estimate of $\theta : E\hat{\theta}_\Delta = \theta$, and its correlation matrix is of the form:

$$\sigma_\Delta^2 = V_\Delta^{-1} \left(\int_\Delta \int_\Delta B(x, y) g(x) g'(y) dx dy \right) V_\Delta^{-1}. \tag{3.1.3}$$

We introduce the diagonal matrix $d_\Delta = \mathrm{diag}(d_{i\Delta})_{i=1,\ldots,q}$ and the matrix measure $\mu_\Delta(d\lambda)$ on $(\mathbf{R}^n, \mathcal{B}^n)$ with density matrix $\mu_{rj}(\Delta))_{r,j=1,\ldots,q}$, where

$$\mu_{rj}(\Delta) = g_\Delta^{(r)}(\lambda) \overline{g_\Delta^{(j)}(\lambda)} \left(\int_{\mathbf{R}^n} |g_\Delta^{(r)}(\lambda)|^2 d\lambda \int_{\mathbf{R}^n} |g_\Delta^{(j)}(\lambda)|^2 d\lambda \right)^{-1/2},$$

$$g_\Delta^{(j)}(\lambda) = \int_\Delta e^{i\langle \lambda, x \rangle} g_j(x) dx, \quad r, j = 1, \ldots, q.$$

Note that $d_{j\Delta}^2 = (2\pi)^{-n} \int_{\mathbf{R}^n} |g_\Delta^{(j)}(\lambda)|^2 d\lambda$. If the field $\epsilon(x)$, $x \in \mathbf{R}^n$, possesses the spectral density function $f(\lambda)$, $\lambda \in \mathbf{R}^n$, then using (1.2.2) we obtain from (3.1.3) that

$$d_\Delta \sigma_\Delta^2 d_\Delta = (2\pi)^n \left(\int_{\mathbf{R}^n} \mu_\Delta(d\lambda) \right)^{-1} \int_{\mathbf{R}^n} f(\lambda) \mu_\Delta(d\lambda) \left(\int_{\mathbf{R}^n} \mu_\Delta(d\lambda) \right)^{-1}. \tag{3.1.4}$$

I. $\mu_\Delta \xrightarrow{\mathfrak{D}} \mu$ as $\Delta \xrightarrow{V.H.} \infty$ and $\mathrm{var}\,\mu = \left(\int_{\mathbf{R}^n} \mu_{rj}(d\lambda) \right)_{r,j=1,\ldots,q}$ is a non-singular matrix.

A measure μ on $(\mathbf{R}^q, \mathbf{B}^q)$ whose existence is postulated in assumption I is called a spectral measure of the regression function $g(x)$.

The matrix measures μ_Δ and μ are, in general, complex-valued. In practice, determination of the measure μ may be based on the relations:

$$d_{r\Delta}^{-1} d_{j\Delta}^{-1} \int_{\Delta \cap (\Delta - h)} g_r(x + h) g_j(x) dx \rightarrow \int_{\mathbf{R}^n} e^{i\langle \lambda, h \rangle} \mu_{rj}(d\lambda), \quad \Delta \xrightarrow{V.H.} \infty,$$

$r, j = 1, \ldots, q$, which are analogous to those in [174, 52].

If $f(\lambda)$, $\lambda \in \mathbf{R}^n$, is a continuous bounded function, then from (3.1.4) and I we obtain that

$$\sigma_1^1 = \lim d_\Delta \sigma_\Delta^2 d_\Delta = (2\pi)^n \left(\int_{\mathbf{R}^n} \mu(d\lambda) \right)^{-1} \int_{\mathbf{R}^n} f(\lambda) \mu(d\lambda) \left(\int_{\mathbf{R}^n} \mu(d\lambda) \right)^{-1}$$

as $\Delta \xrightarrow{V.H.} \infty$.

An unbiased estimator $\hat{\theta}_\Delta$ of the parameter θ is called mean-square consistent, if its correlation matrix tends to the zero matrix as $\Delta \rightarrow \infty$.

Let $\mathfrak{M} = \{\Delta\}$ be a system of linearly ordered sets with ordering relation "$\subseteq$", $\Delta \rightarrow \infty$ and suppose that the function $f(\lambda)$ is bounded and almost everywhere positive. Following A.S. Kholevo's arguments [127], we conclude that the least-squares estimators $\hat{\theta}_\Delta$ are mean square consistent if and only if for any vector $\theta = (\theta_1, \ldots, \theta_q)' \neq 0$ the function $g(x, \theta) = \sum_{j=1}^q \theta_j g_j(x)$ satisfies the condition $\int_{\mathbf{R}^n} g^2(x, \theta) dx = \infty$.

Theorem 3.1.1. Suppose that the vector-function $g(x)$ satisfies assumption I and condition (1.8.10) coordinatewise. Suppose also that the homogeneous random field $\epsilon(x)$, $x \in \mathbf{R}^n$, satisfies the strong mixing condition with mixing rate $\alpha(\rho)$, so that

$$E|\epsilon(0)|^{2+\delta} < \infty, \quad \alpha(\rho) \leq k_1 \rho^{-n-\gamma}, \quad \gamma\delta > 2n, \quad \delta > 0. \qquad (3.1.5)$$

Then there exists a continuous bounded spectral density function $f(\lambda)$, $\lambda \in \mathbf{R}^n$, of the field $\epsilon(x)$, $x \in \mathbf{R}^n$, and if the matrix $\int_{\mathbf{R}^n} f(\lambda) \mu(d\lambda)$ is non-singular, then

$$d_\Delta(\hat{\theta}_\Delta - \theta) \xrightarrow{\mathfrak{D}} N_q(0, \sigma_1^2).$$

as $\Delta \xrightarrow{V.H.} \infty$.

Proof. In view of Remark 1.6.1 this assertion follows from Theorem 1.7.5, if one chooses functions $g_\Delta^{(j)}(x, \theta) = g_j(x) d_{j\Delta}^{-1}$, $j = 1, \ldots, q$, not depending on θ to be the components of the weight function $g_\Delta(x, \theta)$. ∎

The mixing condition in (3.1.5) may be relaxed for collections $\mathfrak{M}$ consisting of rectangles or spheres (cf. §1.7). If $\Delta \xrightarrow{F} \infty$, then in place of (3.1.5) one may utilize restrictions on the mixing rate $\alpha(\rho, d)$ from Theorem 1.7.2 or 1.7.3.

If in (3.1.2) $g(x) \equiv 1$, $f(0) > 0$, then the assertion of Theorem 3.1.1 means that the estimator $\hat{\theta}_\Delta = \int_\Delta \xi(x)dx/|\Delta|$ of the parameter θ is asymptotically normal as $\Delta \xrightarrow{V.H.} \infty : |\Delta|^{1/2}(\hat{\theta}_\Delta - \theta) \xrightarrow{\mathcal{D}} N(0, (2\pi)^n f(0))$.

Using Corollary 1.8.1, Theorem 3.1.1 may be extended to a functional central limit theorem following the proof of Theorem 1.9.2. To illustrate the above, consider the field (3.1.2) and assume that $\mathfrak{M}$ consists of expanding balls $v(r)$. Denote $\hat{\theta}(r) = \hat{\theta}_{v(r)}$. Let the function $a(r) = d^2_{v(r)}$ be monotonically increasing. Replacing condition (3.1.5) by condition (1.9.2) we obtain that the probability measures P_r in $C[0, 1]$ corresponding to the random processes

$$X_r(t) = ta^{1/2}(r)[\hat{\theta}(a^{-1}(ta(r))) - \theta], \ t \in [0, 1], \tag{3.1.6}$$

converge weakly, as $r \to \infty$, to a measure w_b in $C[0, 1]$ corresponding to a Brownian motion process $w_b(t)$, $t \in [0, 1]$, having local variance $b = (2\pi)^n \int_{\mathbf{R}^n} f(\lambda) \times \mu(d\lambda)$ (in (3.1.6) $a^{-1}(\cdot)$ is the inverse of $a(\cdot)$).

Thus, if in (3.1.2) $g(x) = |x|^\nu$, $\nu \geq 0$ (a radial polynomial regression), then

$$a(r) = (n + \nu)^{-1}|s(1)|r^{n+\nu}, \ a^{-1}(u) = ((n + \nu)|s(1)|^{-1}u)^{1/(n+\nu)},$$

$$X_r(t) = (n + \nu)^{1/2}|s(1)|^{-1/2}r^{-(n+\nu)/2} \int_{v(rt^{1/(n+\nu)})} |x|^\nu \epsilon(x)dx, \ t \in [0, 1].$$

Whence, for $\nu = 0$ we obtain the random processes constructed in §1.9.

When $\mathfrak{M}$ is a system of expanding balls and the field $\epsilon(x)$, $x \in \mathbf{R}^n$, is homogeneous and isotropic, the notion of a spectral measure may be modified so that its calculation is reduced to an analysis of single integrals.

To simplify matters, consider the model (3.1.2) only with $\Delta = v(r)$. Then the least squares estimator of the parameter θ becomes

$$\hat{\theta}(r) = \int_{v(r)} \xi(x)g(x)dx/d^2_r, \quad d^2_r = d^2_{v(r)} = \int_{v(r)} g^2(x)dx.$$

We proceed to polar coordinates (1.1.1). Let $g^l_m(r) = \int_{s(1)} g(r, u)S^l_m(u)dm(u)$ be the Fourier coefficients of the function $g(r, u)$ with respect to an orthonormal system of spherical harmonics $\{S_m^l(u)\}$, $(m,l) \in T \equiv \{0,1,\ldots\} \times \{0,1,\ldots$ $\ldots, h(n,m)\}$ (cf. §1.2). Then

$$d_r^2 = \int_0^r \int_{s(1)} g^2(\rho,u)\rho^{n-1}d\rho dm(u) = \sum_{m=0}^{\infty} \sum_{l=1}^{h(n,m)} \int_0^r [g_m^l(\rho)]^2 \rho^{n-1}d\rho, \quad (3.1.7)$$

and by virtue of the Parseval equality for the Hankel transform, we have

$$\int_0^r \rho[g_m^l(\rho)\rho^{(n-2)/2}]^2 d\rho = \int_0^{\infty} \lambda \left[\int_0^r g_m^l(\rho) J_{m+(n-2)/2}(\lambda\rho)\rho^{n/2}d\rho \right]^2 d\lambda.$$
$$(3.1.8)$$

Utilizing (3.1.7), (3.1.8) we introduce the set of normalized measures on $(\mathbf{R}_+^1, \mathcal{B}(\mathbf{R}_+^1))$:

$$\mu_r^{m,l}(d\lambda) = \left[\int_0^r g_m^l(\rho) J_{m+(n-2)/2}(\lambda\rho)\rho^{n/2}d\rho \right]^2 \lambda d\lambda \left[\int_0^r [g_m^l(\rho)]^2 \rho^{n-1}d\rho \right]^{-1}.$$

Let

$$A_{m,l}(r) = a_{m,l}(r) \Big/ \sum_{m=0}^{\infty} \sum_{l=1}^{h(n,m)} a_{m,l}(r),$$

where $a_{m,l}(r) = \int_0^r [g_m^l(\rho)]^2 \rho^{n-1}d\rho$, $(m,l) \in T$. Clearly,

$$A_{m,l}(r) \geq 0, \quad \sum_{m=0}^{\infty} \sum_{l=1}^{h(n,m)} A_{m,l}(r) = 1;$$

thus for every r, a probability distribution on the set T is defined.

Assume that the field $\epsilon(x)$, $x \in \mathbf{R}^n$ has an isotropic spectral density function $\tilde{g}(\lambda)$, $\lambda \in \mathbf{R}_+^1$. Utilizing (1.2.11) and the addition theorem for Bessel functions we obtain a representation for the variance σ_r^2 of the estimator $\hat{\theta}_r$

$$\sigma_r^2 d_r^2 = \int_{v(r)} \int_{v(r)} B(|x-y|)g(x)g(y)dxdy/d_r^2 =$$

$$= d_r^{-2} \int_0^r \int_0^r \int_{s(1)} \int_{s(1)} g(\rho_1, u_1)g(\rho_2, u_2) \int_0^{\infty} c_1^2(n) \sum_{m=0}^{\infty} \sum_{l=1}^{h(n,m)} S_m^l(u_1) \times$$

$$\times S_m^l(u_2) J_{m+(n-2)/2}(\lambda\rho_1)(\lambda\rho_1)^{-(n-2)/2} J_{m+(n-2)/2}(\lambda\rho_2)(\lambda\rho_2)^{-(n-2)/2} \times$$

$$\times dG(\lambda)(\rho_1\rho_2)^{n-1}d\rho_1 d\rho_2 dm(u_1)dm(u_2) =$$

$$= d_r^{-2} c_1^2(n) \sum_{m=0}^{\infty} \sum_{l=1}^{h(n,m)} \int_0^{\infty} \left[\int_0^r g_m^l(\rho) J_{m+(n-2)/2}(\lambda\rho)(\lambda\rho)^{-(n-2)/2} \rho^{n-1} d\rho \right]^2 \times$$

$$\times G(d\lambda) = (2\pi)^n \sum_{m=0}^{\infty} \sum_{l=1}^{h(n,m)} A_{m,l}(r) \int_0^{\infty} \tilde{g}(\lambda) \mu_r^{m,l}(d\lambda). \qquad (3.1.9)$$

We introduce the following assumption:

II. As $r \to \infty$ $\mu_r^{m,l} \xrightarrow{\mathcal{D}} \mu_{m,l}$;

$$\lim_{r \to \infty} A_{m,l}(r) = A_{m,l} \in [0, \infty), \ (m, l) \in T; \qquad (3.1.10)$$

in (3.1.9) the term-by-term proceeding to the limit is legal as $r \to \infty$ and

$$\sigma_2^2 = (2\pi)^n \sum_{m=0}^{\infty} \sum_{l=1}^{h(n,m)} A_{m,l} \int_0^{\infty} \tilde{g}(\lambda) \mu_{m,l}(d\lambda) \in (0, \infty). \qquad (3.1.11)$$

The set of measures $\{\mu_{m,l}\}$, $(m, l) \in T$, whose existence is postulated by assumption II will be called an isotropic spectral measure of the regression function. Let $g(r, u) = g(r)$ be a radial function. Then $g_0^1(r) = g(r)|s(1)|^{1/2}$, the remaining $g_m^l(r)$ vanishing. Hence, $A_{0,1}(r) = A_{0,1} = 1$ the remaining $A_{m,l}(r) = A_{m,l} = 0$ and (3.1.10) is valid. In Assumption II we merely require that $\mu_r^{0,1} \xrightarrow{\mathcal{D}} \mu_{0,1}$ as $r \to \infty$. Then in place of (3.1.11) we obtain:

$$\sigma_2^2 = (2\pi)^n \int_0^{\infty} \tilde{g}(\lambda) \mu_{0,1}(d\lambda) \in (0, \infty). \qquad (3.1.12)$$

Choosing for $g(r, u)$ the sum of a radial function and a finite number of distinct spherical harmonics, we obtain for $n = 2$, 3 the quantities (3.1.11), where $A_{m,l}$ are non-zero for finite subsets of the set T.

If the function $\tilde{g}(\lambda)$ is continuous and bounded and assumption II holds, we obtain from (3.1.9) $\lim_{r \to \infty} \sigma_r^2 d_r^2 = \sigma_2^2$, where σ_2^2 is defined by relation (3.1.11). If the function g is radial, then relation (3.1.12) is valid for σ_2^2.

Theorem 3.1.2. Let the random field (3.1.2) be observed on $\Delta = v(r)$, and let the homogeneous isotropic field $\epsilon(x)$, $x \in \mathbf{R}^n$, satisfy condition (3.1.5) and the regression function $g(x) = g(r, u)$ satisfy condition (1.8.10) with $\Delta = v(r)$. Suppose also that Assumption II holds. Then there exists a continuous bounded isotropic spectral density $\tilde{g}(\lambda)$, $\lambda \in \mathbf{R}_+^1$ and as $r \to \infty$, $d_r(\hat{\theta}(r) - \theta) \xrightarrow{\mathcal{D}} N(0, \sigma_2^2)$.

The assertion of the theorem follows from Theorem 1.7.3 and Remark 1.6.1.

Evidently, the notion of an isotropic spectral measure may be introduced based on decompositions (1.2.12) and (1.2.13) with $n = 2, 3$.

Assume now that the field (3.1.2) is observed on a sphere $s(r)$ and that $D_r^2 = \int_{s(r)} g^2(x)dm(x) < \infty$ for any $r > 0$ and let the field $\epsilon(x)$, $x \in \mathbf{R}^n$, be homogeneous and isotropic. The least squares estimator of the parameter θ is of the form

$$\hat{\theta}^*(r) = \int_{s(r)} g(x)\xi(x)dm(x) \, / \, D_r^2.$$

This is an unbiased estimator, its variance being

$$\sigma_*^2(r) = c_1^2(n) \sum_{m=0}^{\infty} \sum_{l=1}^{h(n,m)} \bar{A}_{m,l}(r) b_{m,n}(r) \Big[\sum_{m=0}^{\infty} \sum_{l=1}^{h(n,m)} [g_m^l(r)]^2 \Big]^{-1}, \qquad (3.1.13)$$

where

$$\bar{A}_{m,l}(r) = [g_m^l(r)]^2 \, / \, \Big[\sum_{m=0}^{\infty} \sum_{l=1}^{h(n,m)} [g_m^l(r)]^2 \Big],$$

$$b_{m,n}(r) = \int_0^{\infty} J_{m+(n-2)/2}(\lambda r)(\lambda r)^{2-n} G(d\lambda);$$

$G(d\lambda)$ appears in the representation (1.2.4).

For example, let

$$B(r) = Y_{n+2}(ar), \ a > 0, \ G(\lambda) = \lambda^n a^{-n} \text{ for } \lambda \le a;$$

$G(\lambda) = 1$ for $\lambda > a$. Then

$$b_{m,n}(r) = 2^{n-2}\Gamma(n/2)\pi^{n/2}J_{m+(n-2)/2}^2(ar)(ar)^{2-n}.$$

If

$$B(r) = \exp\{-ar^2\}, \quad G'(\lambda) = (2a)^{-n/2}\lambda^{n-1}\exp\{-\lambda^2/4a\},$$

then

$$b_{m,n}(r) = 2(2/a)^{(n-2)/2}\pi^{n/2}\Gamma(n/2)r^{2-n}\exp\{-2ar^2\}J_{m+(n-2)/2}(2ar^2), \ a > 0.$$

If $\sup_m b_{m,n}(r) \le k_1$ for all r, then $\hat{\theta}^*(r) \xrightarrow{P} \theta$ as $r \to \infty$.

Consider the asymptotic behaviour of $\sigma_*^2(r)$ as $r \to \infty$.

III. For some $\gamma \in (-1, n-2)$, $n \ge 2$, $G'(\lambda) = |s(1)|\lambda^{\gamma}h(\lambda)$, where $h(\lambda)$ is continuous in a neighbourhood of zero, $h(0) \ne 0$, $h(\lambda)$ is bounded on $\mathbf{R}_+^1$.

Let III hold. Setting in Lemma 1.5.4 $K(z) = J^2_{m+(n-2)/2}z^{\gamma+2-n}$, then in view of relation (1.5.16) we obtain as $r \to \infty$

$$b_{m,n}(r) = |s(1)|h(0)c_2(n,m,\gamma)r^{-1-\gamma}(1+o(1)), \qquad (3.1.14)$$

where

$$c_2(n,m,\gamma) = \Gamma(n-2-\gamma)\Gamma((2m+\gamma+1)/2)\times$$

$$\times[2^{n-2-\gamma)}\Gamma^2((n-1-\gamma)/2)\Gamma((2m+2n-3-\gamma)/2)]^{-1}.$$

Let assumption III hold. Then the limit

$$\lim_{r\to\infty} \bar{A}_{m,l}(r) = \bar{A}_{m,l} \in [0,\infty), \ (m,l) \in T \qquad (3.1.15)$$

exists and the termwise approach to the limit as $r \to \infty$ in (3.1.13) is justified and

$$\sigma_3^2 = (2\pi)^n h(0) \sum_{m=0}^{\infty} \sum_{l=1}^{h(n,m)} \bar{A}_{m,l}c_2(n,m,\gamma) \in (0,\infty). \qquad (3.1.16)$$

Formulas (3.1.13), (3.1.14), (3.1.16) then imply that

$$\sigma_*^2(r) = \sigma_3^2 \Big\{ r^{\gamma+1} \sum_{m=0}^{\infty} \sum_{l=1}^{h(n,m)} [g_m^l(r)]^2 \Big\}^{-1} (1+o(1))$$

as $r \to \infty$.

In particular, if the function $g(r,u) = g(r)$ is radial, condition (3.1.15) holds and as $r \to \infty$,

$$\sigma_*^2(r) = c_3(n,\gamma)h(0)r^{-\gamma-1}g^{-2}(r)(1+o(1)),$$

where

$$c_3(n,\gamma) = c_1^2(n)\Gamma(n-2-\gamma)\Gamma((\gamma+1)/2)\div$$

$$\div[2^{n-2-\gamma}\Gamma^2((n-\gamma-1)/2)\Gamma((2n-3-\gamma)/2)].$$

Note that for a radial function $g(r)$, the estimator $\hat{\theta}^*(r)$ of the parameter θ has minimal variance in the class of all linear unbiased estimators of the parameter θ [140,Ch.IV,§3]. In particular, for $g(r) \equiv 1$ the least squares estimator $\hat{\theta}^*(r) = |s(r)|^{-1} \int_{s(r)} \xi(x)dm(x)$ will be the optimal estimator of the unknown mean θ. Theorem 1.7.8 provides conditions for the asymptotic normality of this estimator for $n \geq 2$.

We now consider examples of asymptotic distributions of the least squares estimators derived on the basis of the results of Chapter 2.

Assume that in the model (3.1.2), the parameter θ is to be estimated from the observations $\{\xi(x),\ x \in \Delta(\mu)\}$, $\mu \to \infty$. Suppose that the field $\epsilon(x) = \tilde{G}(\eta(x))$, $x \in \mathbf{R}^n$, where the function $\tilde{G} : \mathbf{R}^1 \to \mathbf{R}^1$ satisfies assumption III of §2.1 with the coefficient $C_0 = 0$, $m = \operatorname{rank}\tilde{G}$ and the random field $\eta(x)$, $x \in \mathbf{R}^n$, satisfies assumption XXI of §2.10, that is, its spectral density function is unbounded at zero. Then in view of Lemma 2.10.1, the cor. f. $B(|x|) = B(r)$ of the field $\eta(x)$, as $r \to \infty$, is of the form

$$B(r) = c_4(n,\alpha)r^{-\alpha}(1 + o(1)), \quad \alpha \in (0,(n+1)/2), \quad c_4(n,\alpha) = h(0)c_{21}(n,\alpha),$$

where $c_{21}(n,\alpha)$ is defined in §2.10. As to the regression function $g(x) = g(|x|)$, it is assumed to be locally square integrable and admissible for the cor.f. $B(|x|)$ of the field $\eta(x)$ (cf. §2.5). Thus for $m = \operatorname{rank}\tilde{G} \geq 1$, the following positive limit exists:

$$c_5 = c_5(n,m,\alpha,g,\Delta) = m! \lim_{\mu\to\infty} \int_\Delta \int_\Delta g(\mu|x|)g(\mu|y|) \times$$

$$\times g^{-2}(\mu)|x - y|^{-\alpha m}\,dx\,dy, \quad \alpha \in (0,\min\{n/m,(n+1)/2\}). \qquad (3.1.17)$$

For the parameter θ of the model (3.1.2), consider the estimator

$$\hat{\theta}_{\Delta(\mu)} = d_\mu^{-2} \int_{\Delta(\mu)} g(x)\xi(x)\,dx, \quad d_\mu^2 = \int_{\Delta(\mu)} g^2(x)\,dx.$$

Under the conditions of Theorem 2.5.1, for $\mu \to \infty$ and $\alpha \in (0,\min\{n/m,(n+1)/2\})$

$$d_\mu^4 \operatorname{var}\hat{\theta}_{\Delta(\mu)} = A_m^2(\mu)c_5(1 + o(1)),$$

where $A_m^2(\mu) = C_m^2(m!)^{-2}c_4^m\mu^{2n-m\alpha}g^2(\mu)$. As $\mu \to \infty$, the limiting distributions of the r.v.

$$\zeta_\mu = (\hat{\theta}_{\Delta(\mu)} - \theta)/[A_m(\mu)d_\mu^{-2}c_5^{1/2}]$$

and the r.v.

$$Y_{m,\mu,g}(1) = \operatorname{sgn}\{C_m\}c_4^{-m/2}c_5^{-1/2}\mu^{(m\alpha/2)-n}g^{-1}(\mu) \int_{\Delta(\mu)} g(x)H_m(\xi(x))\,dx$$

coincide. Under the conditions of Theorem 2.10.2, as $\mu \to \infty$ the limiting distributions of the r.v.'s $Y_{m,\mu,g}(1)$ (and hence that of ζ_μ) are defined by the multiple stochastic integrals $\operatorname{sgn}\{C_m\}Y_{m,g}(1)$ (cf. (2.10.7)). In particular, for the estimator $\hat{\theta}(r)$ of the unknown mean θ of the field $\{\xi(x),\ x \in v(r)\}$ under the conditions of Theorem 2.10.1, the limiting relation

$$\zeta_r = m! r^{m\alpha/2} |v(1)|^{1/2} |C_m|^{-1} c_4^{-m/2} c_2^{-1/2} (\hat{\theta}_r - \theta) \xrightarrow{D} X_m(1) \operatorname{sgn}\{C_m\}$$

is valid where $c_2 = c_2(n, m, \alpha, v(1))$ is given by formula (2.1.10) and $X_m(1)$ by the relation (2.10.1).

3.2. Consistency of the Least Squares Estimate under Non-Linear Parametrization

Assume that $\Theta \subset R^q$ is an open set and that the function $g(x, \theta)$ in (3.1.1) depends non-linearly on the parameter $\theta \in \Theta$. §§3.2–3.4 provide certain asymptotic properties of the estimator $\hat{\theta}_\Delta$ for $\Delta \to \infty$.

Let $d_\Delta = d_\Delta(\theta) = \operatorname{diag}(d_{i\Delta}(\theta))_{i=1,\ldots,q}$, $\theta \in \Theta$. The matrix d_Δ will be used to normalize the estimator $\hat{\theta}_\Delta$. If $g(x, \theta)$ is differentiable with respect to θ and its derivatives $g_i(x, \theta) = (\partial/\partial\theta_i)g(x, \theta)$ are locally square integrable with respect to x, it is natural to choose

$$d_{i\Delta}^2(\theta) = \int_\Delta g_i^2(x, \theta) dx, \quad i = 1, \ldots, q.$$

Let $T \subset \Theta$ be a compact set.

Below we shall confine ourselves to an important case when the following relation holds:

$$\lim_{\Delta\to\infty} \inf_{\theta\in T} |\Delta|^{-1/2} d_{i\Delta}(\theta) > 0, \quad i = 1, \ldots, q. \tag{3.2.1}$$

(The left-hand side of (3.2.1) may be infinity.) A family of r.v.'s $\tilde{\theta}_\Delta = \tilde{\theta}_\Delta(\xi_\theta(x), x \in \Delta)$, is said to be a uniformly consistent family of estimators of the parameter θ on the set $T \subset \Theta$ ($\tilde{\theta}_\Delta$ is a uniformly consistent estimator of θ in T) if for any $\rho > 0$,

$$\sup_{\theta\in T} P\{|\tilde{\theta}_\Delta - \theta| \geq \rho\} \xrightarrow[\Delta\to\infty]{} 0. \tag{3.2.2}$$

We set

$$\phi_\Delta(\theta_1, \theta_2) = \int_\Delta (g(x, \theta_1) - g(x, \theta_2))^2 dx,$$

$$w_\Delta(\theta_1, \theta_2) = \int_\Delta \epsilon(x)(g(x, \theta_1) - g(x, \theta_2)) dx, \quad \theta_1, \theta_2 \in \Theta^c;$$

$$z_\Delta(\theta_1, \theta_2) = \phi_\Delta^{-1}(\theta_1, \theta_2) w_\Delta(\theta_1, \theta_2), \theta_1 \neq \theta_2,$$

$$z_\Delta(\theta_1, \theta_2) = 0, \ \theta_1 = \theta_2; \ \gamma(\Delta) = |\Delta|^{-1} \int_\Delta \epsilon^2(x) dx.$$

By definition of $\hat{\theta}_\Delta$, a.s. we have

$$\gamma(\Delta) \geq |\Delta|^{-1} L_\Delta(\hat{\theta}_\Delta) = \gamma(\Delta) - 2|\Delta|^{-1} w_\Delta(\hat{\theta}_\Delta, \theta) + |\Delta|^{-1} \phi_\Delta(\hat{\theta}_\Delta, \theta), \quad (3.2.3)$$

or

$$\phi_\Delta(\hat{\theta}_\Delta, \theta)(z_\Delta(\hat{\theta}_\Delta, \theta) - 1/2) \geq 0. \tag{3.2.4}$$

Let $B_\Delta(\theta) \subset \Theta^c$ be a Borel set and $\theta \notin B_\Delta^c(\theta)$. If $\hat{\theta}_\Delta \in B_\Delta(\theta)$ and for any $\delta > 0$ and $\theta_1, \theta_2 \in \Theta^c$, $\inf_{|\theta_1-\theta_2|>\delta} \phi_\Delta(\theta_1, \theta_2) > 0$, then (3.2.4) implies that

$$P\{\hat{\theta}_\Delta \in B_\Delta(\theta)\} \leq P\{ \sup_{\tau \in B_\Delta^c(\theta)} z_\Delta(\tau, \theta) \geq 1/2\}. \tag{3.2.5}$$

Inequality (3.2.5) is a starting point for obtaining sufficient conditions for the uniform consistency of $\hat{\theta}_\Delta$. In particular, if $B_\Delta(\theta) = \Theta^c \cap \{\tau : |\tau - \theta| \geq \rho\}$, then the convergence of the right-hand side of (3.2.5) to zero uniformly in $\theta \in T$ as $\Delta \to \infty$ implies that (3.2.2) holds for $\hat{\theta}_\Delta$.

Set $\Phi_\Delta(u_1, u_2) = \phi_\Delta(\theta + |\Delta|^{1/2} d_\Delta^{-1}(\theta) u_1, \ \theta + |\Delta|^{1/2} d_\Delta^{-1}(\theta) u_2)$. For a fixed $\theta \in \Theta$, the function $\Phi_\Delta(u_1, u_2)$ is defined on the set $U_\Delta^c(\theta) \times U_\Delta^c(\theta)$, $U_\Delta(\theta) = |\Delta|^{-1/2} d_\Delta(\theta)(\Theta - \theta)$.

Assume the following.

IV. For any $\epsilon > 0$ and $R > 0$ there exists $\delta = \delta(\epsilon, R)$ such that

$$\sup_{\theta \in T} \left[\sup_{\substack{u_1, u_2 \in U_\Delta^c(\theta) \cap V^c(R) \\ |u_1-u_2| \leq \delta}} |\Delta|^{-1} \Phi_\Delta(u_1, u_2) \right] \leq \epsilon. \tag{3.2.6}$$

V. For some $R_0 > 0$ and $\rho \in (0, R_0)$ there exist numbers $a = a(R_0) > 0$ and $b = b(\rho, R_0) > 0$ such that

$$\inf_{\theta \in T} \left[\inf_{u \in U_\Delta^c(\theta) \cap (v^c(R_0) \setminus v(\rho))} |\Delta|^{-1} \Phi_\Delta(u, 0) \right] \geq b; \tag{3.2.7}$$

$$\inf_{\theta \in T} \left[\inf_{u \in U_\Delta^c(\theta) \setminus v(R_0)} |\Delta|^{-1} \Phi_\Delta(u, 0) \right] \geq 4B(0) + a. \tag{3.2.8}$$

Lemma 3.2.1. Let the field $\epsilon(x)$ satisfy the assumption of Lemma 1.8.1. with $m = 2s$, $s \geq 1$, and for any $\theta_1, \theta_2 \in \Theta^c$, $|\Delta|^{1/2} \sup_{x \in \Delta} |g(x, \theta_1) - g(x, \theta_2)| \asymp \phi_\Delta^{1/2}(\theta_1, \theta_2)$ as $\Delta \to \infty$. Then

$$E|w_\Delta(\theta_1, \theta_2)|^{4s} \leq k_1 \phi_\Delta^{2s}(\theta_1, \theta_2); \tag{3.2.9}$$

$$E|\gamma(\Delta) - B(0)|^{2s} \leq k_2 |\Delta|^{-s}. \tag{3.2.10}$$

Proof. Inequality (3.2.9) coincides with (1.8.11), while inequality (3.2.10) follows from the fact that the mixing rate $\tilde{\alpha}(\rho)$ of the field $\epsilon^2(x)$ does not exceed $\alpha(\rho)$. ∎

In the theorem below, the conclusions of Lemma 3.2.1 are used as assumptions. Set $\hat{u}_\Delta(\theta) = |\Delta|^{-1/2}d_\Delta(\theta)(\hat{\theta}_\Delta - \theta)$.

Theorem 3.2.1. If assumptions IV, V, (3.2.9) and (3.2.10) are valid, then for any $\rho > 0$

$$\sup_{\theta \in T} P\{|\hat{u}_\Delta(\theta)| > \rho\} = O(|\Delta|^{-s}) \tag{3.2.11}$$

as $\Delta \to \infty$.

Proof. Let $\rho \in (0, R_0)$ be fixed and let R_0, b and a be numbers whose existence is assured by V. Set $z_\Delta(u) \equiv z_\Delta(\theta, \theta + |\Delta|^{-1/2}d_\Delta(\theta)u)$. By inequality (3.2.5), for any $\theta \in T$ we have

$$P\{|\hat{u}_\Delta(\theta)| \geq \rho\} \leq P\left\{\sup_{u \in U_\Delta^c(\theta)\backslash v(R_0)} z_\Delta(u) \geq 1/2\right\} +$$

$$+P\left\{\sup_{u \in U_\Delta^c(\theta)\cap(v^c(R_0)\backslash v(\rho))} z_\Delta(u) \geq 1/2\right\} = P_1 + P_2.$$

Using the Cauchy-Bunyakovskii inequality, condition (3.2.8), Chebyshev's inequality and (3.2.10), we find that

$$P_1 \leq P\left\{\gamma(\Delta) \geq 1/4 \inf_{u \in U_\Delta^c(\theta)\backslash v(R_0)} |\Delta|^{-1}\Phi_\Delta(u,0)\right\} \leq$$

$$\leq P\{\gamma(\Delta) - B(0) \geq a/4\} = O(|\Delta|^{-s}).$$

Let $F^{(1)}, \ldots, F^{(m)} \subset v^c(R_0)\backslash v(\rho)$ be closed sets, $\bigcup_{i=1}^m F^{(i)} = v^c(R_0)\backslash v(\rho)$, the diameter of each $F^{(i)}$ being less than the δ corresponding to the numbers ϵ and R_0 in assumption IV (the value of ϵ will be chosen below), $u^{(i)} \in F^{(i)} \cap U_\Delta^c(\theta)$, $i = 1, \ldots, m$. Then

$$P_2 \leq \sum_{i=1}^m P\left\{\sup_{u \in F^{(i)}\cap v_\Delta^c(\theta)} z_\Delta(u) \geq \tfrac{1}{2}\right\};$$

$$P\left\{\sup_{u \in F^{(i)}\cap U_\Delta^c(\theta)} z_\Delta(u) \geq \tfrac{1}{2}\right\} \leq P\{|z_\Delta(u^{(i)})| \geq \tfrac{1}{4}\} +$$

$$+P\left\{\sup_{u_1,u_2 \in F^{(i)}\cap U_\Delta^c(\theta)} |z_\Delta(u_1) - z_\Delta(u_2)| \geq \tfrac{1}{4}\right\} = P_3^{(i)} + P_4^{(i)}.$$

In view of condition (3.2.7) and inequality (3.2.9), we have

$$P_3^{(i)} \le k_3 \Phi_\Delta^{-2s}(u^{(i)}, 0) \le k_3 b^{-2s} |\Delta|^{-2s}.$$

Note that

$$|z_\Delta(u_1) - z_\Delta(u_2)| \le |w_\Delta(\theta + |\Delta|^{1/2} d_\Delta^{-1} u_1, \theta)| |\Phi_\Delta^{-1}(u_1, 0) - \Phi_\Delta^{-1}(u_2, 0)| +$$
$$+ \Phi_\Delta^{-1}(u_2, 0)|w_\Delta(\theta + |\Delta|^{1/2} d_\Delta^{-1} u_1, \ \theta + |\Delta|^{1/2} d_\Delta^{-1} u_2)|;$$
$$|\Phi_\Delta^{-1}(u_1, 0) - \Phi_\Delta^{-1}(u_2, 0)| \le 2^{1/2} \Phi_\Delta^{1/2}(u_1, u_2) \times$$
$$\left(\Phi_\Delta^{-1/2}(u_1, 0) \Phi_\Delta^{-1}(u_2, 0) + \Phi_\Delta^{-1}(u_1, 0) \Phi_\Delta^{-1/2}(u_2, 0) \right).$$

Hence for $u_1, u_2 \in F^{(i)} \cap U_\Delta^c(\theta)$, by (3.2.6) and (3.2.7) we have:

$$|z_\Delta(u_1) - z_\Delta(u_2)| \le \gamma^{1/2}(\Delta) \Phi_\Delta^{1/2}(u_1, u_2)((1 + 2^{1/2}) \Phi_\Delta^{-1}(u_2, 0) +$$
$$+ 2^{1/2} \Phi_\Delta^{-1/2}(u_1, 0) \Phi_\Delta^{-1/2}(u_2, 0)) |\Delta|^{1/2} \le (1 + 2^{3/2}) \epsilon^{1/2} b^{-1} \gamma^{1/2}(\Delta).$$

Therefore $P_4^{(i)} = O(|\Delta|^{-s})$ provided ϵ is chosen to satisfy $b^2/16(9 + 4\sqrt{2})\epsilon > B(0)$. Combining the bounds for P_1, P_2, $P_3^{(i)}$ and $P_4^{(i)}$ we arrive at the assertion of the theorem. The case $F^{(i)} \cap U_\Delta^c(\theta) = \emptyset$ for some i is handled in the same manner. ∎

Theorem 3.2.2. Let $\epsilon(x)$, $x \in \mathbf{R}^n$, be a homogeneous isotropic random field having a bounded spectral density function $f(\lambda)$ and $\gamma(\Delta) \xrightarrow[\Delta \to \infty]{P} B(0)$. Then if assumptions IV and V hold, for any $\rho > 0$ we have

$$\sup_{\theta \in T} P\{|\hat{u}_\Delta(\theta)| > \rho\} \xrightarrow[\Delta \to \infty]{} 0. \qquad (3.2.12)$$

Proof. The proof proceeds along the lines of the preceding one. By Chebyshev's inequality

$$P_3^{(i)} \le 16 E|z_\Delta(u^{(i)})|^2 \le 16 \sup_{\lambda \in \mathbf{R}^n} f(\lambda) \Phi_\Delta^{-1}(u^{(i)}, 0) = O(|\Delta|^{-1}).$$

The convergence of P_1 and $P_4^{(i)}$ to zero is obvious. ∎

If $|\Delta|^{-1/2} d_{i\Delta}(\theta) \xrightarrow[\Delta \to \infty]{} \infty$ for some or all indices $i = 1, \dots, q$, then (3.2.12) is a stronger property of the estimator $\hat{\theta}_\Delta$ as compared with uniform consistency (3.2.2).

Strengthening assumption IV and the condition for distinguishability of the parameters (contrast assumption V), one can obtain an assertion on large deviations of the estimator $\hat{\theta}_\Delta$ that is more general than Theorem 3.2.1 . We set

$$\tilde{U}_\Delta(\theta) = d_\Delta(\theta)(\Theta - \theta),$$

$$\tilde{\Phi}_\Delta(u_1, u_2) = \phi_\Delta(\theta + d_\Delta^{-1}(\theta)u_1, \ \theta + d_\Delta^{-1}(\theta)u_2), \ u_1, u_2 \in \tilde{U}_\Delta(\theta).$$

VI. For some $\alpha \in (0,1]$ there exist constants $\tilde{k} = \tilde{k}(T) > 0$ and $a = a(T) \geq 0$ such that

$$\sup_{\theta \in T} \left[\sup_{u_1, u_2 \in \tilde{U}_\Delta^c(\theta) \cap v^c(Q)} \tilde{\Phi}_\Delta^{1/2}(u_1, u_2)|u_1 - u_2|^{-\alpha} \right] \leq \tilde{k}(1 + Q^a), Q \geq 0; \quad (3.2.13)$$

VII.

$$\inf_{\theta \in T} \left[\inf_{u \in U_\Delta^c(\theta)} \tilde{\Phi}_\Delta^{1/2}(u, 0)\Psi_\Delta^{-1}(|u|) \right] \geq 1, \quad (3.2.14)$$

where the function $\Psi_\Delta(x)$, $x \geq 0$, is monotonically non-decreasing in Δ and x, $\Psi_\Delta(0) = 0$, $x^{a+\alpha}\Psi_\Delta^{-2}(x) \downarrow 0$ as $\Delta \to \infty$, $x \to \infty$.

We state an assertion on the modulus of continuity of a random field.

Theorem 3.2.3 [53, 140]. Let $\eta(u)$ be a separable measurable random field defined on a closed set $F \subseteq R^q$ and for any u, $\tilde{u}, u + \tilde{u} \in F$ let

$$E|\eta(u + \tilde{u}) - \eta(u)|^m \leq l(u)|\tilde{u}|^p \quad (3.2.15)$$

for some $m \geq p > q$ and locally bounded function $l(u) : R^q \to R_+^1$. Then for any Q, h and $C > 0$

$$P\left\{ \sup_{u', u'' \in F \cap v^c(Q); |u'-u''| \leq h} |\eta(u') - \eta(u'')| > C \right\} \leq$$

$$\leq k_0 \left(\sup_{u \in F \cap v^c(Q)}^\bullet l(u) \right) Q^q h^{p-q} C^{-m}, \quad (3.2.16)$$

where the constant k_0 depends on p, m and q and does not depend on Q, h, C or the set F. In particular, provided the above conditions hold,

$$P\left\{ \sup_{u', u'' \in F \cap v^c(Q)} |\eta(u') - \eta(u'')| > C \right\} \leq \tilde{k}_0 \left(\sup_{u \in F \cap v^c(Q)} l(u) \right) Q^p C^{-m}, \quad (3.2.17)$$

$\tilde{k}_0$ is independent of Q, C and F.

Theorem 3.2.3 will be used to prove the following assertion.

Theorem 3.2.4. Let assumptions VI, VII hold and suppose that for m such that $\alpha m > q$,

$$E|w_\Delta(\theta_1, \theta_2)|^m \leq k_4 \phi_\Delta^{m/2}(\theta_1, \theta_2), \ \theta_1, \theta_2 \in \Theta^c. \quad (3.2.18)$$

Then for some $k_5 > 0$ and $H > 0$

$$\sup_{\theta \in T} P\{|d_\Delta(\theta)(\hat{\theta}_\Delta - \theta)| \geq H\} \leq k_5 H^{(a+\alpha)m} \Psi_\Delta^{-2m}(H) +$$

$$+ \frac{k_5}{H} \int_H^\infty x^{(a+\alpha)m} \Psi_\Delta^{-2m}(x) dx. \qquad (3.2.19)$$

Proof. Denote $U^{(\rho)} = (v^c(H(\rho+2)) \backslash v(H(\rho+1))) \cap \tilde{U}_\Delta^c(\theta), \quad \rho = 0, 1, \ldots$.
For any $\theta \in T$

$$P\{|d_\Delta(\theta)(\hat{\theta}_\Delta - \theta)| > H\} \leq \sum_{\rho=0}^\infty P\left\{ \sup_{u \in U^{(\rho)}} z_\Delta(\theta + d_\Delta^{-1}(\theta)u, \theta) \geq \tfrac{1}{2} \right\}.$$

By condition (3.2.14)

$$P\left\{ \sup_{u \in U^{(\rho)}} z_\Delta(\theta + d_\Delta^{-1}(\theta)u, \theta) \geq \tfrac{1}{2} \right\} \leq$$

$$\leq P\left\{ \sup_{u \in \tilde{U}_\Delta^c(\theta) \cap v^c(H(\rho+2))} |w_\Delta(\theta + d_\Delta^{-1}(\theta)u, \theta)| \geq 2^{-1} \Psi_\Delta^2(H(\rho+1)) \right\}.$$

We shall apply Theorem 3.2.3 to the random field $\eta(u) = \int_\Delta \epsilon(x) g(x, \theta + d_\Delta^{-1}(\theta)u) dx$, $u \in \tilde{U}_\Delta^c(\theta)$. We note that under the conditions (3.2.18) and (3.2.13), the inequality (3.2.15) is valid for $l(u) \asymp (1 + |u|^a)^m$ and $p = \alpha m$. Therefore for $u_1, u_2 \in \tilde{U}_\Delta^c(\theta) \cap v^c(H(\rho+2))$

$$E|w_\Delta(\theta + d_\Delta^{-1}(\theta)u_1, \ \theta + d_\Delta^{-1}(\theta)u_2)|^m \leq$$

$$\leq \tilde{k}^m k_4 (1 + (H(\rho+2))^a)^m |u_1 - u_2|^{\alpha m}$$

and hence in view of (3.2.17),

$$P\left\{ \sup_{u \in U^{(\rho)}} z_\Delta(\theta + d_\Delta^{-1}(\theta)u, \theta) \geq \tfrac{1}{2} \right\} \leq$$

$$\leq 2^m \tilde{k}_0 \tilde{k}^m k_4 (1 + (H(\rho+2))^a)^m (H(\rho+2))^{\alpha m} \Psi_\Delta^{-2m}(H(\rho+1)).$$

Thus

$$\sup_{\theta \in T} P\{|d_\Delta(\theta)(\hat{\theta}_\Delta - \theta)| \geq H\} \leq k_5 \sum_{\rho=0}^\infty (H(\rho+1))^{(a+\alpha)m} \Psi_\Delta^{-2m}(H(\rho+1)) \leq$$

$$\leq k_5 H^{(a+\alpha)m} \Psi_\Delta^{-2m}(H) + k_5 H^{-1} \int_H^\infty x^{(a+\alpha)m} \Psi_\Delta^{-2m}(x)dx. \quad \blacksquare$$

Corollary 3.2.1. Let $\Psi_\Delta(x) = kx^\beta$, $0 < \beta \leq \alpha$, $2\beta - \alpha - a > 0$. Then the right-hand side of (3.2.19) is a quantity of order $H^{-(2\beta - \alpha - a)m}$. In particular, with $\alpha = \beta = 1$, $a = 0$, we have

$$\sup_{\theta \in T} P\{|d_\Delta(\theta)(\hat{\theta}_\Delta - \theta)| \geq H\} \leq k^* H^{-m}. \tag{3.2.20}$$

Setting in (3.2.20) $H = |\Delta|^{1/2}\rho$, $\rho > 0$, $m = 4s$, one obtains a stronger result than Theorem 3.2.1: for any $\rho > 0$

$$\sup_{\theta \in T} P\{|\hat{u}_\Delta(\theta)| \geq \rho\} = O(|\Delta|^{-m/2}), \quad \Delta \to \infty. \tag{3.2.21}$$

A family of r.v.'s $\tilde{\theta}_\Delta = \tilde{\theta}_\Delta(\xi_\theta(x), x \in \Delta)$ is said to be a strongly consistent family of estimators of the parameter $\theta \in \Theta$ ($\tilde{\theta}_\Delta$ is a strongly consistent estimator of $\theta \in \Theta$) if $\tilde{\theta}_\Delta \xrightarrow[\Delta \to \infty]{} \theta$ a.s.

We provide simple sufficient conditions for strong consistency of $\tilde{\theta}_\Delta$ in the case when Θ^c is a compact set and the matrix $d_\Delta(\theta)$ is proportional to the matrix $|\Delta|^{1/2}\mathbf{1}_q$.

VIII. $\mathfrak{M}$ is a linearly ordered set with the order relation "$\subseteq$" and there exists a sequence of sets $\Delta_t \in \mathfrak{M}$ such that: 1)$\Delta_t \subset \Delta_{t+1}$; 2) $\lim_{t \to \infty}\{|\Delta_{t+1}| \div |\Delta_t|\} = 1$; 3) $\sum_t |\Delta_t|^{-1/2} < \infty$.

An example of a sequence satisfying conditions 2) and 3) is the sequence of sets Δ_t possessing volumes $|\Delta_t| = t^{2+\delta}$, $\delta > 0$.

IX.

1) $\quad \overline{\lim_{\Delta \to \infty}} \sup_{\theta_1, \theta_2 \in \Theta^c} (|\Delta|^{-1}\phi_\Delta(\theta_1, \theta_2) - \phi(\theta_1, \theta_2)) \leq 0, \tag{3.2.22}$

where the function $\phi(\theta_1, \theta_2) \geq 0$ is continuous on $\Theta^c \times \Theta^c$ and $\phi(\theta_1, \theta_2) = 0$ if and only if $\theta_1 = \theta_2$;

2) $\quad \overline{\lim_{\Delta \to \infty}} \sup_{\theta \in \Theta^c} |\Delta|^{-1} \int_\Delta g^2(x, \theta)dx < \infty.$

X.   For any $\rho > 0$ there exists a number $k_6 = k_6(\rho) > 0$ such that for any $\theta \in \Theta$

$$\inf_{u \in (\Theta^c - \theta)\backslash v(\rho)} |\Delta|^{-1}\phi_\Delta(\theta + u, \theta) \geq k_6. \tag{3.2.23}$$

The contrast assumption X is a simplified version of V.

Theorem 3.2.5. Let assumptions VIII–X hold and suppose that the homogeneous field $\epsilon(x)$, $x \in \mathbf{R}^n$, possesses the following properties:

1) $\gamma(\Delta) \xrightarrow[\Delta \to \infty]{} B(0)$ a.s.; 2) $\int_{\mathbf{R}^n} B^2(x)dx < \infty$. Then $\hat{\theta}_\Delta$ is a strongly consistent estimator of the parameter $\theta \in \Theta$.

Proof. Set $\eta_\Delta(\tau) = |\Delta|^{-1} \int_\Delta \epsilon(x)g(x,\tau)dx$, $\tau \in \Theta^c$. We shall prove that

$$\sup_{\tau \in \Theta^c} |\eta_\Delta(\tau)| \xrightarrow[\Delta \to \infty]{} 0 \text{ a.s.} \tag{3.2.24}$$

and therefore, in view of (3.2.3),

$$|\Delta|^{-1} w_\Delta(\hat{\theta}_\Delta, \theta) \xrightarrow[\Delta \to \infty]{} 0 \text{ a.s.} \tag{3.2.25}$$

Fix $\tau \in \Theta^c$. Then

$$E\eta_{\Delta_t}^2(\tau) \leq |\Delta_t|^{-1/2} \left(|\Delta_t|^{-1} \int_{\Delta_t} g^2(x,\tau)dx \right) \left(\int_{\mathbf{R}^n} B^2(x)dx \right)^{1/2}. \tag{3.2.26}$$

Under the conditions of the theorem and in view of (3.2.26), $\sum_t E\eta_{\Delta_t}^2(\tau) < \infty$ and $\eta_{\Delta_t}(\tau) \xrightarrow[t \to \infty]{} 0$ a.s. Consider the quantity

$$\eta_t = \sup_{\Delta_t \subseteq \Delta \subset \Delta_{t+1}} |\eta_\Delta(\tau) - \eta_{\Delta_t}(\tau)| \leq (|\Delta_{t+1}||\Delta_t|^{-1} - 1)|\eta_{\Delta_t}(\tau)| +$$

$$+ (|\Delta_{t+1}||\Delta_t|^{-1}\gamma(\Delta_{t+1}) - \gamma(\Delta_t))^{1/2} \times$$

$$\times (|\Delta_t|^{-1} \int_{\Delta_{t+1} \backslash \Delta_t} g^2(x,\tau)dx)^{1/2} \xrightarrow[t \to \infty]{} 0 \text{ a.s.},$$

that is, $\eta_\Delta(\tau) \xrightarrow[\Delta \to \infty]{} 0$ a.s.

Relation (3.2.24) is now readily derived using (3.2.22) and the continuity of the function $\phi(\theta_1, \theta_2)$. From (3.2.25) and (3.2.23) using a proof by contradiction we obtain the assertion of the theorem. ∎

For example, the following conditions are sufficient for IX to hold.

IX$_1$. $|\Delta|^{-1}\phi_\Delta(\theta_1, \theta_2)$ converges uniformly on $\Theta^c \times \Theta^c$ as $\Delta \to \infty$ to the function $\phi(\theta_1, \theta_2)$ having the properties stipulated in IX;

IX$_2$. For some $\alpha > 0$ and $k_7 < \infty$

$$\varlimsup_{\Delta \to \infty} \sup_{\theta_1, \theta_2 \in \Theta^c} |\Delta|^{-1}\phi_\Delta(\theta_1, \theta_2)|\theta_1 - \theta_2|^{-\alpha} \leq k_7.$$

Note that IX_1 implies X.

Inequality (3.2.3) shows that under assumption X, strong consistency of $\hat{\theta}_\Delta$ is implied by (3.2.24). Theorem 3.2.5 provides only an example of fulfillment of (3.2.24) which does not exclude the possibility of proving relation (3.2.24) under assumptions on the function $g(x,\theta)$ different from IX. The conditions for $\gamma(\Delta)$ to converge to $B(0)$ stated in Theorems 3.2.2 and 3.2.5 will be discussed in §4.2.

3.3.   Asymptotic Expansion of Least Squares Estimators

This section presents an asymptotic expansion of the normalized estimator $\hat{\theta}_\Delta$ (Theorem 3.3.1) for a smooth function $g(x,\theta)$. Assume that (3.2.1) holds, $\Theta \subset R^q$ is a convex open set, $T \subset \Theta$ is a compact set. We introduce the following notation.

Let $\alpha = (\alpha_1,\ldots,\alpha_q)$ be a vector with the integer non-negative coordinates, $|\alpha| = \alpha_1 + \ldots + \alpha_q$, $\alpha! = \alpha_1!\ldots\alpha_q!$, $\theta^\alpha = \theta_1^{\alpha_1}\ldots\theta_q^{\alpha_q}$, $\theta \in R^q$. For a smooth function $a(\theta)$ set $a^{(\alpha)} = (\partial^{|\alpha|}/(\partial\theta)^\alpha)a$, $a_i^{(\alpha)} = (\partial/\partial\theta_i)a^{(\alpha)}$, $a_{ij}^{(\alpha)} = (\partial^2/\partial\theta_i\partial\theta_j)a^{(\alpha)}$, and so on.

Suppose that all the partial derivatives of the function $g(x,\theta)$ with respect to the variables $\theta = (\theta_1,\ldots,\theta_q)$ exist up to order $r \geq 2$ inclusive for each $x \in R^n$, and that they are continuous; suppose further that the functions $g^{(\alpha)}(x,\theta)$, $|\alpha| = 1,\ldots,r$, are locally square integrable in x for any $\theta \in \Theta^c$.

Set $s^{(\alpha)}(x,u) = g^{(\alpha)}(x,\theta + d_\Delta^{-1}(\theta)|\Delta|^{1/2}u)$, $|\alpha| \geq 0$.

Similarly, $s_i^{(\alpha)}$, $s_{ij}^{(\alpha)}$ will denote $g_i^{(\alpha)}$, $g_{ij}^{(\alpha)}$ of the same composite argument. We set

$$\Phi_{\alpha\Delta}(u_1,u_2) = \int_\Delta (s^{(\alpha)}(x,u_1) - s^{(\alpha)}(x,u_2))^2 dx,$$

$$d_\Delta^2(\alpha,\theta) = \int_\Delta (g^{(\alpha)}(x,\theta))^2 dx,$$

$$d_\Delta^{(\alpha)}(\theta) = d_{1\Delta}^{\alpha_1}(\theta)\ldots d_{q\Delta}^{\alpha_q}(\theta),$$

$$\Im_\Delta(\theta) = (\Im_\Delta^{ij}(\theta))_{i,j=1,\ldots,q}, \quad \Im_\Delta^{ij}(\theta) = d_{i\Delta}^{-1}(\theta)d_{j\Delta}^{-1}(\theta)\int_\Delta g_i(x,\theta)g_j(x,\theta)dx,$$

$$\Lambda_\Delta(\theta) = (\Lambda_\Delta^{ij}(\theta))_{i,j=1,\ldots,q} = \Im_\Delta^{-1}(\theta).$$

Let $e_i = (0,\ldots,1,\ldots,0)$ be a vector, the ith coordinate of which is 1 and the others are zero. Assume that

$$b_\Delta(\alpha, u) = |\Delta|^{(|\alpha|-1)/2}(d_\Delta^\alpha(\theta))^{-1}\int_\Delta \epsilon(x)s^{(\alpha)}(x, u)dx,$$

$$b_\Delta^i(d, u) = |\Delta|^{|\alpha|/2}(d_\Delta^{\alpha+e_i}(\theta))^{-1}\int_\Delta \epsilon(x)s_i^{(\alpha)}(x, u)dx,$$

$$b_\Delta^{ij}(\alpha, u) = |\Delta|^{(|\alpha|+1)/2}(d_\Delta^{\alpha+e_i+e_j}(\theta))^{-1}\int_\Delta \epsilon(x)s_{ij}^{(\alpha)}(x, u)dx,$$

$$a_\Delta(\alpha, u) = |\Delta|^{-1}EL_\Delta^{(\alpha)}(\theta + |\Delta|^{1/2}d_\Delta^{-1}(\theta)u).$$

Similarly, denote by a_Δ^i, a_Δ^{ij}, a_Δ^{ijk} the mathematical expectations of the derivatives $L_{i\Delta}^{(\alpha)}$, $L_{ij\Delta}^{(\alpha)}$, $L_{ijk\Delta}^{(\alpha)}$.

To state the main result of §3.3, the following assumptions will be needed.

XI. For any $R > 0$ there exist constants $k(\alpha, R)$, $k_r(\alpha, R) < \infty$ such that

$$\sup_{\theta\in T}\left[\sup_{u\in U_\Delta^c(\theta)\cap v^c(R)} |\Delta|^{(|\alpha|-1)/2}(d_\Delta^\alpha(\theta))^{-1}d_\Delta(\alpha, \theta + |\Delta|^{1/2}d_\Delta^{-1}(\theta)u)\right] \le$$

$$\le k(\alpha, R), \quad |\alpha| = 1, \ldots, r; \tag{3.3.1}$$

$$\sup_{\theta\in T}\left[\sup_{u_1,u_2\in U_\Delta^c(\theta)\cap v^c(R)} |\Delta|^{(|\alpha|-1)}(d_\Delta^\alpha(\theta))^{-2}\Phi_{\alpha\Delta}(u_1, u_2)|u_1 - u_2|^{-2}\right] \le$$

$$\le k_r(\alpha, R), \quad |\alpha| = r. \tag{3.3.2}$$

Denote by $\lambda_{\min}(A)$ the smallest eigenvalue of the positive definite $q \times q$ matrix A.

XII. For $\lambda_* = \lambda_*(T) > 0$,

$$\inf_{\theta\in T} \lambda_{\min}(\Im_\Delta(\theta)) \ge \lambda_*. \tag{3.3.3}$$

XIII. For some $m \ge 2$ and any locally square integrable function $s(x)$, $x \in \mathbf{R}^n$

$$E\left|\int_\Delta \epsilon(x)s(x)dx\right|^{2m} \le k_1\left[\int_\Delta s^2(x)dx\right]^m; \tag{3.3.4}$$

$$E|\gamma(\Delta) - B(0)|^m \le k_2|\Delta|^{-m/2}. \tag{3.3.5}$$

Sufficient conditions for (3.3.4) and (3.3.5) to hold are given in §1.8.

It is assumed below that differentiation under the integral sign with respect to the variables $\tau_1, \ldots, \tau_q$ in the expression for $L_\Delta(\tau)$ is justified without qualifying it as a separate requirement for the model (3.1.1).

Suppose that $\tau_\Delta \xrightarrow[\Delta \to \infty]{} \infty$, $\tau_\Delta t^r \xrightarrow[\Delta \to \infty]{} 0$, $t = |\Delta|^{-1/2}$.

Theorem 3.3.1. Let assumptions XI–XIII hold and let the estimator $\hat\theta_\Delta$ satisfy the property (3.2.21). Then

$$\sup_{\theta \in T} P\{|d_\Delta(\theta)(\hat\theta_\Delta - \theta) - \sum_{\nu=1}^{r-1} t^{\nu-1} h_{\nu\Delta}(\theta)| \geq \tau_\Delta t^{r-1}\} =$$

$$= O(|\Delta|^{-m/2}) + O(\tau_\Delta^{-2m/r}) \qquad (3.3.6)$$

as $\Delta \to \infty$; the $h_{\nu\Delta}(\theta)$ are homogeneous vector-valued polynomials of degree ν in the random variables $b_\Delta(\alpha, 0)$, $|\alpha| = 1, \ldots, \nu$, with coefficients uniformly bounded in Δ and $\theta \in T$. We shall prove several lemmas.

Lemma 3.3.1. Condition (3.3.2) for the vector α, $0 \leq |\alpha| \leq r - 1$, is implied by relations (3.3.1) which hold for $\alpha + e_i$, $i = 1, \ldots, q$.

Proof. The proof consists in applying the finite increments formula to the function $|\Delta|^{|\alpha|-1}\Phi_{\alpha\Delta}(u_1, u_2)(d_\Delta^\alpha(\theta))^{-1}$, $u_1, u_2 \in U_\Delta^c(\theta) \cap v^c(R)$. ∎

Denote $A_\Delta(\alpha, \theta) = (a_\Delta^i(\alpha, 0))_{i=1,\ldots,q}$, $B_\Delta(\alpha, \theta) = (b_\Delta^i(\alpha, 0))_{i=1,\ldots,q}$, $B_\Delta^{(2)}(\theta) = (b_\Delta^{ij}(0, 0))_{i,j=1,\ldots,q}$ and write down the McLaurin expansion in the variable u for the gradient of the function $|\Delta|^{-1}L_\Delta(\theta + |\Delta|^{1/2}d_\Delta^{-1}(\theta)u)$:

$$t^2 \nabla L_\Delta(\theta + t^{-1}d_\Delta^{-1}(\theta)u) = -2tB_\Delta(0, \theta) + 2\Im_\Delta(\theta)u - 2tB_\Delta^{(2)}(\theta)u +$$

$$+ \sum_{2 \leq |\alpha| \leq r-1} (1/\alpha!)(A_\Delta(\alpha, \theta) - 2tB_\Delta(\alpha, \theta))u^\alpha + \zeta_\Delta(u); \qquad (3.3.7)$$

$$\zeta_\Delta(u) = (\zeta_{i\Delta}(u))_{i=1,\ldots,q},$$

$$\zeta_{i\Delta}(u) = \sum_{|\alpha|=r-1} \left(\frac{1}{\alpha!}t^2 \int_\Delta (G_i^{(\alpha)}(x, \theta + t^{-1}d_\Delta^{-1}(\theta)u_i^*) - \right.$$
$$\left. - G_i^{(\alpha)}(x, \theta))dx\right)u^\alpha, \quad |u_i^*| \leq |u|. \qquad (3.3.8)$$

The analogous expansion for the function $t^2 L_{ij\Delta}(\theta + t^{-1}d_\Delta^{-1}(\theta)u)$ is of the form:

$$t^2 L_{ij\Delta}(\theta + t^{-1}d_\Delta^{-1}(\theta)u) = 2\Im_\Delta^{ij}(\theta) - 2tb_\Delta^{ij}(0, 0) +$$

$$+ \sum_{1 \leq |\alpha| \leq r-2} \frac{1}{\alpha!}(a_\Delta^{ij}(\alpha, 0) - 2tb_\Delta^{ij}(\alpha, 0))u^\alpha + \zeta_{ij\Delta}(u); \qquad (3.3.9)$$

$$\zeta_{ij\Delta}(u) = \sum_{|\alpha|=r-2} \left(\frac{1}{\alpha!}\int_\Delta (G_{ij}^{(\alpha)}(x, \theta + t^{-1}d_\Delta^{-1}(\theta)u_{ij}^*) - \right.$$
$$\left. - G_{ij}^{(\alpha)}(x, \theta))dx\right)u^\alpha, \quad |u_{ij}^*| \leq |u|. \qquad (3.3.10)$$

If $r = 2$, then the sums in (3.3.7) and (3.3.9) are absent and the remainder in (3.3.10) is given by

$$\zeta_{ij\Delta}(u) = t^2 L_{ij\Delta}(\theta + t^{-1}d_\Delta^{-1}(\theta)u) - t^2 L_{ij\Delta}(\theta).$$

Lemma 3.3.2. Let $|u| \leq \delta < 1$ and the event $\{\gamma(\Delta) \leq B(0) + 1\}$ occur. If XI is valid, then

$$\sup_{\theta \in T} |\zeta_{i\Delta}(u)| \leq k_3 \delta^r, \quad \sup_{\theta \in T} |\zeta_{ij\Delta}(u)| \leq k_4 \delta^{r-1}, \quad i, j = 1, \ldots, q.$$

Proof. We shall show that for a fixed α, $|\alpha| = r$,

$$|\zeta_\Delta(\alpha, u)| = t^2 \left| \int_\Delta (G^{(\alpha)}(x, \theta + t^{-1} d_\Delta^{-1}(\theta) u) - G^{(\alpha)}(x, \theta)) dx \right| \leq k_5 \delta.$$

Note that $\zeta_\Delta(\alpha, u) = \zeta_\Delta^{(1)}(\alpha, u) + \zeta_\Delta^{(2)}(\alpha, u)$, where

$$\zeta_\Delta^{(1)}(\alpha, u) = 2|\Delta|^{|\alpha|/2-1} (d_\Delta^\alpha(\theta))^{-1} \int_\Delta [\epsilon(x)(s^{(\alpha)}(x, 0) - s^{(\alpha)}(x, u)) +$$

$$+ s^{(\alpha)}(x, u)(s(x, u) - s(x, 0))] dx;$$

$$\zeta_\Delta^{(2)}(\alpha, u) = \sum_{|\beta| \geq 1, \beta < \alpha} k(\alpha, \beta) |\Delta|^{(|\alpha|/2)-1} (d_\Delta^\alpha(\theta))^{-1} \int_\Delta (s^{(\alpha-\beta)}(x, u) s^{(\beta)}(x, u) -$$

$$- s^{(\alpha-\beta)}(x, 0) s^{(\beta)}(x, 0)) dx,$$

$k(\alpha, \beta)$ are integer constants. By virtue of the conditions of this lemma and the conclusion of Lemma 3.3.1 we obtain

$$|\zeta_\Delta^{(1)}(\alpha, u)| \leq 2|\Delta|^{(|\alpha|-1)/2} (d_\Delta^\alpha(\theta))^{-1} [\Phi_{\alpha\Delta}^{1/2}(0, u) \gamma^{1/2}(\Delta) +$$

$$+ d_\Delta(\alpha, \theta + |\Delta|^{1/2} d_\Delta^{-1}(\theta) u) |\Delta|^{-1/2} \Phi_\Delta^{1/2}(0, u)] \leq$$

$$\leq 2(k_r^{1/2}(\alpha, 1)(B(0) + 1)^{1/2} + k(\alpha, 1) k_6) \delta, \qquad (3.3.11)$$

and one may choose $k_6 = 2(\sum_{i=1}^q k^2(e_i, 1))^{1/2}$. Similarly, for fixed α and β,

$$|\Delta|^{(|\alpha|/2)-1} (d_\Delta^\alpha(\theta))^{-1} \left| \int_\Delta (s^{(\alpha-\beta)}(x, u) s^{(\beta)}(x, u) -\right.$$

$$\left. - s^{(\alpha-\beta)}(x, 0) s^{(\beta)}(x, 0)) dx \right| \leq (|\Delta|^{(|\alpha-\beta|-1)/2} (d_\Delta^{(\alpha-\beta)}(\theta))^{-1} \Phi_{\alpha-\beta,\Delta}^{1/2}(0, u)) \times$$

$$\times (|\Delta|^{(|\beta|-1)/2} (d_\Delta^\beta(\theta))^{-1} d_\Delta(\beta, \theta + |\Delta|^{1/2} d_\Delta^{-1}(\theta) u)) +$$

$$+(|\Delta|^{(|\beta|-1)/2}(d_\Delta^\beta(\theta))^{-1}\Phi_{\beta\Delta}^{1/2}(0,u))(|\Delta|^{(|\alpha-\beta|-1)/2}\times$$

$$\times(d_\Delta^{\alpha-\beta}(\theta))^{-1}d_\Delta(\alpha-\beta,\theta)) \le (k_7 k(\beta,1) + k_8 k(\alpha-\beta,1))\delta, \qquad (3.3.12)$$

where one may set $k_7 = 2(\sum_{i=1}^q k^2(\alpha-\beta+e_i,1))^{1/2}$ and $k_8 = 2(\sum_{i=1}^q k^2(\beta + e_i,1))^{1/2}$.

The assertion of the lemma follows from (3.3.8), (3.3.10), (3.3.11), (3.3.12) and the obvious inequality $|u^\alpha| \le |u|^{|\alpha|}$. ∎

Denote by $L_\Delta^{(2)} = (L_{ij\Delta})_{i,j=1,\dots,q}$ the Hessian of the function L_Δ.

Lemma 3.3.3. Let $\theta \in T$, suppose that the events $\{\gamma(\Delta) \le B(0) + 1\}$ and $\{i|b_\Delta(\alpha,0)| < \delta\}$, $|\alpha| = 2,\dots,r$, occur for sufficiently small $\delta > 0$ and that assumptions XI and VII are valid. Then there exists a number $\kappa_0 = \kappa_0(T) > 0$ such that

$$\inf_{u \in U_\Delta^c(\theta) \cap v^c(\kappa_0)} \lambda_{\min}(t^2 L_\Delta^{(2)}(\theta + t^{-1}d_\Delta^{-1}(\theta)u)) \ge \lambda_*.$$

Proof. We use the relation [231]

$$|\lambda_{\min}(t^2 L_\Delta^{(2)}(\theta + t^{-1}d_\Delta^{-1}(\theta)u)) - \lambda_{\min}(2\Im_\Delta(\theta))| \le$$

$$\le q \max_{i,j} |t^2 L_{ij\Delta}(\theta + t^{-1}d_\Delta^{-1}(\theta)u) - 2\Im_\Delta^{ij}(\theta)|. \qquad (3.3.13)$$

Since the $t|b_\Delta^{ij}(0,0)|$ in right-hand side of expansion (3.3.9) are less than δ by the condition of the lemma and the sum of the subsequent terms does not exceed $k_9\delta$, $k_9 = k_9(T)$, by the assumption of this lemma and the assertion of Lemma 3.3.2, the right-hand side (3.3.13) does not exceed $q(2 + k_9)\delta$. Consequently, the lemma is valid for $\kappa_0 \le \delta < \lambda_*(2 + k_9)^{-1}q^{-1}$. ∎

If the event $\{|\hat{u}_\Delta(\theta)| \le \kappa_0\}$ occurs and the assumptions of Lemma 3.3.3 hold, the mapping $u \to t^2 L_\Delta(\theta + t^{-1}d_\Delta^{-1}(\theta)u)$ is convex in the ball $v(\kappa_0)$ and the system of equations

$$t^2 \nabla L_\Delta(\theta) = 0 \qquad (3.3.14)$$

has a unique solution identical to $\hat{\theta}_\Delta$. Thus,

$$0 = t^2 \nabla L_\Delta(\hat{\theta}_\Delta) = -2tB_\Delta(0,\theta) + t^2 L_\Delta^{(2)}(\theta + t^{-1}d_\Delta^{-1}(\theta)u^*)\hat{u}_\Delta(\theta),$$

$$|u^*| \le |\hat{u}_\Delta(0)|, \qquad (3.3.15)$$

and by Lemma 3.3.3,

$$|\hat{u}_\Delta(\theta)| = 2t|(t^2 L_\Delta^{(2)}(\theta + t^{-1}d_\Delta^{-1}(\theta)u^*))^{-1}B_\Delta(0,\theta)| \le$$

$$\le 2t\lambda_*^{-1}|B_\Delta(0,\theta)|. \tag{3.3.16}$$

Denote by $\mathcal{L}_{r-1}(u,t)$ the expansion (3.3.7) without the remainder and by $\mathcal{L}_\infty(u,t)$ the series obtained from $\mathcal{L}_{r-1}(u,t)$ by a formal extension of summation to infinity:

$$\mathcal{L}_\infty(u,t) = \sum_{|\alpha|=0}^{\infty} \frac{1}{\alpha!}(A_\Delta(\alpha,\theta) - 2tB_\Delta(\alpha,\theta))u^\alpha. \tag{3.3.17}$$

Let $u(t) = th_1 + \ldots + t^{r-1}h_{r-1} + \ldots$ be the formal expansion into a series of the solution of the equation $\mathcal{L}_\infty(u,t) = 0$. On substituting its initial segment

$$u^{(r-1)}(t) = th_1 + \ldots + t^{r-1}h_{r-1} \tag{3.3.18}$$

into $\mathcal{L}_\infty(u,t)$, the terms containing t^i, $i = 1,\ldots,r-1$ will vanish. Therefore one can write

$$\mathcal{L}_\infty(u^{(r-1)}(t),t) = \sum_{i \ge r} t^i h_{i,r-1}. \tag{3.3.19}$$

Lemma 3.3.4. All the h_i, $i = 1,\ldots,r-1$; $h_{i,r-1}$, $i \ge r$, from representations (3.3.18) and (3.3.19) are vectors whose coordinates are homogeneous polynomials in $b_\Delta(\alpha,0)$ of degree i, $|\alpha| = 1,\ldots,i$.

Proof. The proof of the lemma is by induction on r. If $r = 2$, then $h_1 = \Lambda_\Delta(\theta)B_\Delta(0,\theta)$ and the assertion concerning h_{i1}, $i \ge 2$, is verified directly. Let the assertion be valid for some $r \ge 2$. Then h_r is obtained from the condition that the coefficients of t^r appearing in

$$\mathcal{L}_\infty(u^{(r-1)}(t) + t^r h_r, t) = \mathcal{L}_\infty(u^{(r-1)}(t),t) - 2t^r \Im_\Delta(\theta)h_r + \ldots \tag{3.3.20}$$

vanish. In (3.3.20) the omitted terms contain powers t^m, $m \ge r$. From (3.3.19) we obtain $h_r = \frac{1}{2}\Lambda_\Delta(\theta)h_{r,r-1}$ and the assertion about h_r is implied by the induction assumption. The quantity $b_\Delta(\alpha,0)$ with $|\alpha| \ge r+1$ appears in $\mathcal{L}_\infty(u^{(r)}(t),t)$ only in the terms of the form $tB_\Delta(\beta,0)(u^{(r)}(t))^\beta/\beta!$, $|\beta| = |\alpha|-1$, that contain t in powers not smaller than $|\alpha|$. Therefore $b_\Delta(\alpha,0)$ does not appear in $h_{i,r}$ for $|\alpha| > i$. We note that upon replacing the quantities $b_\Delta(\alpha,0)$ in $\mathcal{L}_\infty(u^{(r-1)}(t),t)$ by $t^{-1}b_\Delta(\alpha,0)$, the series $\mathcal{L}_\infty(u^{(r-1)}(t),t)$ becomes independent of t. This follows from (3.3.17) and the property of $u^{(r-1)}(t)$ established above. Hence the $h_{i,r-1}$ are homogeneous polynomials of degree i in the variables $b_\Delta(\alpha, 0)$, $|\alpha| = 1, \ldots, i$. ∎

Clearly the function $\mathcal{L}_{r-1}(u, t)$ is obtained from $\mathcal{L}_\infty(u, t)$ when $a_\Delta(\alpha, 0) = b_\Delta(\alpha, 0) = 0$, $|\alpha| \geq r$. We retain the notation $u^{(r-1)}(t)$, h_i, $h_{i,r-1}$ with reference to $\mathcal{L}_{r-1}(u, t)$. Then in place of (3.3.19) we obtain the relation

$$\mathcal{L}_{r-1}(u^{(r-1)}(t), t) = \sum_{i \geq r} t^i h_{i,r-1}, \qquad (3.3.21)$$

where the sum contains a finite number of terms.

Proof. We now prove Theorem 3.3.1. We shall show that if

$$\hat{u}_\Delta(\theta) \in v(\kappa_0), \quad \gamma(\Delta) \leq B(0) + 1, \quad |b_\Delta(\alpha, 0)| \leq k\tau_\Delta^{1/r}, \ |\alpha| = 1, \ldots, r, \tag{3.3.22}$$

then

$$|\hat{u}_\Delta(\theta) - u_\Delta^{(r-1)}(t)| \leq k_{10} k^r t^r \tau_\Delta. \tag{3.3.23}$$

This will imply the assertion of the theorem. Thus we set $k = k_{10}^{-1/r}$ in (3.3.22). We then obtain from (3.3.23)

$$\sup_{\theta \in T} P\{|\hat{u}_\Delta(\theta) - u_\Delta^{(r-1)}(t)| \geq \tau_\Delta t^r\} \leq \sup_{\theta \in T} P\{|\hat{u}_\Delta(\theta)| \geq \kappa_0\} +$$

$$+ P\{\gamma(\Delta) - B(0) \geq 1\} + \sum_{|\alpha|=1}^{r} \sup_{\theta \in T} P\{|b_\Delta(\alpha, 0)| \geq k\tau_\Delta^{1/r}\} =$$

$$= \pi_1 + \pi_2 + \pi_3. \tag{3.3.24}$$

In view of (3.2.21), $\pi_1 = O(|\Delta|^{-m/2})$; by (3.3.5), $\pi_2 = O(|\Delta|^{-m/2})$. Using Chebyshev's inequality, (3.3.4) and (3.3.1), we obtain

$$\sup_{\theta \in T} P\{|b_\Delta(\alpha, 0)| \geq k\tau_\Delta^{1/r}\} = O(\tau_\Delta^{-2m/r}), \ |\alpha| = 1, \ldots, r.$$

Hence $\pi_3 = O(\tau_\Delta^{-2m/r})$.

Inequality (3.3.22) and the conditions of the theorem permit us to use Lemma 3.3.3 to deduce that $\hat{\theta}_\Delta$ is a solution of the system of equations (3.3.14) and (3.3.16) is fulfilled. Set $y_\Delta = kt\tau_\Delta^{1/r}$. By assumption, $y_\Delta \xrightarrow[\Delta \to \infty]{} 0$; moreover (3.3.16) and (3.3.22) imply that

$$|\hat{u}_\Delta(\theta)| \leq k_{11} y_\Delta. \tag{3.3.25}$$

Since the terms $t^i h_i$ and $t^i h_{i,r-1}$ in (3.3.18) and (3.3.21) are homogeneous in $t b_\Delta(\alpha, 0)$, it follows that $t^i |h_i| \leq k_{12} y_\Delta^i$, $t^i |h_{i,r-1}| \leq k_{13} y_\Delta^i$. Whence

$$|u^{(r-1)}(t)| \leq k_{14} y_\Delta, \quad |\mathcal{L}_{r-1}(u^{(r-1)}(t)t)| \leq k_{15} y_\Delta^r. \tag{3.3.26}$$

Lemma 3.3.2, (3.3.25) and the first of inequalities (3.3.26) imply that

$$|\zeta_\Delta(\hat{u}_\Delta(\theta))| \leq k_{16} y_\Delta^r, \quad |\zeta_\Delta(u^{(r-1)}(t))| \leq k_{17} y_\Delta^r. \tag{3.3.27}$$

The second of inequalities (3.3.26) and the second of inequalities (3.3.27) yield

$$t^2 |\nabla L_\Delta(\theta + t^{-1} d_\Delta^{-1}(\theta) u^{(r-1)}(t))| \leq |\mathcal{L}_{r-1}(u^{(r-1)}(t), t)| +$$

$$+ |\zeta_\Delta(u^{(r-1)}(t))| \leq (k_{15} + k_{17}) y_\Delta^r. \tag{3.3.28}$$

Utilizing (3.3.7), (3.3.27) and inequality $|u_1^\alpha - u_2^\alpha| \leq k(\alpha)|u_1 - u_2|(|u_1| \vee |u_2|)^{|\alpha|-1}$, $u_1, u_2 \in R^q$, we have the estimate

$$t^2 |\nabla L_\Delta(\hat{\theta}_\Delta) - \nabla L_\Delta(\theta + t^{-1} d_\Delta^{-1}(\theta) u^{(r-1)}(t))| \geq |\mathcal{L}_{r-1}(\hat{u}_\Delta(\theta), t) -$$

$$- \mathcal{L}_{r-1}(u^{(r-1)}(t), t)| - (|\zeta_\Delta(\hat{u}_\Delta(\theta))| + |\zeta_\Delta(u^{(r-1)}(t))|) \geq$$

$$\geq |\hat{u}_\Delta(\theta) - u^{(r-1)}(t)|(2\lambda_* - 2qt \max_{i,j} |b_\Delta^{ij}(0,0)| -$$

$$- \sum_{j=1}^{r-2} k^{(j)}(|\hat{u}_\Delta(\theta)|^j \vee |u^{(r-1)}(t)|^j) - (k_{16} + k_{17}) y_\Delta^r \geq$$

$$\geq \lambda_* |\hat{u}_\Delta(\theta) - u^{(r-1)}(t)| - (k_{16} + k_{17}) y_\Delta^r. \tag{3.3.29}$$

Since $\nabla L_\Delta(\hat{\theta}_\Delta) = 0$, (3.3.23) follows from (3.3.28) and (3.3.29). ∎

The precise form of the polynomials $h_{i\Delta}$ may be obtained by substituting $u^{(r-1)}(t)$ in (3.3.7) and equating the coefficients of the powers t^i to zero (see Lemma 3.3.4). Using this method, we have

$$h_{1\Delta}(\theta) = \left(\sum_{i_1=1}^{q} \Lambda_\Delta^{ii_1}(\theta) b_\Delta^{i_1}(0,0) \right)_{i=1,\dots,q};$$

$$h_{2\Delta}(\theta) = \Big(\sum_{i_1,\dots,i_5=1}^{q} \Lambda_\Delta^{ii_1}(\theta) \Lambda_\Delta^{i_2 i_3}(\theta) (b_\Delta^{i_1 i_2}(0,0) b_\Delta^{i_3}(0,0) -$$

$$- \tfrac{1}{4} \Lambda_\Delta^{i_4 i_5}(\theta) a^{i_1 i_2 i_4}(0,0) b_\Delta^{i_3}(0,0) b_\Delta^{i_5}(0,0) \Big)_{i=1,\dots,q}.$$

It is easy to verify that in the expression for $h_{2\Delta}(\theta)$,

$$a_\Delta^{ijk}(0,0) = 2(\Pi_\Delta^{(ij)(k)}(\theta) + \Pi_\Delta^{(ik)(j)}(\theta) + \Pi_\Delta^{(jk)(i)}(\theta)),$$

$$\Pi_\Delta^{(ij)(k)}(\theta) = |\Delta|^{1/2} d_{i\Delta}^{-1}(\theta) d_{j\Delta}^{-1}(\theta) d_{k\Delta}^{-1}(\theta) \int_\Delta g_{ij}(x,\theta) g_k(x,\theta) dx.$$

Corollary 3.3.1. If in Theorem 3.3.1 assumption XI is fulfilled for $r = 2$, then for any $\epsilon > 0$

$$\pi_\Delta = \sup_{\theta \in T} P\{|d_\Delta(\theta)(\hat{\theta}_\Delta - \theta) - h_{1\Delta}(\theta)| > \epsilon\} = O(|\Delta|^{-m/2}) \qquad (3.3.30)$$

as $\Delta \to \infty$.

To prove this assertion, it suffices to set in (3.3.6): $r = 2$, $\tau_\Delta = \epsilon |\Delta|^{1/2}$, $\epsilon > 0$.

Corollary 3.3.2. Let assumption XI for $r = 2$, assumption XII, (3.3.4) for $m = 1$, and (3.2.12) all hold and suppose that $\gamma(\Delta) \xrightarrow[\Delta \to \infty]{P} B(0)$. Then for any $\epsilon > 0$,

$$\pi_\Delta \to 0, \quad \Delta \to \infty. \qquad (3.3.31)$$

This fact follows from (3.3.24) and Corollary 3.3.1.

If $\epsilon(x)$, $x \in \mathbf{R}^n$, is a Gaussian homogeneous field, Theorem 3.3.1 may be refined. Firstly in the case of a Gaussian field $\epsilon(x)$ and a function $g(x, \theta)$ satisfying assumptions IV, V of Theorem 3.2.1, relation (3.2.21) holds for any $m \geq 2$). Moreover, under assumptions VI and VII, which appear in Theorem 3.2.4, an exponential bound in $|\Delta|^{1/2}$ for the rate of decrease of the quantity $\sup_{\theta \in T} P\{|\hat{u}_\Delta(\theta)| > \rho\}$ can be achieved instead of the polynomial bound (3.2.21). Set $\tau_\Delta = \log^{r/2}|\Delta|$. Using the inequality

$$\int_\epsilon^\infty e^{-x^2/2} dx \leq \frac{1}{\epsilon} e^{-\epsilon^2/2}, \qquad (3.3.32)$$

conditions (3.3.1), and assuming that the field $\epsilon(x)$ possesses a bounded spectral density function $f(\lambda)$, $\lambda \in \mathbf{R}^n$, it is easy to derive the bound

$$\sup_{\theta \in T} P\{|b_\Delta(\alpha,0)| \geq k \log^{1/2}|\Delta|\} \leq \left(\frac{2}{\pi}\right)^{1/2} f_0^{1/2} k_0 k^{-1} \log^{-1/2}|\Delta| \times$$

$$\times \exp\{-k^2 \log|\Delta|/2 f_0 k_0^2\},$$

where $f_0 = \sup_{\lambda \in \mathbf{R}^n} f(\lambda)$, $k_0 = \max_{|\alpha|=1,\ldots,r} k(\alpha,0)$, the $k(\alpha,0)$ being the constants appearing in (3.3.1). Set $k = f_0^{1/2} k_0 m^{1/2}$. Then

$$\max_{|\alpha|=1,\ldots,r} \sup_{\theta \in T} P\{|b_\Delta(\alpha,0)| \geq k \log^{1/2}|\Delta|\} = o(|\Delta|^{-m/2}). \tag{3.3.33}$$

Inequalities (3.3.24) and (3.3.33) imply the following assertion.

Theorem 3.3.2. Let $\epsilon(x)$, $x \in \mathbf{R}^n$ be a homogeneous Gaussian field having cor. f. $B(x) \in L_1(\mathbf{R}^n)$. Suppose that the estimator $\hat{\theta}_\Delta$ possesses the property (3.2.21), $m \geq 2$ and that assumptions XI and XII hold. Then for some constant k_{18} depending on m,

$$\sup_{\theta \in T} P\{|d_\Delta(\theta)(\hat{\theta}_\Delta - \theta) - \sum_{\nu=1}^{r-1} t^{\nu-1} h_{\nu\Delta}(\theta)| \geq k_{18}(\log|\Delta|)^{r/2} t^{r-1}\} =$$

$$= O(|\Delta|^{-m/2}). \tag{3.3.34}$$

3.4. Asymptotic Normality and Convergence of Moments for Least Squares Estimators

Corollaries 3.3.1 and 3.3.2 allow us to derive sufficient conditions for asymptotic normality of the estimator $\hat{\theta}_\Delta$.

Theorem 3.4.1. Let g be a non-negative differentiable function on $[0,\infty)$ such that: 1) $b = \int_0^\infty |g'(\rho)| \rho^{q-1} d\rho < \infty$; 2) $\lim_{t\to\infty} g(t) = 0$. Then for any convex set $A \in \mathcal{B}^q$ and any ϵ, $\delta > 0$, the inequality

$$\int_{A_\epsilon \backslash A_{-\delta}} g(|x|)dx \leq b|s_{q-1}(1)|(\epsilon+\delta) \tag{3.4.1}$$

is valid.

The proof of this remarkable theorem is presented in reference [149].

We say that a vector-function $(g_1(x,\theta),\ldots,g_q(x,\theta))$ has the (matrix) spectral measure $\mu(d\lambda,\theta)$ uniformly in T if for the collection of (matrix) measures $\mu_\Delta(d\lambda,\theta)$ constructed from the functions $g_1,\ldots,g_q$ as in §3.1, $\mu_\Delta(\cdot,\theta) \xrightarrow{\mathcal{D}} \mu(\cdot,\theta)$ holds uniformly in T as $\Delta \to \infty$.

XIV. The vector-function $\nabla g(x,\theta)$ possesses the spectral measure $\mu(d\lambda,\theta)$ uniformly in T, the matrix $\int_{\mathbf{R}^n} f(\lambda)\mu(d\lambda,\theta)$ is positive definite uniformly in

$\theta \in T$ and the spectral density $f(\lambda)$ of the homogeneous field $\epsilon(x)$ is a continuous bounded function on $\mathbf{R}^n$.

Assumption XIV assures the fulfillment of assumption IV in §1.7. Let $_1F_\Delta(y,\theta)$ be the d.f. of the vector $d_\Delta(\theta)(\hat{\theta}_\Delta - \theta)$.

Theorem 3.4.2. Assume that the random field $\epsilon(x)$, $x \in \mathbf{R}^n$, satisfies the conditions of Theorem 1.7.5 and that $\nabla g(x,\theta)$ satisfies assumption XIV and condition (1.8.10) coordinatewise uniformly in $\theta \in T$. Let XII and (3.3.31) hold. Then uniformly in T,

$$_1F_\Delta(\cdot,\theta) \to \Phi^{(q)}_{0,K_1(\theta)}(\cdot) \quad \text{as } \Delta \xrightarrow{V.H.} \infty, \tag{3.4.2}$$

where

$$K_1(\theta) = (2\pi)^n \left(\int_{\mathbf{R}^n} \mu(d\lambda,\theta) \right)^{-1} \int_{\mathbf{R}^n} f(\lambda)\mu(d\lambda,\theta) \left(\int_{\mathbf{R}^n} \mu(d\lambda,\theta) \right)^{-1}. \tag{3.4.3}$$

Proof. The uniform positive definiteness of the matrix $\int_{\mathbf{R}^n} \mu(d\lambda,\theta)$ in $\theta \in T$ is assured by XII.

We set

$$\Pi(-\infty, y \pm \bar{\epsilon}) = (-\infty, y_1 \pm \epsilon) \times \ldots \times (-\infty, y_q \pm \epsilon), \quad \epsilon > 0.$$

In view of (3.3.31) we obtain for $\theta \in T$

$$-\pi_\Delta + P\{h_{1\Delta}(\theta) \in \Pi(-\infty, y - \bar{\epsilon})\} \leq {}_1F_\Delta(y,\theta) \leq$$

$$\leq P\{h_{1\Delta}(\theta) \in \Pi(-\infty, y + \bar{\epsilon})\} + \pi_\Delta. \tag{3.4.4}$$

By the uniform central limit theorem 1.7.5 we have as $\Delta \xrightarrow{V.H.} \infty$,

$$\sup_{\theta \in T} |P\{h_{1\Delta}(\theta) \in \Pi(-\infty, y \pm \bar{\epsilon})\} - \Phi^{(q)}_{0,K_1(\theta)}(y + \bar{\epsilon})| \to 0. \tag{3.4.5}$$

Let $\phi(y,\theta)$ be the Gaussian density function corresponding to the d.f. $\Phi^{(q)}_{0,K_1(\theta)}(y)$. Since $\inf_{\theta \in T} \lambda_{\min}(K_1(\theta)) = \underline{\lambda}_T > 0$, $\sup_{\theta \in T} \lambda_{\max}(K_1(\theta)) = \bar{\lambda}_T < \infty$,

$$\phi(y,\theta) \leq (2\pi\underline{\lambda}_T)^{-q/2} \exp\{-|y|^2/2\bar{\lambda}_T\} = g_T(|y|).$$

If $A = \Pi(-\infty, y)$, then $A_{-\epsilon} = \Pi(-\infty, y - \bar{\epsilon}]$, $(\Pi(-\infty, y + \bar{\epsilon}))_{-\epsilon} = A^c$. Application of Theorem 3.4.1 to the function g_r yields

$$\sup_{\theta \in T} |\Phi^{(q)}_{0,K_1(\theta)}(y) - \Phi^{(q)}_{0,K_1(\theta)}(y - \bar{\epsilon})| =$$

$$= \sup_{\theta \in T} \int_{A \backslash A_{-\epsilon}} \phi(y, \theta) dy \le b |s_{q-1}(1)| \epsilon; \qquad (3.4.6)$$

$$\sup_{\theta \in T} |\Phi^{(q)}_{0, K_1(\theta)}(y + \vec{\epsilon}) - \Phi^{(q)}_{0, K_1(\theta)}(y)| =$$

$$\sup_{\theta \in T} \int_{\Pi} \phi(y, \theta) dy \le b |s_{q-1}(1)| \epsilon, \quad \Pi = \Pi(-\infty, y + \vec{\epsilon}) \backslash A^c. \qquad (3.4.7)$$

Thus, (3.4.2) follows from (3.4.4)–(3.4.7) since $\epsilon > 0$ is arbitrary. ■

Let $\lambda \in s_{q-1}(1)$. Theorem 3.4.2 implies in particular that the d.f. of the r.v. $\langle \lambda, d_\Delta(\theta)(\hat{\theta}_\Delta - \theta) \rangle$ converges uniformly in $\theta \in T$ to the d.f. of the r.v. $\langle \lambda, \xi \rangle$ as $\Delta \to \infty$, where ξ is a Gaussian vector with parameters 0 and $K_1(\theta)$.

If the estimator $\hat{\theta}_\Delta$ satisfies inequality (3.2.20) for some $m > 0$, then also $\sup_{\theta \in T} P\{|\langle \lambda, d_\Delta(\theta)(\hat{\theta}_\Delta - \theta) \rangle| \ge H\} \le k^* H^{-m}$. Hence for any $0 < m' < m$,

$$\sup_{\theta \in T} E |\langle \lambda, d_\Delta(\theta)(\hat{\theta}_\Delta - \theta) \rangle|^{m'} \le k < \infty. \qquad (3.4.8)$$

Theorem 3.4.2 and inequality (3.4.8) yield the following result.

Theorem 3.4.3. If the estimator $\hat{\theta}_\Delta$ possesses properties (3.2.20) and (3.4.2), then for any $0 < m' < m$, $\lambda \in s_{q-1}(1)$,

$$\sup_{\theta \in T} |E|\langle \lambda, d_\Delta(\theta)(\hat{\theta}_\Delta - \theta) \rangle|^{m'} - E|\langle \lambda, \xi \rangle|^{m'}| \to 0, \quad \Delta \xrightarrow{V.H.} \infty.$$

Proof. The assertion follows from the uniform version in $\theta \in T$ of the convergence theorem in [191,Chap.IV,§11.4]. ■

Let $K_\Delta(\theta)$ be the correlation matrix of the vector $d_\Delta(\theta)(\hat{\theta}_\Delta - \theta)$.

Corollary 3.4.1. If in Theorem 3.4.3 $m > 2$, then

$$\sup_{\theta \in T} |K_\Delta(\theta) - K_1(\theta)| \to 0, \quad \Delta \xrightarrow{V.H.} \infty.$$

Let $q = 1$ and $\sup_{\theta \in T} \int_{\mathbf{R}^n} f^{-1}(\lambda) \mu(d\lambda, \theta) < \infty$. Since

$$1 = \int_{\mathbf{R}^n} \mu(d\lambda, \theta) \le \left(\int_{\mathbf{R}^n} f^{-1}(\lambda) \mu(d\lambda, \theta) \right)^{1/2} \left(\int_{\mathbf{R}^n} f(\lambda) \mu(d\lambda, \theta) \right)^{1/2},$$

then in accordance with (3.4.3),

$$K_1(\theta) \ge (2\pi)^n \left(\int_{\mathbf{R}^n} f^{-1}(\lambda) \mu(d\lambda, \theta) \right)^{-1}. \qquad (3.4.9)$$

Under certain additional requirements on the functions $f(\lambda)$ and $g(x,\theta)$, the bound (3.4.9) may be derived for $q > 1$ as well [52,128].

In this case inequality (3.4.9) is interpreted in the sense of non-negative definiteness of the matrices.

The lower bound (3.4.9) has an important optimality property. This property is that if $g_i(x,\theta)$ are known functions, then the right-hand side of (3.4.9) yields an asymptotically minimal value for the correlation matrix of estimators for the coefficients θ_j of the function $\sum_{j=1}^{q} \theta_j g_j(x,\theta)$ in the class of linear unbiased estimators [52]. Although $\hat{\theta}_\Delta$ does not belong, in general, to this class, (3.4.9) may be used to determine the efficiency of estimating the parameter θ of a non-linear function $g(x,\theta)$ by the least squares method. Specifically, if $f(\lambda) = \mathrm{const} > 0 \pmod{\mu(d\lambda,\theta)}$, then in (3.4.9) equality is attained and naturally the estimator may be called asymptotically efficient.

3.5.   Consistency of the Least Moduli Estimators

In §§3.5–3.6 the same non-linear model of observation (3.1.11) as in §§3.2–3.4 will be considered but to estimate the parameter θ, a different estimator will be used. Set $Q(x,\tau) = |\xi_\theta(x) - g(x,\tau)|$, $R_\Delta(\tau) = \int_\Delta Q(x,\tau)dx$.

The least moduli estimator of the parameter $\theta \in \Theta$ obtained from observations on the random field $\xi(x)$, $x \in \Delta$ is defined as a r.v. $\check{\theta}_\Delta = (\check{\theta}_{1\Delta},\ldots,\check{\theta}_{q\Delta})$ for which $R_\Delta(\check{\theta}_\Delta) = \inf_{\tau \in \Theta^c} R_\Delta(\tau)$.

Setting $s(x,u) = g(x,\theta + |\Delta|^{1/2}d_\Delta^{-1}(\theta)u)$, we write

$$\Phi_\Delta^{(t)}(u_1,u_2) = \int_\Delta |s(x,u_1) - s(x,u_2)|^t dx, \quad u_1,u_2 \in U_\Delta^c(\theta), \ t = 1,2.$$

Thus, $\Phi_\Delta^{(2)}(u_1,u_2)$ coincides with the function $\Phi_\Delta(u_1,u_2)$ defined in §3.2.

XV.   For any $\epsilon > 0$ and $R > 0$, a $\delta = \delta(\epsilon, R) > 0$ exists such that

$$\sup_{\theta \in T}\left[\sup_{u_1,u_2 \in U_\Delta^c(\theta) \cap v^c(R); |u_1-u_2|<\delta} |\Delta|^{-1}\Phi_\Delta^{(1)}(u_1,u_2)\right] \le \epsilon. \qquad (3.5.1)$$

XVI.   For any $R > 0$ a constant $k(R) < \infty$ exists such that

$$\sup_{\theta \in T}\left[\sup_{u \in U_\Delta^c(\theta) \cap v^c(R)} |\Delta|^{-1}\Phi_\Delta^{(2)}(u,0)\right] \le k(R). \qquad (3.5.2)$$

The following assumption (a contrast assumption) requires an appropriate coordination of properties of the functions $g(x,\theta)$ and $\epsilon(x)$. We set

$$\mu_t = E|\epsilon(0)|^t, \ t > 0, \quad \tilde{R}_\Delta(u) = R_\Delta(\theta + |\Delta|^{1/2}d_\Delta^{-1}(\theta)u).$$

XVII. For any $\rho > 0$, a number $a(\rho) > 0$ exists such that

$$\inf_{\theta \in T} \left[\inf_{u \in U_\Delta^c(\theta) \backslash v(\rho)} |\Delta|^{-1} E \widetilde{R}_\Delta(u) \right] \geq \mu_1 + a(\rho), \qquad (3.5.3)$$

and there exists $\rho_0 > 0$ such that $a(\rho_0) = q_0 \mu_1 + a_0$, where $q_0 > 2$ and $a_0 > 0$ are certain numbers.

We set $\check{u}_\Delta(\theta) = |\Delta|^{-1/2} d_\Delta(\theta)(\check{\theta}_\Delta - \theta)$.

Theorem 3.5.1. If assumptions XV–XVII are valid and the field $\epsilon(x)$, $x \in R^n$, satisfies the assumption of Lemma 1.8.2 for $m = 1$, then for any $\rho > 0$

$$\sup_{\theta \in T} P\{|\check{u}_\Delta(\theta)| \geq \rho\} = O(|\Delta|^{-1}), \quad \Delta \to \infty. \qquad (3.5.4)$$

Proof. Choose $\theta \in T$ and set $b_\Delta(\theta, u) = \widetilde{R}_\Delta(u) - E \widetilde{R}_\Delta(u)$. Clearly, $b_\Delta(\theta, 0) = \int_\Delta |\epsilon(x)| dx - \mu_1 |\Delta|$. By definition of the estimator $\check{\theta}_\Delta$, $R_\Delta(\check{\theta}_\Delta) \leq b_\Delta(\theta, 0) + \mu_1 |\Delta|$ a.s. Hence, by assumption XVII,

$$P\{|\check{u}_\Delta(\theta)| \geq \rho\} \leq P\{|\Delta|^{-1} b_\Delta(\theta, 0) + \inf_{u \in U_\Delta^c(\theta) \backslash v(\rho)} |\Delta|^{-1} E \widetilde{R}_\Delta(u) -$$

$$-a(\rho) \geq \inf_{u \in U_\Delta^c(\theta) \backslash v(\rho)} \widetilde{R}_\Delta(u)\} \leq P\{|\Delta|^{-1} b_\Delta(\theta, 0) > (1 - \gamma)a(\rho)\} +$$

$$+ P\Big\{ \inf_{u \in U_\Delta^c(\theta) \backslash v(\rho)} |\Delta|^{-1} \widetilde{R}_\Delta(u) - \inf_{u \in U_\Delta^c(\theta) \backslash v(\rho)} |\Delta|^{-1} E \widetilde{R}_\Delta(u) \leq$$

$$\leq -\gamma a(\rho) \Big\} = P^{(1)} + P^{(2)},$$

where $\gamma \in (0, 1)$.

Applying Chebyshev's inequality and inequality (1.8.1) to the field $|\epsilon(x)| - \mu_1$, we obtain $P^{(1)} = O(|\Delta|^{-1})$. On the other hand,

$$P^{(2)} \leq P\Big\{ \inf_{u \in U_\Delta^c(\theta) \backslash v(\rho)} |\Delta|^{-1} b_\Delta(\theta, u) \leq -\gamma a(\rho) \Big\}. \qquad (3.5.5)$$

Since obviously,

$$\Phi_\Delta^{(1)}(u, 0) - \int_\Delta |\epsilon(x)| dx \leq \widetilde{R}_\Delta(u) \leq \Phi_\Delta^{(1)}(u, 0) + \int_\Delta |\epsilon(x)| dx,$$

it follows that

$$|\Delta|^{-1} b_\Delta(\theta, u) \geq -|\Delta|^{-1} \int_\Delta |\epsilon(x)| dx - \mu_1. \qquad (3.5.6)$$

Set $\rho = \rho_0$ and $\gamma = \frac{2}{q_0}$ where ρ_0 and q_0 are the numbers in assumption XVII. Then in view of assumption XVII and inequalities (3.5.6) and (1.8.1), the probability (3.5.5) is a quantity of order $O(|\Delta|^{-1})$. Thus, it remains to evaluate the probability

$$P\{\rho_0 > |\breve{u}_\Delta(\theta)| \geq \rho\} \leq P\{|\Delta|^{-1}b_\Delta(\theta, 0) \geq (1 - \gamma')a(\rho)\} +$$

$$+ P\left\{ \inf_{u \in U^c_\Delta(\theta) \cap (v^c(\rho_0) \backslash v(\rho))} |\Delta|^{-1}b_\Delta(\theta, u) \leq -\gamma'a(\rho) \right\} \leq$$

$$\leq P\left\{ \sup_{u \in U^c_\Delta(\theta) \cap v^c(\rho_0)} |\Delta|^{-1}|b_\Delta(\theta, u)| \geq \gamma'a(\rho) \right\} + O(|\Delta|^{-1}),$$

where $\gamma' \in (0, 1)$.

Let $F^{(1)}, \ldots, F^{(s)} \in v^c(\rho_0)$ be closed sets whose diameters do not exceed the quantity δ corresponding to the numbers $R = \rho_0$ and $\epsilon = \beta a(\rho)\gamma'/2$ in assumption XV, where $\beta \in (0, 1)$. Choose $u^{(i)} \in F^{(i)} \cap U^c_\Delta(\theta)$, $i = 1, \ldots, s$. Then

$$P^{(3)} = P\left\{ \sup_{u \in U^c_\Delta(\theta) \cap v^c(\rho_0)} |\Delta|^{-1}|b_\Delta(\theta, u)| \geq \gamma'a(\rho) \right\} \leq$$

$$\leq \sum_{i=1}^s P\left\{ \sup_{u_1, u_2 \in U^c_\Delta(\theta) \cap F^{(i)}} |\Delta|^{-1}|b_\Delta(\theta, u_1) - b_\Delta(\theta, u_2)| + \right.$$

$$\left. + |\Delta|^{-1}|b_\Delta(\theta, u^{(i)})| \geq \gamma'a(\rho) \right\}.$$

Note that

$$|b_\Delta(\theta, u_1) - b_\Delta(\theta, u_2)| \leq |\tilde{R}_\Delta(u_1) - \tilde{R}_\Delta(u_2)| +$$

$$+ E|\tilde{R}_\Delta(u_1) - \tilde{R}_\Delta(u_2)| \leq 2\Phi^{(1)}_\Delta(u_1, u_2).$$

Therefore by assumption XV,

$$P^{(3)} \leq \sum_{i=1}^s P\{|\Delta|^{-1}|b_\Delta(\theta, u^{(i)})| \geq (1 - \beta)\gamma'a(\rho)\}. \qquad (3.5.7)$$

We shall estimate each term of the last sum separately. Denote $\tilde{Q}(x, u) = Q(x, \theta + |\Delta|^{1/2}d_\Delta^{-1}(\theta)u)$. By Lemma 1.6.2 and assumption XVI,

$$|\Delta|^{-2} \operatorname{var} b_\Delta(\theta, u^{(i)}) = |\Delta|^{-2} \int_\Delta \int_\Delta \operatorname{cov}(\tilde{Q}(x, u^{(i)}), \tilde{Q}(y, u^{(i)})) dx dy \leq$$

$$\leq 10|\Delta|^{-2} \int_\Delta \int_\Delta (E\tilde{Q}^{2+\delta}(x, u^{(i)}))^{1/(2+\delta)} \times$$

$$\times (E\widetilde{Q}^{2+\delta}(y,u^{(i)}))^{1/(2+\delta)}\alpha^{\delta/(2+\delta)}(|x-y|)dxdy \le 20 \cdot 2^{\delta/(2+\delta)}|\Delta|^{-2}\times$$

$$\times \int_\Delta \int_\Delta \alpha^{\delta/(2+\delta)}(|x-y|)(\mu_{2+\delta}^{2/(2+\delta)} + |s(x,0) - s(x,u^{(i)})|^2)dxdy \le$$

$$\le 20 \cdot 2^{\delta/(2+\delta)}|s(1)|\alpha_1(\mu_{2+\delta}^{2/(2+\delta)} + |\Delta|^{-1}\Phi_\Delta^{(2)}(u^{(i)},0))|\Delta|^{-1} \le$$

$$\le 20 \cdot 2^{\delta/(2+\delta)}|s(1)|\alpha_1(\mu_{2+\delta}^{2/(2+\delta)} + k(\rho_0))|\Delta|^{-1},$$

$$\alpha_1 = \int_0^\infty \rho^{n-1}\alpha^{\delta/(2+\delta)}(\rho)d\rho, \quad i=1,\ldots,s. \tag{3.5.8}$$

Thus in view of (3.5.7), Chebyshev's inequality and (3.5.8), we obtain $P^{(3)} = O(|\Delta|^{-1})$. The case when some sets $F^{(i)} \cap U_\Delta^c(\theta)$ are empty is handled in a similar manner. ∎

We present a sufficient condition for the validity of XVII. Suppose that the r.v. $\epsilon(0)$ is symmetric and has an absolutely continuous d.f. F, $F'(x) = \tau(x)$. Suppose also that the conditions

$$\sup_{x \in \mathbf{R}^n} \sup_{\theta_1,\theta_2 \in \Theta^c} |g(x,\theta_1) - g(x,\theta_2)| = g_0 < \infty, \tag{3.5.9}$$

$$\inf_{|x| \le g_0} \tau(x) = \tau^* > 0 \tag{3.5.10}$$

hold. Since for any $g \in [-g_0, g_0]$

$$E|\epsilon(0) + g| - \mu_1 = \int_{-|g|}^{|g|} (|g| - |x|)F(dx) \ge \tau^* g^2,$$

it follows that for $u \in U_\Delta^c(\theta)$

$$|\Delta|^{-1}E\widetilde{R}_\Delta(u) \ge \mu_1 + \tau^*|\Delta|^{-1}\Phi_\Delta^{(2)}(u,0).$$

Therefore XVII is a corollary of the following assumption.

XVII$_1$. For any $\rho > 0$ there exists a number $a(\rho) > 0$ such that

$$\inf_{\theta \in T}\left[\inf_{u \in U_\Delta^c(\theta)\backslash v(\rho)} |\Delta|^{-1}\Phi_\Delta^{(2)}(u,0)\right] \ge a(\rho)$$

and there exists $\rho_0 > 0$ such that $\tau^* a(\rho_0) = q_0\mu_1 + a_0$; $q_0 > 2$, $a_0 > 0$ are certain numbers. Assumption XVII is similar to V presented in §3.2. Clearly, (3.5.9) ensures the validity of assumption XVI.

Theorem 3.5.2. Let $\epsilon(x)$, $x \in \mathbf{R}^n$, be a homogeneous Gaussian field with cor. f. $B(x) \in L_1(\mathbf{R}^n)$. If assumptions XV–XVII hold, the estimator $\check{\theta}_\Delta$ satisfies (3.5.4).

Proof. The proof of this theorem differs from that of the preceding one only in the evaluation of the quantity $|\Delta|^{-2} \operatorname{var} b_\Delta(\theta, u)$, $u \in U_\Delta^c(\theta) \cap v^c(\rho_0)$. Since $B = \int_{\mathbf{R}^n} |B(x)| dx < \infty$, $\epsilon(x)$ possesses a spectral density function. Thus by the Riemann-Lebesgue theorem, $\lim_{|x| \to \infty} B(x) = 0$. Let $\epsilon > 0$ be a number such that $k_0 = \sup_{|x| > \epsilon} (B(x)/B(0)) < 1$. Then

$$|\Delta|^{-2} \operatorname{var} b_\Delta(\theta, u) \le |\Delta|^{-2} \iint_{\Delta^2 \cap \{|x-y| \le \epsilon\}} (B(0) + |s(x,0) - s(x,u)|^2) dx dy +$$

$$+ |\Delta|^{-2} \iint_{\Delta^2 \cap \{|x-y| > \epsilon\}} \operatorname{cov}(\widetilde{Q}(x,u), \widetilde{Q}(y,u)) dx dy = D_1 + D_2.$$

By assumption XVI,

$$D_1 \le (B(0) + |\Delta|^{-1} \Phi_\Delta^{(2)}(u,0)) |v(1)| \epsilon^n |\Delta|^{-1} \le$$

$$\le (B(0) + k(\rho_0)) |v(1)| \epsilon^n |\Delta|^{-1}. \tag{3.5.11}$$

To evaluate D_2, we utilize expansion (2.1.5) of the bivariate Gaussian density function:

$$D_2 = |\Delta|^{-2} \sum_{m=1}^\infty \int_{\mathbf{R}^2} \frac{1}{m!} H_m(t_1) H_m(t_2) \frac{1}{2\pi} e^{-(t_1^2 + t_2^2)/2} \times$$

$$\times \left(\iint_{\Delta^2 \cap \{|x-y| > \epsilon\}} |B^{1/2}(0) t_1 + s(x,0) - s(x,u)| |B^{1/2}(0) t_2 + \right.$$

$$\left. + s(y,0) - s(y,u)| \frac{B^m(x-y)}{B^m(0)} dx dy \right) dt_1 dt_2 \le$$

$$\le |\Delta|^{-2} \sum_{m=1}^\infty \int_{\mathbf{R}^2} \frac{1}{m!} |H_m(t_1) H_m(t_2)| \frac{1}{2\pi} e^{-(t_1^2 + t_2^2)/2} \iint_{\Delta^2 \cap \{|x-y| > \epsilon\}} \left(B(0) |t_1 t_2| + \right.$$

$$+ 2 B^{1/2}(0) |t_1| \, |s(y,0) - s(y,u)| + |s(x,0) - s(x,u)|^2 \right) \frac{|B(x-y)|^m}{B^m(0)} dx dy \, dt_1 dt_2. \tag{3.5.12}$$

We estimate the series corresponding to the term $2B^{1/2}(0)|t_1|\,|s(y,0) - s(y,u)|$ in the inner integral of (3.5.12). Note that by the Cauchy-Bunyakovskii inequality

$$(2\pi)^{-1/2} \int_{-\infty}^{\infty} |t|^a |H_m(t)| e^{-t^2/2} dt \le (m!)^{1/2}, \quad a = 0, 1.$$

Therefore

$$2B^{1/2}(0)|\Delta|^{-2} \sum_{m=1}^{\infty} \int_{R^2} \frac{1}{m!} |t_1 H_m(t_1) H_m(t_2)| \frac{1}{2\pi} e^{-(t_1^2 + t_2^2)/2} dt_1 dt_2 \times$$

$$\times \iint_{\Delta^2 \cap \{|x-y|>\epsilon\}} |s(y,0) - s(y,u)| \frac{|B(x-y)|^m}{B^m(0)} dx\,dy \le 2B^{-1/2}(0)(1-k_0)^{-1} \times$$

$$\times |\Delta|^{-2} \int_\Delta \int_\Delta |s(y,0) - s(y,u)|\,|B(x-y)| dx\,dy \le \frac{2B\,k^{1/2}(\rho_0)}{B^{1/2}(0)(1-k_0)} |\Delta|^{-1}.$$

In the same way, the series corresponding to the term $B(0)|t_1 t_2|$ is bounded by the quantity $B(1-k_0)^{-1}|\Delta|^{-1}$ and the series corresponding to the term $|s(y,0) - s(y,u)|^2$ by the quantity $B^{-1}(0)B(1-k_0)^{-1}k(\rho_0)|\Delta|^{-1}$. Hence

$$D_2 \le B(1-k_0)^{-1}(2B^{-1/2}(0)k^{1/2}(\rho_0) + B^{-1}(0)k(\rho_0) + 1)|\Delta|^{-1}. \quad (3.5.13)$$

The bounds (3.5.11) and (3.5.13) show that $|\Delta|^{-2} \operatorname{var} b_\Delta(\theta, u) = O(|\Delta|^{-1})$. ∎

Since (3.5.10) holds for a Gaussian density, assumption XVII will hold, for example, for a bounded function $g(x,\theta)$ provided assumption XVII$_1$ is valid.

3.6.   Asymptotic Normality of the Least Moduli Estimators

The principal difficulty when proving the asymptotic normality of the estimator $\check\theta_\Delta$ is that the functional $R_\Delta(\tau)$ is not differentiable and thus one cannot use the standard arguments equating the gradient of $R_\Delta(\tau)$ to zero at the point $\check\theta_\Delta$ (cf §§3.3–3.4). On the other hand, the functional $R_\Delta(\tau)$ has derivatives in all directions provided function $g(x,\theta)$ is differentiable with respect to $\theta \in \Theta^c$ and passage to the limit under the integral sign is justified. Indeed, if $\tau \in \Theta$ and l is an arbitrary direction in R^q, then

$$\frac{\partial}{\partial l} R_\Delta(\tau) = \int_\Delta \langle \nabla g(x,\tau), l \rangle (2\chi\{\xi_\theta(x)\} * g(x,\tau)\} - 1) dx, \quad (3.6.1)$$

where "*" denotes $\leq$ if $\langle \nabla g(x,\tau), l \rangle > 0$ and $<$ if $\langle \nabla g(x,\tau), l \rangle < 0$.

Let $\tilde{\rho} > 0$ be the distance between the compact set T and the set $R^q \backslash \Theta$. If $\rho < \tilde{\rho}$ and event $|\breve{\theta}_\Delta - \theta| < \rho$ occurs, then for any direction l, $\frac{\partial}{\partial l} R_\Delta(\breve{\theta}_\Delta) \geq 0$. This simple remark allows us to obtain the asymptotic normality of the estimator $\breve{\theta}_\Delta$.

We shall impose a number of requirements on the observations model (3.1.1) using the notation of the preceding sections and retaining the requirements of §3.1 with regard to function $g(x,\tau)$.

Assume that Θ is a convex set and suppose that all the derivatives of function $g(x,\theta)$ with respect to the variables $\theta = (\theta_1, \ldots, \theta_q)$ up to second order inclusive, exist for each $x \in \mathbf{R}^n$ and are continuous; $g_i(x,\theta)$ and $g_{ij}(x,\theta)$, $i, j = 1, \ldots, q$, are locally square integrable in x for any $\theta \in \Theta^c$.

XVIII.   For any $\rho > 0$,

$$\sup_{\theta \in T} \left[\sup_{u \in U^c_\Delta(\theta) \cap v^c(\rho)} |\Delta|^{1/2} d^{-1}_{i\Delta}(\theta) \sup_{x \in \Delta} |s_i(x,u)| \right] \leq k^{(i)}(\rho), \quad i = 1, \ldots, q; \quad (3.6.2)$$

$$\sup_{\theta \in T} \left[\sup_{u \in U^c_\Delta(\theta) \cap v^c(\rho)} |\Delta|^{1/2} d^{-1}_{i\Delta}(\theta) d^{-1}_{j\Delta}(\theta) d_{ij\Delta}(\theta + \right.$$

$$\left. + |\Delta|^{1/2} d^{-1}_\Delta(\theta) u) \right] \leq k^{(ij)}(\rho), \quad i, j = 1, \ldots, q. \quad (3.6.3)$$

From (3.6.2) and (3.6.3) and by Lemma 3.3.1, we have the inequalities:

$$\sup_{\theta \in T} \left[\sup_{u_1, u_2 \in U^c_\Delta(\theta) \cap v^c(\rho)} |\Delta|^{-1} \Phi_\Delta(u_1, u_2) |u_1 - u_2|^{-2} \right] \leq k(\rho); \quad (3.6.4)$$

$$\sup_{\theta \in T} \left[\sup_{u_1, u_2 \in U^c_\Delta(\theta) \cap v^c(\rho)} d^{-2}_{i\Delta}(\theta) \Phi_{i\Delta}(u_1, u_2) |u_1 - u_2|^{-2} \right] \leq k_i(\rho), \quad i = 1, \ldots, q.$$

$$(3.6.5)$$

These inequalities will be utilized below. The assumption (3.6.3) is assumption (3.3.1) for $|\alpha| = 2$ and (3.6.2) is sufficient for the validity of (3.3.1) with $|\alpha| = 1$.

XIX.   The r.v. $\epsilon(0)$ is symmetric. Its d.f. $F(x)$ is absolutely continuous: $F'(x) = \tau(x)$ and   1) $\tau_0 = \sup_{x \in \mathbf{R}^n} \tau(x) < \infty$;   2) $\tau(0) > 0$;   3) $|\tau(x) - \tau(0)| \leq H|x|$, where H is constant.

Consider along with the strictly homogeneous random field $\epsilon(x)$, $x \in \mathbf{R}^n$, the strictly homogeneous random field $\eta(x) = 2\chi\{\epsilon(x) < 0\} - 1$, $x \in \mathbf{R}^n$. Clearly, $E\eta(x) = 0$ for the symmetric r.v. $\epsilon(x)$; $B_\eta(x) = E\eta(x)\eta(0) = 4P\{\epsilon(x) < 0, \ \epsilon(0) < 0\} - 1$.

If $\epsilon(x)$ is a Gaussian field, then by the Stieltjes-Sheppard formula [160] we have

$$B_\eta(x) = 4 \int_{-\infty}^{0} \int_{-\infty}^{0} \phi\left(t, s, \frac{B(x)}{B(0)}\right) dt\,t\,ds - 1 =$$
$$= \frac{2}{\pi} \sin^{-1} \frac{B(x)}{B(0)}.$$

Let $\alpha'(\rho)$ be the mixing rate (see §1.6) of the field $\eta(x)$. Then $\alpha'(\rho) \leq \alpha(\rho)$, $\alpha(\rho)$ is the mixing rate of $\epsilon(x)$. If for $\alpha(\rho)$, assumption IX of §1.7 holds, then $B_\eta(x)$ is integrable over $\mathbf{R}^n$ (see Remark 1.6.1) and thus, $\eta(x)$ has a continuous bounded spectral density function $f_\eta(\lambda)$, $\lambda \in \mathbf{R}^n$.

XX. The vector-function $\nabla g(x, \theta)$ has the spectral measure $\mu(d\lambda, \theta)$ uniformly in T and the matrix $\int_{\mathbf{R}^n} f_\eta(\lambda)\mu(d\lambda, \theta)$ is positive definite uniformly in $\theta \in T$.

Theorem 3.6.1. Assume that the mixing rate $\alpha(\rho)$ satisfies assumption IX given in §1.7, $g(x, \tau)$ satisfies assumptions XVIII, XX and XII given in §3.3 and that the r.v. $\epsilon(0)$ satisfies assumption XIX, $\Delta \xrightarrow{V.H.} \infty$. Let the estimator $\check{\theta}_\Delta$ possess property (3.5.4). Then uniformly in T,

$$_2F_\Delta(\cdot, \theta) \xrightarrow{\mathcal{D}} \Phi_{0, K_2(\theta)}^{(q)}(\cdot), \quad \Delta \xrightarrow{V.H.} \infty, \qquad (3.6.6)$$

where $_2F_\Delta(y, \theta)$ is the d.f. of the vector $d_\Delta(\theta)(\check{\theta}_\Delta - \theta)$,

$$K_2(\theta) = \frac{(2\pi)^n}{4\tau^2(0)} \left(\int_{\mathbf{R}^n} \mu(d\lambda, \theta)\right)^{-1} \int_{\mathbf{R}^n} f_\eta(\lambda)\mu(d\lambda, \theta)\left(\int_{\mathbf{R}^n} \mu(d\lambda, \theta)\right)^{-1}.$$
$$(3.6.7)$$

We shall subdivide the proof into several stages. Let $l_1, \ldots, l_q$ be the positive directions of the coordinate axes. Consider the vectors $R_\Delta^\pm(\tau)$ with coordinates $R_{i\Delta}^\pm(\tau) = d_{i\Delta}^{-1}(\theta)\frac{\partial}{\partial(\pm l_i)}R_\Delta(\tau)$, $i = 1, \ldots, q$ and the vectors $ER_\Delta^\pm(\tau)$ with coordinates

$$ER_{i\Delta}^\pm(\tau) = \pm d_{i\Delta}^{-1}(\theta) \int_\Delta g_i(x, \tau)(2F(g(x, \tau) - g(x, \theta)) - 1)dx, \quad i = 1, \ldots, q.$$
$$(3.6.8)$$

Evidently, $ER_\Delta^\pm(\theta) = 0$ in view of the symmetry of the r.v. $\epsilon(0)$. We set

$$z_\Delta^\pm(\theta, u) = \left| R_\Delta^\pm(\theta + |\Delta|^{1/2}d_\Delta^{-1}(\theta)u) - R_\Delta^\pm(\theta) - \right.$$

$$\left. - ER_\Delta^\pm(\theta + |\Delta|^{1/2}d_\Delta^{-1}(\theta)u)\right|(1 + |ER_\Delta^\pm(\theta + |\Delta|^{1/2}d_\Delta^{-1}(\theta)u)|)^{-1}.$$

Lemma 3.6.1. Under conditions of Theorem 3.6.1, for any $\epsilon > 0$ and sufficiently small $\rho > 0$,

$$\sup_{\theta \in T} P\left\{ \sup_{u \in U_\Delta^c(\theta) \cap v^c(\rho)} z_\Delta^\pm(\theta, u) > \epsilon \right\} \to 0, \quad \Delta \xrightarrow{V.H.} \infty. \qquad (3.6.9)$$

Proof. We shall carry out the proof for the quantity $z_\Delta^+(\theta, u)$. To simplify matters, assume that $\rho = 1$ and the supremum under the probability sign in (3.6.9) is defined in the cube $C_0 = \{u : |u|_0 = \max_{1 \le i \le q} |u_i| \le 1\} \supset v(1)$. Cover the cube C_0 by $N_0 = O(\ln |\Delta|)$ cubes $C_{(1)}, \ldots, C_{(N_0)}$ in the following manner. Let $p \in (0, 1)$. We construct a concentric system of sets $C^{(m)} = \{u : |u|_0 \in [(1-p)^{m+1}, (1-p)^m]\}$, $m = 0, \ldots, m_0 - 1$, $C^{(m_0)} = \{u : |u|_0 \le (1-p)^{m_0}\}$. Cover each one of the sets $C^{(m)}$ by identical cubes with side $a_m = (1-p)^m - (1-p)^{m+1} = p(1-p)^m$ and enumerate them. They will form a covering $C_{(1)}, \ldots, C_{(N_0-1)}, C_{(N_0)} = C^{(m_0)}$. Choose $m_0 = m_0(\Delta)$ from the condition $(1-p)^{\tilde{m}_0} = |\Delta|^{-\gamma}$, $m_0 = [\tilde{m}_0]$, $\gamma \in (\frac{1}{2}, 1)$. Note that the $|\cdot|_0$-distance from $C_{(j)}$ to 0 equals $\rho(j) = (1-p)|\Delta|^{-\gamma m / \tilde{m}_0}$ and the $|\cdot|_0$-diameter of $C_{(j)}$ equals $a(j) = p|\Delta|^{-\gamma m / \tilde{m}_0}$ for some $m = m(j)$, $j = 1, \ldots, N_0 - 1$. Indeed, let the cube $C_{(j)}$ be an element of the covering of the set $C^{(m)}$. Then $a(j) = a_m$, $\rho(j) = p(1-p)^{m+1} + \ldots + p(1-p)^{m_0-1} + (1-p)^{m_0}$. The number of cubes $C_{(j)}$ covering each set $C^{(m)}$ can be made independent of m and, hence, of Δ. To verify this, consider any octant in R^q. The volume of the subset of $C^{(m)}$ falling into this octant is $(1-p)^{mq} - (1-p)^{(m+1)q}$ and $|C_{(j)}| = a^q(j) = p^q(1-p)^{mq}$. Thus, this octant contains at most $((1-p)^{mq} - (1-p)^{(m+1)q})p^{-q}(1-p)^{-mq} = (1-(1-p)^q)p^{-q}$ cubes. Since $m_0 = O(\ln |\Delta|)$, we have $N_0 = O(\ln |\Delta|)$. Choose $\theta \in T$. Then

$$P\left\{ \sup_{u \in C_0} z_\Delta^+(\theta, u) > \epsilon \right\} \le \sum_{j=1}^{N_0} P\left\{ \sup_{u \in C_{(j)}} z_\Delta^+(\theta, u) > \epsilon \right\}. \qquad (3.6.10)$$

Evaluate each term in (3.6.10). A common element of the matrix-derivative $D_\Delta(u)$ of the mapping $u \to ER_\Delta^+(\theta + |\Delta|^{1/2} d_\Delta^{-1}(\theta)u)$ is of the form

$$D_\Delta^{ij}(u) = \frac{\partial}{\partial u_j} ER_{i\Delta}^+(\theta + |\Delta|^{1/2} d_\Delta^{-1}(\theta)u) =$$

$$= |\Delta|^{1/2} d_{i\Delta}^{-1}(\theta) d_{j\Delta}^{-1}(\theta) \int_\Delta s_{ij}(x, u)(2F(s(x, u) - s(x, 0)) - 1) dx +$$

$$+ 2|\Delta|^{1/2} d_{i\Delta}^{-1}(\theta) d_{j\Delta}^{-1}(\theta) \int_\Delta s_i(x, u) s_j(x, u) \tau(s(x, u) - s(x, 0)) dx =$$

$$= {}_1 D_\Delta^{ij}(u) + {}_2 D_\Delta^{ij}(u).$$

In view of (3.6.3) (3.6.4) and the boundedness of $\tau(x)$, for $|u| < \rho$ we obtain

$$|\Delta|_1^{-1/2} D_\Delta^{ij}(u) \le 2|\Delta|^{1/2} d_{i\Delta}^{-1}(\theta) d_{j\Delta}^{-1}(\theta) d_{ij\Delta}(\theta + |\Delta|^{1/2} d_\Delta^{-1}(\theta)u) \times$$

$$\times \left(|\Delta|^{-1} \int_\Delta (F(s(x,u) - s(x,0)) - F(0))^2 dx\right)^{1/2} \le$$

$$\le 2k^{(ij)}(\rho)k^{1/2}(\rho)\tau_0|u|. \tag{3.6.11}$$

On the other hand,

$$|\tfrac{1}{2}|\Delta|^{-1/2}{}_2 D_\Delta^{ij}(u) - \tau(0)\Im_\Delta^{ij}(\theta)| \le \tau_0[d_{i\Delta}^{-1}(\theta)d_{i\Delta}(\theta +$$

$$+|\Delta|^{1/2}d_\Delta^{-1}(\theta)u)d_{j\Delta}^{-1}(\theta)\Phi_{j\Delta}^{1/2}(u,0) + d_{i\Delta}^{-1}(\theta)\Phi_{i\Delta}^{1/2}(u,0)]+$$

$$+d_{i\Delta}^{-1}(\theta)d_{j\Delta}^{-1}(\theta)\left|\int_\Delta g_i(x,\theta)g_j(x,\theta)(\tau(s(x,u) - s(x,0)) - \tau(0))dx\right|. \tag{3.6.12}$$

In view of (3.6.2) and (3.6.5) the terms in the square brackets are bounded by the quantity $\tau_0(k^{(i)}(\rho)k^{1/2}(\rho) + k_i^{1/2}(\rho))|u|$. From assumption XIX and (3.6.2) with $u = 0$, we find the following majorant for the last term:

$$|\Delta|^{1/2}d_{i\Delta}^{-1}(\theta) \sup_{x \in \Delta} |g_i(x,\theta)| \left(|\Delta|^{-1} \int_\Delta (\tau(s(x,u) - s(x,0)) - \tau(0))^2 dx\right)^{1/2} \le$$

$$\le k^{(i)}(0)H|\Delta|^{-1/2}\Phi_\Delta^{1/2}(u,0) \le k^{(i)}(0)k^{1/2}(\rho)H|u|. \tag{3.6.13}$$

Since by assumption XII, the matrix $|\Delta|^{-1/2}D_\Delta(0) = 2\tau(0)\Im_\Delta(\theta)$ is positive definite, the above arguments show that for a sufficiently small u (assume for simplicity that $u \in C_0$) and some $k_0 > 0$

$$\inf_{\theta \in T} |ER_\Delta^+(\theta + |\Delta|^{1/2}d_\Delta^{-1}(\theta)u)| \ge k_0|\Delta|^{1/2}|u|_0. \tag{3.6.14}$$

Let $j \ne N_0$, $v \in C_{(j)}$ be an arbitrary point. In view of (3.6.14) one can write

$$\sup_{u \in C_{(j)}} z_\Delta^+(\theta, u) \le \left(\sup_{u \in C_{(j)}} W_\Delta^{(j)}(\theta, u, v) + Y_\Delta^{(j)}(\theta, v)\right)(1 + k_0|\Delta|^{1/2}\rho(j))^{-1};$$

$$W_\Delta^{(j)}(\theta, u, v) = \sum_{\lambda=1}^4 W_{\lambda\Delta}^{(j)}(\theta, u, v),$$

$$W_{1\Delta}^{(j)}(\theta, u, v) = 2 \left| d_{\Delta}^{-1}(\theta) \int_{\Delta} \nabla s(x, u)(\chi\{\xi_{\theta}(x) * s(x, u)\} - \chi\{\xi_{\theta}(x) < s(x, v)\}) dx \right|;$$

$$W_{2\Delta}^{(j)}(\theta, u, v) = \left| d_{\Delta}^{-1}(\theta) \int_{\Delta} (\nabla s(x, u) - \nabla s(x, v))(2\chi\{\xi_{\theta}(x) < s(x, v)\} - 1) dx \right|;$$

$$W_{3\Delta}^{(j)}(\theta, u, v) = 2 \left| d_{\Delta}^{-1}(\theta) \int_{\Delta} \nabla s(x, u)(F(s(x, u) - s(x, 0)) - F(s(x, v) - s(x, 0))) dx \right|;$$

$$W_{4\Delta}^{(j)}(\theta, u, v) = \left| d_{\Delta}^{-1}(\theta) \int_{\Delta} (\nabla s(x, u) - \nabla s(x, v))(2F(s(x, v) - s(x, 0)) - 1) dx \right|;$$

$$Y_{\Delta}^{(j)}(\theta, v) = \left| d_{\Delta}^{-1}(\theta) \int_{\Delta} (\nabla s(x, v)(2\chi\{\xi_{\theta}(x) < s(x, v)\} - 1) - \right.$$
$$\left. - \nabla s(x, 0)(2\chi\{\epsilon(x) * 0\} - 1) - \nabla s(x, v)(2F(s(x, v) - s(x, 0)) - 1)) dx \right|.$$

From (3.6.5) we obtain for $u, v \in C_{(j)}$ that

$$|\Delta|^{-1/2} W_{2\Delta}^{(j)}(\theta, u, v) \leq \left(\sum_{i=1}^{q} d_{i\Delta}^{-2}(\theta) \Phi_{i\Delta}(u, v) \right)^{1/2} \leq k_1 a(j). \qquad (3.6.15)$$

Note further that

$$|\Delta|^{-1/2} W_{3\Delta}^{(j)}(\theta, u, v) \leq 2\tau_0 |\Delta|^{-1/2} \Phi_{\Delta}^{1/2}(u, v) \times$$

$$\times \left(\sum_{i=1}^{q} d_{i\Delta}^{-2}(\theta) d_{i\Delta}^{2}(\theta + |\Delta|^{1/2} d_{\Delta}^{-1}(\theta) u) \right)^{1/2} \leq k_2 a(j) \qquad (3.6.16)$$

in view of (3.6.2), (3.6.4) and XIX. By analogy

$$|\Delta|^{-1/2} W_{4\Delta}^{(j)}(\theta, u, v) \leq 2\tau_0 \left(\sum_{i=1}^{q} d_{i\Delta}^{-2}(\theta) \Phi_{i\Delta}(u, v) \right)^{1/2} \times$$

$$\times |\Delta|^{-1/2} \Phi_{\Delta}^{1/2}(v, 0) \leq k_3 a(j). \qquad (3.6.17)$$

We evaluate $W_{1\Delta}^{(j)}(\theta, u, v)$. For any $u, v \in C_{(j)}$, we have a.s.

$$|\chi\{\xi_{\theta}(x) * s(x, u)\} - \chi\{\xi_{\theta}(x) < s(x, v)\}| \leq \chi\{\inf_{u \in C_{(j)}} s(x, u) -$$

$$-g(x,\theta) \le \epsilon(x) \le \sup_{u \in C_{(j)}} s(x,u) - g(x,\theta)\} = \chi(x).$$

Hence from (3.6.2),

$$|\Delta|^{-1/2}W_{1\Delta}^{(j)}(\theta, u, v) \le |\Delta|^{-1/2}\Big(\sum_{i=1}^{q}(d_{i\Delta}^{-1}(\theta)\sup_{x \in \Delta}|s_i(x,u)|)^2\Big)^{1/2} \times$$

$$\times \int_{\Delta}\chi(x)dx \le k_4|\Delta|^{-1}\int_{\Delta}\chi(x)dx. \qquad (3.6.18)$$

Utilizing the formula of finite increments, we find that

$$|\Delta|^{-1}\int_{\Delta}E\chi(x)dx = |\Delta|^{-1}\int_{\Delta}\Big(F\Big(\sup_{u \in C_{(j)}} s(x,u) - g(x,\theta)\Big) -$$

$$-F\Big(\inf_{u \in C_{(j)}} s(x,u) - g(x,\theta))\Big)dx \le \tau_0|\Delta|^{-1}\int_{\Delta}\sup_{u_1,u_2 \in C_{(j)}}|s(x,u_1) - s(x,u_2)|dx \le$$

$$\le \tau_0\Big(\sum_{i=1}^{q}(|\Delta|^{1/2}d_{i\Delta}^{-1}(\theta)\sup_{x \in \Delta, u \in C_{(j)}}|s_i(x,u)|)^2\Big)^{1/2}a(j)q^{1/2} \le k_5a(j). \quad (3.6.19)$$

The estimates (3.6.15)–(3.6.19) show that constants k_6 and k_7 exist such that

$$P\Big\{\sup_{u \in C_{(j)}} W_{\Delta}^{(j)}(\theta, u, v)(1 + k_0|\Delta|^{1/2}\rho(j))^{-1} \ge \frac{\epsilon}{2}\Big\} \le$$

$$\le P\{k_6|\Delta|^{-1}\int_{\Delta}(\chi(x) - E\chi(x))dx > \frac{\epsilon}{2}\rho(j) - k_7a(j)\}. \qquad (3.6.20)$$

The quantity $\frac{\epsilon}{2}\rho(j) - k_7a(j) = (\frac{\epsilon}{2}(1-p) - k_7p)|\Delta|^{-\gamma m/\bar{m}_0} > 0$ if p is chosen sufficiently small.

Apply the inequality of Lemma 1.6.2 to the r.v.'s $\chi(x)$ and $\chi(y)$. In view of (3.6.19) we have for $t > 2$

$$|\mathrm{cov}(\chi(x), \chi(y))| \le 10(E\chi(x))^{1/t}(E\chi(y))^{1/t}\alpha^{1-2/t}(|x - y|) \le$$

$$\le k_8a^{2/t}(j)\alpha^{1-2/t}(|x - y|). \qquad (3.6.21)$$

Hence, by Chebyshev's inequality the probability (3.6.20) is bounded by the quantity

$$k_9\alpha^{(t)}|\Delta|^{-1+\gamma m(2-2/t)/\bar{m}_0}, \qquad \alpha^{(t)} = \int_{0}^{\infty}\alpha^{1-2/t}(\rho)\rho^{n-1}d\rho. \qquad (3.6.22)$$

Since $\alpha(\rho) = O(\rho^{-n-\epsilon})$ for some $\epsilon > 0$, $\alpha^{(t)} < \infty$ provided $t > 2 + \frac{2n}{\epsilon}$.

In order for quantity (3.6.22) to converge to zero as $|\Delta| \to \infty$ it suffices to choose $\gamma < \frac{1}{2} + \frac{1}{2t-2}$. Denote $t_{1i}(x) = (s_i(x,v) - s_i(x,0))(2\chi\{\xi_\theta(x) < s(x,v)\} - 1)$, $t_{2i}(x) = 2s_i(x,0)(\chi\{\xi_\theta(x) < s(x,v)\} - \chi\{\epsilon(x)*0\})$, $i = 1,\ldots,q$. Then

$$P\{Y_\Delta^{(j)}(\theta,v)(1 + k_0|\Delta|^{1/2}\rho(j))^{-1} > \tfrac{1}{2}\epsilon\} \le 8(\epsilon k_0)^{-2}|\Delta|^{-1}\rho^{-2}(j) \times$$

$$\times \sum_{i=1}^{q} d_{i\Delta}^{-2}(\theta) \sum_{j=1}^{2} E\left(\int_\Delta (t_{ji}(x) - Et_{ji}(x))dx\right)^2. \tag{3.6.23}$$

We apply the inequality of Lemma 1.6.2 to the r.v.'s $t_{ji}(x)$ and $t_{ji}(y)$:

$$\mathrm{var}\left(\int_\Delta t_{1i}(x)dx\right) \le 10\int_\Delta\int_\Delta \alpha^{1-2/t}(|x-y|)(E|t_{1i}(x)|^t)^{2/t}dxdy \le$$

$$\le 10|s(1)|\alpha^{(t)}\Phi_{i\Delta}(v,0); \tag{3.6.24}$$

$$\mathrm{var}\left(\int_\Delta t_{2i}(x)dx\right) \le 40|s(1)|\alpha^{(t)}\int_\Delta s_i^2(x,0)|F(s(x,v)-s(x,0))-$$

$$-F(0)|^{2/t}dx \le 40|s(1)|\alpha^{(t)}\tau_0 d_{i\Delta}^2(\theta)\Big(\sum_{i=1}^{q}(|\Delta|^{1/2}d_{i\Delta}^{-1}(\theta)\times$$

$$\times \sup_{x\in\Delta,v\in C_0}|s_i(x,v)|)^2\Big)^{1/t}(a(j)+\rho(j))^{2/t}q^{1/t}. \tag{3.6.25}$$

It follows from the conditions of the theorem, (3.6.24) and (3.6.25) that the right-hand side of (3.6.23) is bounded by the quantity

$$k_{10}|\Delta|^{-1}\rho^{-2}(j)((a(j)+\rho(j))^2 + (a(j)+\rho(j))^{2/t}) =$$

$$= O(|\Delta|^{-1+2\gamma m(1-1/t)/\tilde{m}_0}).$$

The bounds obtained show that for $j = 1,\ldots,N_0 - 1$,

$$\sup_{\theta\in T} P\Big\{\sup_{u\in C_{(j)}} z_\Delta^+(\theta,u) > \epsilon\Big\} = O(|\Delta|^{-1+2m\gamma(1-1/t)/\tilde{m}_0}) \tag{3.6.26}$$

for some $m = m(j) < m_0$.

Consider the case $j = N_0$. Clearly,

$$P\Big\{\sup_{u\in C_{(N_0)}} z_\Delta^+(\theta,u) > \epsilon\Big\} \le P\Big\{\sup_{|u|_0\le|\Delta|^{-\gamma m_0/\tilde{m}_0}} |R_\Delta^+(\theta+|\Delta|^{1/2}d_\Delta^{-1}(\theta)u)-$$

$$-R_\Delta^+(\theta) - ER_\Delta^+(\theta + |\Delta|^{1/2}d_\Delta^{-1}(\theta)u)| > \epsilon\Big\}. \tag{3.6.27}$$

We write the expression appearing under the norm sign in (3.6.27) as the sum of vectors $\beta_{1\Delta}(\theta, u) + \beta_{2\Delta}(\theta, u) + \beta_{3\Delta}(\theta, u)$, where

$$\beta_{1\Delta}(\theta, u) = d_\Delta^{-1}(\theta) \int_\Delta (\nabla s(x, u) - \nabla s(x, 0))(2\chi\{\xi_\theta(x) * s(x, u)\} - 1)dx,$$

$$\beta_{2\Delta}(\theta, u) = 2d_\Delta^{-1}(\theta) \int_\Delta \nabla s(x, 0)(\chi\{\xi_\theta(x) * s(x, u)\} - \chi\{\epsilon(x) * 0\})dx,$$

$$\beta_{3\Delta}(\theta, u) = d_\Delta^{-1}(\theta) \int_\Delta \nabla s(x, u)(2F(s(x, u) - s(x, 0)) - 1)dx.$$

It is easy to show that for $|u|_0 \leq |\Delta|^{-\gamma m_0/\tilde{m}_0}$,

$$|\beta_{1\Delta}(\theta, u)| \leq |\Delta|^{1/2}\Big(\sum_{i=1}^q d_{i\Delta}^{-2}(\theta)\Phi_{i\Delta}(u, 0)\Big)^{1/2} \leq k_{11}|\Delta|^{1/2-\gamma m_0/\tilde{m}_0}; \tag{3.6.28}$$

$$|\beta_{3\Delta}(\theta, u)| \leq \Big(\sum_{i=1}^q d_{i\Delta}^{-2}(\theta)d_{i\Delta}^2(\theta + |\Delta|^{1/2}d_\Delta^{-1}(\theta)u)\Big)^{1/2} \times$$

$$\times 2\tau_0\Phi_\Delta^{1/2}(u, 0) \leq k_{12}|\Delta|^{1/2-\gamma m_0/\tilde{m}_0} \tag{3.6.29}$$

If $\gamma > \frac{1}{2}$, then for $|\Delta|$ sufficiently large, the exponents in (3.6.28) and (3.6.29) are negative. Thus, it remains to evaluate the probability

$$P\Big\{\sup_{|u|_0 \leq |\Delta|^{-\gamma m_0/\tilde{m}_0}} |\beta_{2\Delta}(\theta, u)| > \epsilon\Big\} \leq$$

$$\leq P\Big\{2\Big(\sum_{i=1}^q (d_{i\Delta}^{-1}(\theta)\sup_{x \in \Delta} |g_i(x, \theta)|^2\Big)^{1/2} \int_\Delta \tilde{\chi}(x)dx > \epsilon\Big\} \leq$$

$$\leq P\Big\{k_{13}|\Delta|^{-1/2} \int_\Delta \tilde{\chi}(x)dx > \epsilon\Big\},$$

$$\tilde{\chi}(x) = \chi\Big\{\inf_{|u|_0 \leq |\Delta|^{-\gamma m_0/\tilde{m}_0}} s(x, u) - g(x, \theta) \leq$$

$$\leq \epsilon(x) \leq \sup_{|u|_0 \leq |\Delta|^{-\gamma m_0/\tilde{m}_0}} s(x, u) - g(x, \theta)\Big\}. \tag{3.6.30}$$

Since under the assumption of the theorem, $E\tilde{\chi}(x) \leq k_{14}|\Delta|^{-\gamma m_0/\tilde{m}_0}$, in place of (3.6.30) it suffices to estimate the probability $P\{|\Delta|^{-1/2}\int_\Delta(\tilde{\chi}(x) - E\tilde{\chi}(x))dx > \epsilon\}$ for any $\epsilon > 0$. Chebyshev's inequality and a bound on $\text{cov}(\tilde{\chi}(x), \tilde{\chi}(y))$ analogous to (3.6.21) imply that this probability is bounded by the quantity $k_{15}|\Delta|^{-2\gamma m_0/\tilde{m}_0 t}$. Since all of the bounds are uniform in $\theta \in T$, Lemma 3.6.1 is proved. ■

Set $ER^{\pm}_{\Delta}(\breve{\theta}_{\Delta}) = (ER^{\pm}_{\Delta}(\tau))|_{\tau=\breve{\theta}_{\Delta}}$.

Lemma 3.6.2. Under the conditions of Theorem 3.6.1, for any $\epsilon > 0$

$$\sup_{\theta \in T} P\{|R^+_{\Delta}(\theta) + ER^+_{\Delta}(\breve{\theta}_{\Delta})| > \epsilon\} \to 0, \quad |\Delta| \xrightarrow{V.H.} \infty \qquad (3.6.31)$$

Proof. We introduce the events $A^{\pm}_{i\Delta}(\theta) = \{R^{\pm}_{i\Delta}(\theta) + ER^{\pm}_{i\Delta}(\breve{\theta}_{\Delta}) - R^{\pm}_{i\Delta}(\breve{\theta}_{\Delta}) \geq -\epsilon(1 + |ER^{\pm}_{\Delta}(\breve{\theta}_{\Delta})|)\}$, $i = 1, \dots, q$. From (3.5.4) and the preceding lemma it follows that

$$\inf_{\theta \in T} P\left\{A^{\pm}_{i\Delta}(\theta)\right\} \to 1, \quad \Delta \xrightarrow{V.H.} \infty, \ i = 1, \dots, q. \qquad (3.6.32)$$

For the event $|\breve{\theta}_n - \theta| < \rho$, $\rho < \tilde{\rho}$, $R^{\pm}_{i\Delta}(\breve{\theta}_{\Delta}) \geq 0$, therefore relations (3.6.32) are valid for the events $B^{\pm}_{\Delta}(\theta) = \{R^{\pm}_{i\Delta}(\theta) + ER^{\pm}_{i\Delta}(\breve{\theta}_{\Delta}) \geq -\epsilon(1 + |ER^{\pm}_{\Delta}(\breve{\theta}_{\Delta})|)\} \supseteq A^{\pm}_{i\Delta}(\theta)$ as well. On the other hand,

$$R^+_{i\Delta}(\theta) + R^-_{i\Delta}(\theta) = 2\int_{\Delta} |g_i(x, \theta)|\chi\{\epsilon(x) = 0\}dx = 0 \ a.s.$$

and the events $B^-_{i\Delta}(\theta)$ are equiprobable to the events

$$C^+_{i\Delta}(\theta) = \{R^+_{i\Delta}(\theta) + ER^+_{i\Delta}(\breve{\theta}_{\Delta}) \leq \epsilon(1 + |ER^{\pm}_{\Delta}(\breve{\theta}_{\Delta})|)\}.$$

Furthermore for $\epsilon < \frac{1}{q}$,

$$D^+_{i\Delta}(\theta) = B^+_{i\Delta}(\theta) \cap C^+_{i\Delta}(\theta) = \{|ER^+_{i\Delta}(\breve{\theta}_{\Delta}) + R^+_{i\Delta}(\theta)| \leq$$

$$\leq \epsilon(1 + |ER^+_{\Delta}(\breve{\theta}_{\Delta})|)\}, \ i = 1, \dots, q; \qquad (3.6.33)$$

$$\bigcap_{i=1}^{q} D^+_{i\Delta}(\theta) \subseteq \{|ER^+_{\Delta}(\breve{\theta}_{\Delta}) + R^+_{\Delta}(\theta)| \leq q\epsilon(1 + |ER^+_{\Delta}(\breve{\theta}_{\Delta})|)\} \subseteq$$

$$\subseteq \{|ER^+_{\Delta}(\breve{\theta}_{\Delta})| \leq (1 - q\epsilon)^{-1}(q\epsilon + |R^+_{\Delta}(\theta)|)\} = E^+_{\Delta}(\theta),$$

that is, $\inf_{\theta \in T} P\{E^+_{\Delta}(\theta)\} \to 1$, $\Delta \xrightarrow{V.H.} \infty$. Since

$$P\{|ER^+_{\Delta}(\breve{\theta}_{\Delta})| > \rho\} \leq P\{\widehat{E^+_{\Delta}(\theta)}\} + P\{|R^+_{\Delta}(\theta)| > \rho(1 - q\epsilon) - q\epsilon\},$$

the r.v. $|ER_\Delta^+(\check{\theta}_\Delta)|$ is bounded in probability uniformly in $\theta \in T$. In accordance with (3.6.33),

$$\sup_{\theta \in T} P\{|R_{i\Delta}^+(\theta) + ER_{i\Delta}^+(\check{\theta}_\Delta)| > \epsilon(1 + |ER_\Delta^+(\check{\theta}_\Delta)|)\} \to 0, \ \Delta \xrightarrow{V.H.} \infty;$$

thus (3.6.31) holds. Note that boundedness in probability of the r.v. $|ER_\Delta^+(\check{\theta}_\Delta)|$ follows directly from (3.6.8) and the assumption of the theorem. ∎

Lemma 3.6.3. Under the conditions of Theorem 3.6.1, for any $\epsilon > 0$

$$\sup_{\theta \in T} P\{|ER_\Delta^+(\check{\theta}_\Delta) - 2\tau(0)\Im_\Delta(\theta)d_\Delta(\theta)(\check{\theta}_\Delta - \theta)| > \epsilon\} \to 0, \Delta \xrightarrow{V.H.} \infty. \quad (3.6.34)$$

Proof. If the quantity $|\check{u}_\Delta(\theta)|$ is small, inequality (3.6.14) and the boundedness in probability of the r.v. $|ER_\Delta^+(\check{\theta}_\Delta)|$ imply boundedness in probability of the norm of the vector $d_\Delta(\theta)(\check{\theta}_\Delta - \theta)$. The assertion of the lemma thus follows from (3.5.4) and inequalities (3.6.11)–(3.6.13). ∎

We shall prove Theorem 3.6.1. Relations (3.6.31) and (3.6.34) show that for any $\epsilon > 0$

$$\sup_{\theta \in T} P\left\{\left|\frac{1}{2\tau(0)}\Lambda_\Delta(\theta)R_\Delta^+(\theta) + d_\Delta(\theta)(\check{\theta}_\Delta - \theta)\right| > \epsilon\right\} \to 0, \ \Delta \xrightarrow{V.H.} \infty. \quad (3.6.35)$$

Thus, the limiting distributions (if they exist) of the vectors $d_\Delta(\theta)(\check{\theta}_\Delta - \theta)$ and $-\frac{1}{2\tau(0)}\Lambda_\Delta(\theta)R_\Delta^+(\theta)$ coincide. Since

$$R_{i\Delta}^+(\theta) - d_{i\Delta}^{-1}(\theta) \int_\Delta g_i(x, \theta)\eta(x)dx =$$

$$= 2d_{i\Delta}^{-1}(\theta) \int_{\Delta \cap \{x: g_i(x,\theta) > 0\}} g_i(x, \theta)\chi\{\epsilon(x) = 0\}d(x) = 0 \text{ a.s.,}$$

it follows that for $\Delta \xrightarrow{V.H.} \infty$, the distribution of the vector $d_\Delta(\theta)(\check{\theta}_\Delta - \theta)$ coincides with the distribution of the vector $-(2\tau(0))^{-1}\Lambda_\Delta(\theta)\mathcal{R}_\Delta(\theta)$, where

$$\mathcal{R}_\Delta(\theta) = (d_{i\Delta}^{-1}(\theta) \int_\Delta g_i(x, \theta)\eta(x)dx)_{i=1,\ldots,q}.$$

However, by the uniform central limit theorem 1.7.5, the vectors $\pm \mathcal{R}_\Delta(\theta)$ are uniformly asymptotically normal in T with zero mean and correlation matrix $(2\pi)^n \int_{\mathbf{R}^n} f_\eta(\lambda)\mu(d\lambda, \theta)$. This, together with (3.6.35), implies the assertion of Theorem 3.6.1 in the same manner as the assertion of Theorem 3.4.2. ∎

CHAPTER 4

Estimation of the Correlation Function

4.1. Definition of Estimators

Let $\xi(x)$, $x \in \mathbf{R}^n$ be a homogeneous random field having zero mean and an unknown cor. f. $B(h)$; let $\Delta \in \mathcal{B}^n$ be a bounded set and $K \in \mathcal{B}^n$ a compactum containing the origin. To estimate $B(h)$ at a point $h \in k$ from observations of a sample function of the field $\xi(x)$, we shall use the statistics

$$\hat{B}^{(1)}_{\Delta}(h) = |\Delta|^{-1} \int_{\Delta} \xi(x)\xi(x+h)dx,$$

$$\hat{B}^{(2)}_{\Delta}(h) = |\Delta|^{-1} \int_{\Delta \cap (\Delta - h)} \xi(x)\xi(x+h)dx,$$

which are non-parametric estimators of the cor. f. $B(h)$ analogous to the correlogram of a random process.

The sample spectral density function (periodogram) of the field $\xi(x)$ observed on the set Δ is defined by the equality

$$I_{\Delta}(\lambda) = (2\pi)^{-n}|\Delta|^{-1}\left|\int_{\Delta} e^{i(\lambda,x)}\xi(x)dx\right|^2.$$

The corresponding sample cor. f. is $\hat{B}^{(2)}_{\Delta}$, that is, the estimator $\hat{B}^{(2)}_{\Delta}$, is a direct statistical analogue of $B(h)$. Estimator $\hat{B}^{(1)}_{\Delta}$ is simpler as compared to $\hat{B}^{(2)}_{\Delta}$ but has the drawback that to compute it at $h \in K$, one has to use observations of the field $\xi(x)$ on the set $\Delta + K$, that is, to use more observations than in the case of the estimator $\hat{B}^{(2)}_{\Delta}$.

A different statistical problem arises when the set Δ is fixed but many independent observations of the field $\xi(x)$ on Δ are available. Here, estimators of $B(h)$ may be obtained by averaging the estimators $\hat{B}^{(1)}_{\Delta}$ and $\hat{B}^{(2)}_{\Delta}$ over the number of observed sample functions.

This chapter discusses statistical properties of the estimators $\hat{B}_\Delta^{(1)}$ and $\hat{B}_\Delta^{(2)}$ constructed from both a single sample function of the field $\xi(x)$ as $\Delta \to \infty$ and from m sample functions of the field $\xi(x)$ for a fixed Δ and $m \to \infty$.

Throughout this chapter, $\xi(x)$, $x \in \mathbf{R}^n$, is assumed to be a strictly homogeneous m.s. continuous separable measurable field. Measurability of $\xi(x)$ and the Fubini-Tonelli theorem imply that the integral $\int_{\mathbf{R}^n} \xi^2(x)e^{-|x|}dx$ is a.s. finite. Let $\widetilde{\Omega}$ be the event that $\xi(x)$, $x \in \mathbf{R}^n$ is a measurable function and $\xi^2(x)e^{-|x|}$ is a function integrable over $\mathbf{R}^n$. Thus $P\{\widetilde{\Omega}\} = 1$. Let $D \in \mathcal{B}^n$ be an arbitrary bounded set, $h \in \mathbf{R}^n$, $\omega \in \widetilde{\Omega}$ fixed and $\xi(x)\xi(x+h) = u_h(x)$. Then

$$|u_j(x)|\chi_D(x) \le \tfrac{1}{2}\xi^2(x)\chi_D(x) + \tfrac{1}{2}\xi^2(x+h)\chi_D(x) \le$$

$$\le \tfrac{1}{2}\exp\{r(D) - |x|\}\xi^2(x) + \tfrac{1}{2}\exp\{r(D_{|h|}) - |x+h|\}\xi^2(x+h),$$

where $r(D)$, $r(D_{|h|})$ are the radii of the balls containing the sets D and $D_{|h|}$. Hence

$$\int_D |u_h(x)|dx \le \exp\{r(D_{|h|})\} \int_{\mathbf{R}^n} \xi^2(x)e^{-|x|}dx < \infty.$$

Set $\hat{B}_\Delta^{(1)}(h) = |\Delta|^{-1} \int_\Delta u_h(x)dx$, $\hat{B}_\Delta^{(2)}(h) = |\Delta|^{-1} \int_{\Delta \cap (\Delta - h)} u_h(x)dx$ if $\omega \in \widetilde{\Omega}$, $\hat{B}_\Delta^{(1)}(h) = \hat{B}_\Delta^{(2)}(h) = 0$ if $\omega \in \Omega \backslash \widetilde{\Omega}$. The functions $\hat{B}_\Delta^{(i)}(h)$, $h \in \mathbf{R}^n$, $i = 1,2$, are random fields. Indeed, for Δ and $h \in \mathbf{R}^n$ fixed, by the Fubini-Tonelli theorem, the r.v.'s

$$\tilde{B}_\Delta^{(1)}(h) = |\Delta|^{-1} \int_\Delta u_h(x)dx, \ \tilde{B}_\Delta^{(2)}(h) = |\Delta|^{-1} \int_{\Delta \cap (\Delta - h)} u_h(x)dx$$

assume finite values for $\omega \in \Omega_{\Delta,h}$, $P\{\Omega_{\Delta,h}\} = 1$. Hence, $\hat{B}_\Delta^{(i)}(h)$ are r.v.'s since $\hat{B}_\Delta^{(i)}(h) = \tilde{B}_\Delta^{(i)}(h)$, $i = 1,2$, for $\omega \in \widetilde{\Omega} \cap \Omega_{\Delta,h}$ and the measure P is complete.

The random field $\hat{B}_\Delta^{(1)}(h)$, $h \in K$, is a.s. continuous. This follows from the continuity of the Lebesgue integral $\int_\Delta u_h(x)dx$ in h. A.s. continuity of the field $\hat{B}_\Delta^{(2)}(h)$, $h \in K$, is a somewhat more complicated matter. Let $h, h + \tau$ belong to K, where h is fixed. Then

$$|\hat{B}_\Delta^{(2)}(h + \tau) - \hat{B}_\Delta^{(2)}(h)| \le |\Delta|^{-1} \int_\Delta |u_{h+\tau}(x) - u_h(x)|dx +$$

$$+ |\Delta|^{-1}\Big| \int_{\Delta \cap (\Delta - h - \tau)} u_h(x)dx - \int_{\Delta \cap (\Delta - h)} u_h(x)dx \Big| = I^{(1)} + I^{(2)}. \quad (4.1.1)$$

The integral $I^{(1)}$ converges to zero as $\tau \to 0$.

Denote by $R \ominus Q = (R \cup Q) \backslash (R \cap Q)$ the symmetric difference of the sets R and Q. The quantity $I^{(2)}$ is bounded by the integral

$$|\Delta|^{-1} \int_{\Delta \cap ((\Delta - h - \tau) \ominus (\Delta - h))} |u_h(x)| dx =$$

$$= |\Delta|^{-1} \int_{(\Delta + h) \cap (\Delta \ominus (\Delta - \tau))} |u_{-h}(x)| dx \le |\Delta|^{-1} \int_{\Delta \ominus (\Delta - \tau)} |u_{-h}(x)| dx. \quad (4.1.2)$$

Since for any bounded $\Delta \in \mathbf{B}^n$, $\ |\Delta \ominus (\Delta - \tau)| \underset{\tau \to 0}{\longrightarrow} 0$, $I^{(2)} \underset{\tau \to 0}{\longrightarrow} 0$ also. Thus the fields $\hat{B}_\Delta^{(i)}(h)$, $h \in K$, $i = 1, 2$, generate probability measures $\mu_\Delta^{(i)}$, $i = 1, 2$ on $\mathbf{B}(C(K))$ in the space $C(K)$.

4.2. Consistency

We obtain conditions under which the estimators $\hat{B}_\Delta^{(i)}(h)$, $h \in K$, $i = 1, 2$, consistently in the norm of $C(K)$ estimate the cor.f. $B(h)$, $h \in K$, of the field $\xi(x)$ as $\Delta \to \infty$. Assume that $\mathfrak{M}$ is a linearly ordered set with the order relation "$\subseteq$".

Suppose the following.

I. For any $\alpha > 0$ there exists a sequence $\Delta_t \in \mathfrak{M}$ such that 1) $\Delta_t \subset \Delta_{t+1}$; 2) $\sum_t (|\Delta_{t+1}| |\Delta_t|^{-1} - 1)^2 < \infty$; 3) $\sum_t |\Delta_t|^{-1} < \infty$; 4) for $t > t_0$ and some fixed $k < \infty$, the sets $\Delta_t \subset v(k|\Delta_t|^{1/n})$ are convex.

II. $E|\xi(0)|^4 < \infty$ and for some $\beta \in [0, 1)$

$$\varlimsup_{\Delta \to \infty} \sup_{h \in K} |\Delta|^{-\beta} \int_{\Delta - \Delta} (|s_4^{(\xi)}(x, x + h, h, 0)| +$$

$$+ |B(x - h)B(x + h)|)dx < \infty. \quad (4.2.1)$$

If as $\Delta \to \infty$, $|\Delta| \to \infty$ continuously in the sense that $|\Delta|$ assumes all of the intermediate values on the interval $[a_0, \infty), a_0 > 0$, then requirements 1)–3) of assumption I are always satisfied, since sets having volume $|\Delta_t| = t^{1/\alpha + \delta}$, $\delta > 0$, can be selected for the sets Δ_t. Requirement 4) of assumption I imposed on the sequence Δ_t arises when studying the estimators $\hat{B}_\Delta^{(2)}$ and it holds if the convex sets $\Delta_t \overset{F}{\longrightarrow} \infty$ (cf. §1.1).

If $B(x) \in L_2(\mathbf{R}^n)$ and $\sup_{h \in K} \int_{\mathbf{R}^n} |s_4^{(\xi)}(x, x + h, h, 0)| dx < \infty$, then evidently condition (4.2.1) holds for $\beta = 0$.

Denote $w_h(x) = u_h(x) - B(h)$ and for $i = 1, 2$,

$$z_\Delta^{(i)}(h) = |\Delta|^{-1} \int_{\Delta^{(i)}} w_h(x)dx, \quad \Delta^{(1)} = \Delta, \quad \Delta^{(2)} = \Delta \cap (\Delta - h).$$

Theorem 4.2.1. If conditions 1)–3) of I and II hold, then for any $h \in K$, $\hat{B}_\Delta^{(1)}(h) \underset{\Delta \to \infty}{\longrightarrow} B(h)$ a.s. This assertion holds also for $\hat{B}_\Delta^{(2)}(h)$ provided 4) of I is added to the above stated conditions.

Proof. We shall show that for any $h \in K$ $z_\Delta^{(i)}(h) \underset{\Delta \to \infty}{\longrightarrow} 0$ a.s. In view of assumption II,

$$E(z_\Delta^{(i)}(h))^2 \leq |\Delta|^{-2} \int_\Delta \int_\Delta (|s_4^{(\xi)}(x^{(1)} - x^{(2)}, x^{(1)} - x^{(2)} + h, h, 0)| +$$

$$+ B^2(x^{(1)} - x^{(2)}) + |B(x^{(1)} - x^{(2)} + h)B(x^{(1)} - x^{(2)} - h)|)dx^{(1)}dx^{(2)} =$$

$$= O(|\Delta|^{-1+\beta}).$$

Let Δ_t be a sequence of sets as given in assumption I corresponding to $\alpha = 1 - \beta \leq 1$. Then for every $h \in K$, $z_{\Delta_t}^{(i)} \underset{t \to \infty}{\longrightarrow} 0$ a.s. We consider the quantity

$$\zeta_t^{(i)}(h) = \sup_{\Delta_t \subseteq \Delta \subseteq \Delta_{t+1}} |z_\Delta^{(i)}(h) - z_{\Delta_t}^{(i)}(h)|$$

and show that $\zeta_t^{(i)}(h) \underset{t \to \infty}{\longrightarrow} 0$ a.s. It is easy to observe that

$$\zeta_t^{(i)}(h) \leq (|\Delta_{t+1}| |\Delta_t|^{-1} - 1)|z_{\Delta_t}^{(i)}(h)| + \eta_t^{(i)}(h), \quad i = 1, 2,$$

where

$$\eta_t^{(1)}(h) = |\Delta_t|^{-1} \int_{\Delta_{t+1} \backslash \Delta_t} |w_h(x)|dx, \quad \eta_t^{(2)}(h) = |\Delta_t|^{-1} \int_{\Delta_t(h)} |w_h(x)|dx,$$

$$\Delta_t(h) = (\Delta_{t+1} \cap (\Delta_{t+1} - h)) \backslash (\Delta_t \cap (\Delta_t - h)).$$

Since

$$E(\eta_t^{(1)}(h))^2 \leq E|\xi(0)|^4 (|\Delta_{t+1}| |\Delta_t|^{-1} - 1)^2,$$

$\eta_t^{(1)}(h) \underset{t \to \infty}{\longrightarrow} 0$ a.s. The inclusion

$$\Delta_t(h) \subseteq (\Delta_{t+1} \backslash \Delta_t) \cup ((\Delta_{t+1} \backslash \Delta_t) - h) \tag{4.2.2}$$

is valid and thus

$$E(\eta_t^{(2)}(h))^2 \leq 4E|\xi(0)|^4(|\Delta_{t+1}|\,|\Delta_t|^{-1} - 1)^2,$$

that is, $\eta_t^{(2)}(h) \underset{t \to \infty}{\longrightarrow} 0$ a.s. also. Since $z_\Delta^{(1)}(h) = \hat{B}_\Delta^{(1)}(h) - B(h)$, this completes the proof of the theorem with regard to to the estimator $\hat{B}_\Delta^{(1)}(h)$.

Note further that

$$z_\Delta^{(2)}(h) = \hat{B}_\Delta^{(2)}(h) - B(h) + \left(1 - \frac{|\Delta \cap (\Delta - h)|}{|\Delta|}\right) B(h). \qquad (4.2.3)$$

We shall show that $|\Delta_t|^{-1}|\Delta_t \cap (\Delta_t - h)| \underset{t \to \infty}{\longrightarrow} 1$. Indeed,

$$1 - |\Delta_t|^{-1}|\Delta_t \cap (\Delta_t - h)| = |\Delta_t|^{-1}|\Delta_t \backslash (\Delta_t - h)| \leq$$

$$\leq |\Delta_t|^{-1}|\Delta_t \backslash (\Delta_t)_{-|h|}|. \qquad (4.2.4)$$

We utilize Theorem 3.4.1 and set $g(\rho) = \exp\{-\rho^2|\Delta_t|^{-2/n}\}$, $q = n$. Then in view of requirement 4) in I, it follows that

$$|\Delta_t \backslash (\Delta_t)_{-|h|}|\, e^{-k^2} \leq \int_{\Delta_t \backslash (\Delta_t)_{-|h|}} e^{-|x|^2/|\Delta_t|^{2/n}}\, dx \leq b|s_{n-1}(1)|\,|h|,$$

$$b = b_1|\Delta_t|^{1-1/n}, \quad b_1 = 2\int_0^\infty e^{-\rho^2}\rho^n d\rho.$$

Thus we have uniformly in $h \in K$:

$$|\Delta_t \backslash (\Delta_t)_{-|h|}|\,|\Delta_t|^{-1} \leq b_1|s_{n-1}(1)|e^{k^2}|h|\,|\Delta_t|^{-1/n}. \qquad (4.2.5)$$

For specific sets Δ_t, the constant in the estimate (4.2.5) may be refined. Note that

$$\delta_t(h) = \sup_{\Delta_t \subseteq \Delta \subset \Delta_{t+1}} \left\| |\Delta|^{-1}|\Delta \cap (\Delta - h)| - |\Delta_t|^{-1}|\Delta_t \cap (\Delta_t - h)| \right\| =$$

$$= \delta_t^{(1)}(h) \vee \delta_t^{(2)}(h); \qquad (4.2.6)$$

$$\delta_t^{(1)}(h) = |\Delta_t|^{-1}|\Delta_t(h)| \leq 2(|\Delta_{t+1}|\,|\Delta_t|^{-1} - 1) \underset{t \to \infty}{\longrightarrow} 0; \qquad (4.2.7)$$

$$\delta_t^{(2)}(h) = |\Delta_t \cap (\Delta_t - h)|(|\Delta_t|^{-1} - |\Delta_{t+1}|^{-1}) =$$

$$= |\Delta_t|^{-1}|\Delta_t \cap (\Delta_t - h)|(1 - |\Delta_t|\,|\Delta_{t+1}|^{-1}) \underset{t \to \infty}{\longrightarrow} 0. \qquad (4.2.8)$$

Relations (4.2.3)–(4.2.8) imply that $\hat{B}_\Delta^{(2)}(h) \xrightarrow[\Delta\to\infty]{} B(h)$ a.s. ∎

Remark 4.2.1.   Condition 4) in I has not been utilized to its full extent in the proof of the theorem. The assertion of the theorem remains valid for the estimator $\hat{B}_\Delta^{(2)}(h)$ if $\lim\limits_{t\to\infty} \sup\limits_{h\in K} |\Delta_t|^{-1}|\Delta_t\setminus(\Delta_t - h)| = 0$.

To prove stronger assertions about the consistency of the estimators $\hat{B}_\Delta^{(1)}$ and $\hat{B}_\Delta^{(2)}$ we have to impose more stringent conditions on the field $\xi(x)$.

III.   For some $p > n$, $0 < \beta \leq 1$, $\delta_1, \delta_2 > 0$ and for any $h, h+\tau \in K$,

$$1)\quad E|u_{h+\tau}(0) - u_h(0)|^p \leq k_1|\tau|^{n+\delta_1}; \tag{4.2.9}$$

$$2)\quad |B(h+\tau) - B(h)| \leq k_2|\tau|^\beta, \; p\beta = n + \delta_2. \tag{4.2.10}$$

(Note that 2) follows from 1)).

For p as given in assumption III we set $m = [p]+1$ if $[p]$ is odd; $m = [p]+2$ if p is non-integer and $[p]$ is even; $m = p$ if p is even.

IV.   1) $E|\xi(0)|^{2m} < \infty$;   2)   for $r = 2,\ldots,m$ as $\Delta \to \infty$,

$$\sup_{h\in K} \int_{\Delta^r} |s_r^{(w_h(\cdot))}(x^{(1)} - x^{(r)}, \ldots, x^{(r-1)} - x^{(r)}, 0)|dx^{(1)}\ldots dx^{(r)} = O(|\Delta|^{r/2}).$$
$$\tag{4.2.11}$$

Evidently, (4.2.9) holds if $E|\xi(0)|^{2p} < \infty$ and $E|\xi(\tau) - \xi(0)|^{2p} \leq k_3|\tau|^{2n+\delta}$ for some $\delta > 0$. Integrability of the cumulants $s_r^{(w_h(\cdot))}$ is a sufficient condition in order that condition (4.2.11) be fulfilled: if

$$\sup_{h\in K} \int_{R^{(r-1)n}} |s_r^{w_h(\cdot)}(x^{(1)}, \ldots, x^{(r-1)}, 0)|dx^{(1)}\ldots dx^{(r-1)} < \infty \tag{4.2.12},$$

then

$$\sup_{h\in K} \int_{\Delta^r} |s_r^{(w_h(\cdot))}(x^{(1)} - x^{(r)}, \ldots, x^{(r-1)} - x^{(r)}, 0)|dx^{(1)}\ldots dx^{(r)} = O(|\Delta|).$$

Lemma 4.2.1.   Let $\Delta \to \infty$ and let assumptions III and IV hold. Then for some $\delta' > 0$, $\gamma \in (0,1)$, any $h, h+\tau \in K$ and any Borel set $D \subseteq \Delta$,

$$a_D^{(p)}(h, \tau) = E\left|\int_D (w_{h+\tau}(x) - w_h(x))dx\right|^p \leq k_4|\Delta|^{p(1-\gamma/2)}|\tau|^{n+\delta'}. \tag{4.2.13}$$

Proof.   In assumption III we can suppose that $\delta_1 = \delta_2 = \delta > 0$. In view of Hölder's inequality, (4.2.9) and (4.2.10),

$$a_D^{(p)}(h, \tau) \leq 2^{p-1}\left(E\left(\int_D |u_{h+\tau}(x) - u_h(x)|dx\right)^p +\right.$$

$$+|D|^p|B(h+\tau)-B(h)|^p\Big)\leq 2^{p-1}(k_1+k_2^p)|D|^p\tau^{n+\delta}. \qquad (4.2.14)$$

On the other hand by the Leonov-Shiryaev formula (1.3.2) and (4.2.11),

$$a_D^{(p)}(h,\tau)\leq(a_D^{(m)}(h,\tau))^{p/m}\leq 2^p\sup_{h\in K}\Big(E\Big|\int_D w_h(x)dx\Big|^m\Big)^{p/m}\leq$$

$$\leq k_5|\Delta|^{p/2}. \qquad (4.2.15)$$

Let $\gamma<\delta/(n+\delta)$. Then, utilizing the bounds (4.2.14) and (4.2.15) we obtain uniformly in $h\in K$

$$a_D^{(p)}(h,\tau)=(a_D^{(p)}(h,\tau))^\gamma(a_D^{(p)}(h,\tau))^{1-\gamma}\leq k_4|\Delta|^{p(1-\gamma/2)}|\tau|^{n+\delta'},$$

where

$$k_4=(2^{p-1}(k_1+k_2^p))^{1-\gamma}k_5^\gamma,\ \delta'=\delta-\gamma(n+\delta). \quad\blacksquare$$

Incidentally, (4.2.14) yields for any set $D\in\mathcal{B}^n$ and a.s. continuous field $\int_D|w_h(x)|dx$, the estimate

$$b_D^{(p)}(h,\tau)=E\Big|\int_D(|w_{h+\tau}(x)|-|w_h(x)|)dx\Big|^p\leq k_4|D|^p|\tau|^{n+\delta}. \qquad (4.2.16)$$

Expressing the higher order cumulants of the field $w_h(x)$ in terms of the cumulants of the original field $\xi(x)$ is in general not a simple matter; thus verification of conditions (4.2.11) or (4.2.12) is troublesome. One can avoid this difficulty by assuming that the r.v. $\xi(0)$ possesses moments of sufficiently high orders.

V. $E|\xi(0)|^{4(m-1)}<\infty$ for m as given in assumption IV, and (4.2.11) holds for $r=2$.

Lemma 4.2.2. If in the conditions of Lemma 4.2.1, IV is replaced by V then for any Borel set $D\subseteq\Delta$

$$a_D^{(p)}(h,\tau)\leq k_6|\Delta|^{p(1-\gamma/2m)}|\tau|^{n+\delta'}. \qquad (4.2.17)$$

Proof. Indeed, $a_D^{(p)}\leq(a_D^{(2)})^{p/2m}(a_D^{(2m-2)})^{p/2m}$. By assumption V,

$$a_D^{(2)}(h,\tau)\leq k_7|\Delta|;$$

$$a_D^{(2m-2)}(h,\tau) \le E|w_{h+\tau}(0) - w_h(0)|^{2m-2}|D|^{2m-2} \le$$

$$\le 2^{4m-5}(E|\xi(0)|^{4(m-1)} + B^{2m-2}(0))D^{2m-2} = k_8|D|^{2m-2}.$$

Therefore

$$(a_D^{(p)}(h,\tau))^\gamma \le (k_7k_8)^{\gamma p/2m}|\Delta|^{p\gamma(1-1/2m)}. \tag{4.2.18}$$

The estimate (4.2.18) together with (4.2.14) yields (4.2.17), where $k_6 = (2^{p-1} \times (k_1 + k_2^p)^{1-\gamma}(k_7k_8)^{\gamma p/2m}$. ∎

Evidently, the estimate (4.2.17) is coarser by comparison with (4.2.13). Since $s_2^{(w_h(\cdot))}(x,0) = s_4^{(\xi)}(x, x+h, h, 0) + B^2(x) + B(x-h)B(x+h)$, assumption V differs only sightly from assumption II and one may state it analogously.

Theorem 4.2.2. If assumptions I, III hold and either assumption IV or V holds then for $i = 1,2$,

$$\sup_{h\in K} |\hat{B}_\Delta^{(i)}(h) - B(h)| \xrightarrow[\Delta\to\infty]{} 0 \ \text{a.s.}$$

Proof. For definiteness, let assumption IV be valid. In view of Lemma 4.2.1, we have for the sequence of sets Δ_t in assumption I corresponding to $\alpha = p\gamma/2$:

$$E|z_{\Delta_t}^{(1)}(h+\tau) - z_{\Delta_t}^{(1)}(h)|^p = |\Delta_t|^{-p}a_{\Delta_t}^{(p)}(h,\tau) \le$$

$$\le k_4|\Delta_t|^{-p\gamma/2}|\tau|^{n+\delta'}. \tag{4.2.19}$$

The estimate (4.2.19) and (3.2.16) imply that for any $\epsilon > 0$, some constant k_9 independent of t, and $\phi(z_{\Delta_t}^{(1)};\rho) = \sup_{|\tau|\le\rho;h,h+\tau\in K} |z_{\Delta_t}^{(1)}(h+\tau) - z_{\Delta_t}^{(1)}(h)|$, we have

$$P\{\phi(z_{\Delta_t}^{(1)};\rho) \ge \epsilon\} \le k_9\epsilon^{-p}|\Delta_t|^{-p\gamma/2}\rho^{\delta'}. \tag{4.2.20}$$

Since $\sum_t |\Delta_t|^{-p\gamma/2} < \infty$, $\phi(z_{\Delta_t}^{(1)};\rho) \xrightarrow[t\to\infty]{} 0$ a.s. for any $\rho > 0$. Let $\tilde{K} \subset K$ be a countable everywhere dense subset of K and Ω_1 the event that $z_{\Delta_t}^{(1)}(h) \xrightarrow[t\to\infty]{} 0$ simultaneously for all $h \in \tilde{K}$. In view of Theorem 4.2.1, $P\{\Omega_1\} = 1$. Let $\rho = 1$ and let Ω_2 be the event that $\phi(z_{\Delta_t}^{(1)};1) \xrightarrow[t\to\infty]{} 0$; let $K_1 \subset \tilde{K}$ be a finite 1-net of the compact set K. Then for $\omega \in \Omega_1 \cap \Omega_2$,

$$\sup_{h\in K} |z_{\Delta_t}^{(1)}(h)| \le \phi(z_{\Delta_t}^{(1)};1) + \max_{h^{(j)}\in K} |z_{\Delta_t}^{(1)}(h^{(j)})| \xrightarrow[t\to\infty]{} 0. \tag{4.2.21}$$

We now apply the estimate (4.2.16) to the sets $D = \Delta_{t+1} \backslash \Delta_t$:

$$|\Delta_t|^{-p} b^{(p)}_{\Delta_{t+1} \backslash \Delta_t}(h, \tau) \le k_4 (|\Delta_{t+1}| |\Delta_t|^{-1} - 1)^p |\tau|^{n+\delta}.$$

Since $p > n \ge 2$, we have $\sum_t (|\Delta_{t+1}| |\Delta_t|^{-1} - 1)^p < \infty$ and for the fields $\eta_t^{(1)}(h) = |\Delta_t|^{-1} \int_{\Delta_{t+1} \backslash \Delta_t} |w_h(x)| dx$ we obtain a relation similar to (4.2.21), that is,

$$\sup_{h \in K} \eta_t^{(1)}(h) \xrightarrow[t \to \infty]{} 0 \text{ a.s.} \qquad (4.2.22)$$

Proceeding to the estimator $B^{(2)}_\Delta(h)$ we note that

$$E|z^{(2)}_{\Delta_t}(h + \tau) - z^{(2)}_{\Delta_t}(h)|^p \le 2^{p-1} |\Delta_t|^{-p} \left(a^{(p)}_{\Delta_t \cap (\Delta_t - h - \tau)}(h, \tau) + \right.$$

$$\left. + E \left| \int_{\Delta_t \cap (\Delta_t - h - \tau)} w_h(x) dx - \int_{\Delta_t \cap (\Delta_t - h)} w_h(x) dx \right|^p \right).$$

By Lemma 4.2.1,

$$|\Delta^t|^{-p} a_{\Delta_t \cap (\Delta_t - h - \tau)}(h, \tau) \le k_4 |\Delta_t|^{-p\gamma/2} |\tau|^{n+\delta'}.$$

On the other hand, if Δ_t satisfies condition 4) in I, then by analogy with (4.1.2), and (4.2.4), (4.2.5) we obtain

$$|\Delta_t|^{-p} E \left| \int_{\Delta_t \cap (\Delta_t - h - \tau)} w_h(x) dx - \int_{\Delta_t \cap (\Delta_t - h)} w_h(x) dx \right|^p \le$$

$$\le |\Delta_t|^{-p} E \left| \int_{\Delta_t \ominus (\Delta_t - \tau)} |w_{-h}(x)| dx \right|^p \le E|w_h(0)|^p |\Delta_t|^{-p} |\Delta_t \ominus (\Delta_t - \tau)|^p \le$$

$$\le (2 b_1 |s_{n-1}(1)| e^{k^2})^p E|w_h(0)|^p |\Delta_t|^{-p/n} |\tau|^p. \qquad (4.2.23)$$

Hence if the sequence of sets Δ_t satisfies assumption I with $\alpha = (p\gamma/2) \wedge (p/n)$ and $\delta^* = \delta' \wedge (p - n)$, then for any $\epsilon > 0$ the inequality $P\{\phi(z^{(2)}_{\Delta_t}; \rho) \ge \epsilon\} \le k_{10} \epsilon^{-p} |\Delta_t|^{-\alpha} \rho^{\delta^*}$ analogous to (4.2.20) holds. Utilizing the result of Theorem 4.2.1 we obtain, proceeding as in the case of $z^{(1)}_\Delta(h)$, that

$$\sup_{h \in K} |z^{(2)}_{\Delta_t}(h)| \xrightarrow[t \to \infty]{} 0 \text{ a.s.} \qquad (4.2.24)$$

We shall show that for the sequence of a.s. continuous fields $\eta_t^{(2)}(h)$ introduced in the proof of Theorem 4.2.1, the relation

$$\sup_{h \in K} |\eta_t^{(2)}(h)| \xrightarrow[t \to \infty]{} 0 \text{ a.s.} \qquad (4.2.25)$$

is valid. Indeed,

$$E|\eta_t^{(2)}(h+\tau) - \eta_t^{(2)}(h)|^p \leq 2^{p-1}|\Delta_t|^{-p}\left(b_{\Delta_t(h+\tau)}^{(p)}(h,\tau)+\right.$$

$$\left. +E\left|\int_{\Delta_t(h+\tau)}|w_h(x)|dx - \int_{\Delta_t(h)}|w_h(x)|dx\right|^p.$$

The estimate (4.2.16) and the inclusion (4.2.2) imply that

$$|\Delta_t|^{-p}b_{\Delta_t(h+\tau)}^{(p)}(h,\tau) \leq k_4|\Delta_t|^{-p}|\Delta_t(h+\tau)|^p|\tau|^{n+\delta} \leq$$

$$\leq 2^p k_4(|\Delta_{t+1}||\Delta_t|^{-1} - 1)^p|\tau|^{n+\delta}.$$

Furthermore, analogously to (4.2.23) we obtain

$$w = |\Delta_t|^{-p}E\left|\int_{\Delta_t(h+\tau)}|w_h(x)|dx - \int_{\Delta_t(h)}|w_h(x)|dx\right|^p \leq$$

$$\leq E|w_h(0)|^p|\Delta_t|^{-p}|\Delta_t(h+\tau)\ominus\Delta_t(h)|^p.$$

Since

$$\Delta_t(\tau_1)\backslash\Delta_t(\tau_2) \subseteq ((\Delta_{t+1}-\tau_1)\backslash(\Delta_{t+1}-\tau_2))\cup((\Delta_t-\tau_2)\backslash(\Delta_t-\tau_1)),$$

it follows that

$$|\Delta_t(h+\tau)\ominus\Delta_t(h)|^p \leq 2^{p-1}(|\Delta_{t+1}\ominus(\Delta_{t+1}-\tau)|^p+$$

$$+|\Delta_t\ominus(\Delta_t-\tau)|^p);$$

hence in view of (4.2.4) and (4.2.5),

$$w \leq k_{11}E|w_h(0)|^p((|\Delta_{t+1}|/|\Delta_t|)^p|\Delta_{t+1}|^{-p/n} + |\Delta_t|^{-p/n})|\tau|^p.$$

The validity of (4.2.25) is thus verified. Since for $i = 1, 2$,

$$\zeta_t^{(i)} = \sup_{\Delta_t\subseteq\Delta\subseteq\Delta_{t+1}}\left|\sup_{h\in K}|z_\Delta^{(i)}(h)| - \sup_{h\in K}|z_{\Delta_t}^{(i)}(h)|\right| \leq$$

$$\leq (|\Delta_{t+1}||\Delta_t|^{-1} - 1)\sup_{h\in K}|z_{\Delta_t}^{(i)}(h)| + \sup_{h\in K}\eta_t^{(i)}(h),$$

the assertion of the theorem for the estimator $\hat{B}_\Delta^{(1)}$ follows from (4.2.21) and (4.2.22), while for the estimator $\hat{B}_\Delta^{(2)}$, it follows from (4.2.24), (4.2.25), (4.2.3) and the arguments presented in the proof of Theorem 4.2.1. The use of assumption V in place of IV does not change the proof, provided that Lemma 4.2.1 is replaced by Lemma 4.2.2. ∎

For the Gaussian field $\xi(x)$, Theorems 4.2.1 and 4.2.2 are combined into the following assertion.

Theorem 4.2.3. Let $\mathfrak{M}$ be a system of sets satisfying I and $\xi(x)$, $x \in \mathbf{R}^n$, a Gaussian field with a correlation function $B(x)$ such that:

1) $\overline{\lim_{\Delta \to \infty}} \sup_{h \in K} |\Delta|^{-\beta} \int_{(\Delta - \Delta) \pm h} B^2(x) dx < \infty$ for some $\beta \in [0, 1)$;

2) for some $\epsilon > 0$ and $\delta > 0$ and for $\tau \in v(\epsilon)$, $B(0) - B(\tau) \le k_{12}|\tau|^\delta$.

Then $\sup_{h \in k} |\hat{B}_\Delta^{(i)}(h) - B(h)| \xrightarrow[\Delta \to \infty]{} 0$ a.s., $i = 1, 2$.

Proof. Indeed, condition 1) and I ensure the validity of Theorem 4.2.1. Let m be an integer such that $m\delta > 2n$, and $D \subseteq \Delta$ a Borel set. Then $a_D^{(m+1)} \le (a_D^{(2)})^{1/2}(a_D^{(2m)})^{1/2}$. By condition 1), $a_D^{(2)}(h, \tau) \le k_{13}|\Delta|^{1+\beta}$. Since

$$a_D^{(2m)}(h, \tau) \le 2^{2m-1}|D|^{2m}[(E|\xi(0)|^{4m})^{1/2} \times$$

$$\times (E|\xi(\tau) - \xi(0)|^{4m})^{1/2} + (B(h+\tau) - B(h))^{2m}],$$

$$E|\xi(\tau) - \xi(0)|^{4m} = (4m-1)!! 2^{2m}(B(0) - B(\tau))^{2m}$$

and

$$|B(h+\tau) - B(h)| \le (2B(0))^{1/2}(B(0) - B(\tau))^{1/2},$$

we have, by the assumption of the theorem:

$$a_D^{(m+1)}(h, \tau) \le k_{14}|\Delta|^{m+1/2+\beta/2}|\tau|^{m\delta/2}, \tag{4.2.26}$$

where a number $k_{14} \ge 2^{m-1/2}k_{12}^{m/2}k_{13}^{1/2}(2B(0))^{m/2}((4m-1)!! + 1)^{1/2}$ can be chosen for the constant k_{14}. The bound (4.2.26) is a slight modification of the bound (4.2.17) and the bound $b_D^{(m)}(h, \tau) \le k_{15}|D|^m \tau^{m\delta/2}$, $D \in \mathcal{B}^n$, which readily follows from the above, is analogous to the bound (4.2.16). ∎

Remark 4.2.2. The strict homogeneity of the field $\xi(x)$, $x \in \mathbf{R}^n$, has not been utilized to its full extent in the proofs of the theorems of §4.2. The results remain valid if one assumes in Theorem 4.2.1 that $\xi(x) \in S_1^{(4)}$, that $\xi(x) \in S^{(2m)}$ in Theorem 4.2.2 provided assumption IV is valid, and that $\xi(x) \in S_1^{(4(m-1))}$ under assumption V.

4.3. Asymptotic Normality

Let $K = \Pi(H)$, $H = (H_1, \ldots, H_n)$. We introduce the random field $X_\Delta^{(1)}(h) = |\Delta|^{1/2}(\hat{B}_\Delta^{(1)}(h) - B(h))$, $h \in \Pi(H)$. We shall show that under certain conditions on the field $\xi(x)$ and on the mode of approach of the sets Δ to infinity, the measures $\mu_\Delta^{(1)}$ in $C(\Pi(H))$ corresponding to the fields $X_\Delta^{(1)}(h)$ converge weakly to a Gaussian measure μ. In the case when the Δ are parallelepipeds, an identical result will be derived for measures $\mu_\Delta^{(2)}$ corresponding to the fields $X_\Delta^{(2)}(h) = |\Delta|^{1/2}(\hat{B}_\Delta^{(2)}(h) - E\hat{B}_\Delta^{(2)}(h))$, $h \in \Pi(H)$. Theorem 1.1.4 is the basis for the arguments in this section. Set $\Psi_\Delta(\lambda) = |\Delta|^{-1}\left|\int_\Delta e^{i\langle\lambda,x\rangle}dx\right|^2$. Let the following assumption hold.

VI. $\Delta \xrightarrow{V.H.} \infty$ and the function $\Psi_\Delta(\lambda)$ possesses the kernel property, that is, for any $\epsilon > 0$,

$$\int_{R^n \setminus v(\epsilon)} \Psi_\Delta(\lambda)d\lambda \to 0, \quad \Delta \xrightarrow{V.H.} \infty.$$

If the random field $\xi(x)$, $x \in R^n$, satisfies the mixing condition stipulated in Remark 1.6.1, it then has a continuous bounded spectral density $f(\lambda)$, $\lambda \in R^n$.

VII.   For some $\beta \in (0,1]$,

$$I_1(\beta) = \int_{R^n} \prod_{i=1}^n (1 + |\lambda_i|^{1+\beta})f(\lambda)d\lambda < \infty.$$

Below we shall assume (cf. §1.3) that the field $\xi(x) \in \Xi_1^{(4)}$, that is, $\xi(x)$ possesses the spectral density of the 4th order $f_4(\lambda^{(1)}, \lambda^{(2)}, \lambda^{(3)})$. Suppose that the following assumption holds.

VIII.   For some $\beta \in (0,1]$

$$\sup_{\lambda \in R^n} \int_{R^{2n}} \prod_{i=1}^n (1 + |\lambda_i^{(1)}|^{(1+\beta)/2})(1 + |\lambda_i^{(2)}|^{(1+\beta)/2}) \times$$

$$\times |f_4(\lambda^{(1)}, \lambda - \lambda^{(1)}, \lambda^{(2)})|d\lambda^{(1)}d\lambda^{(2)} < \infty.$$

Let $X(h)$, $h \in K$, be a separable Gaussian field with zero mean and cor. f.

$$\rho(h^{(1)}, h^{(2)}) = (2\pi)^n \left\{ \int_{R^{2n}} \exp\{i(\langle\lambda^{(1)}, h^{(1)}\rangle + \langle\lambda^{(2)}, h^{(2)}\rangle)\} \right\} \times$$

$$\times f_4(\lambda^{(1)}, -\lambda^{(1)}, \lambda^{(2)})d\lambda^{(1)}d\lambda^{(2)} + \int_{R^n} (\exp\{i\langle\lambda, h^{(1)} - h^{(2)}\rangle\} +$$

$$+ \exp\{i\langle\lambda, h^{(1)} + h^{(2)}\rangle\})f^2(\lambda)d\lambda = \rho_1(h^{(1)}, h^{(2)}) + \rho_2(h^{(1)}, h^{(2)}). \quad (4.3.1)$$

Below we will show that $\rho(h^{(1)}, h^{(2)})$ is the pointwise limit of the cor. f. $\rho_\Delta^{(1)}(h^{(1)}, h^{(2)})$ of the random fields $X_\Delta^{(1)}(h)$ as $\Delta \to \infty$.

Lemma 4.3.1. Assume that for some $\beta \in (0,1]$ the integrals

$$\int_{R^{2n}} (\ln(e + |\lambda^{(1)}|) \ln(e + |\lambda^{(2)}|))^{(1+\beta)/2} |f_4(\lambda^{(1)}, -\lambda^{(1)}, \lambda^{(2)})| d\lambda^{(1)} d\lambda^{(2)},$$

$$\int_{R^n} \ln^{1+\beta}(e + |\lambda|) f^2(\lambda) d\lambda$$

are finite. Then the Gaussian field $X(h)$, $h \in K$ is a.s. continuous.

Proof. Let $t \geq 1$, $s \in (0,1)$ and $\gamma \in [0,1]$. Then [72]

$$\sin ts \leq (\ln et / \ln s^{-1})^\gamma. \tag{4.3.2}$$

Let $\gamma = 1$, $t \geq e$, $0 < s \leq e^{-1}$. If $ts \geq 1$, $\ln t / \ln s^{-1} \geq 1$. If $ts < 1$, then $\sin ts \leq ts < \ln t / \ln s^{-1}$ since the function $u^{-1} \ln u$, $u > 0$, attains its maximum at the point $u = e$ and decreases monotonically in the interval $[e, \infty)$. Next note that for $t \geq 1$ and $s \in (0,1)$, $\sin ts \leq \ln et / \ln es^{-1} \leq \ln et / \ln s^{-1}$. Using inequality (4.3.2) we verify the validity of the conditions of R.M. Dudley's theorem [166] (this theorem is presented in §4.5):

$$E|X(h^{(1)}) - X(h^{(2)})|^2 = \rho(h^{(1)}, h^{(1)}) - \rho(h^{(1)}, h^{(2)}) - \rho(h^{(2)}, h^{(1)}) + \rho(h^{(2)}, h^{(2)}) \leq$$

$$\leq (2\pi)^n \Big\{ \int_{R^{2n}} |(e^{i\langle \lambda^{(1)}, h^{(2)} \rangle} - e^{i\langle \lambda^{(1)}, h^{(1)} \rangle})(e^{i\langle \lambda^{(2)}, h^{(2)} \rangle} - e^{i\langle \lambda^{(2)}, h^{(1)} \rangle})| \times$$

$$\times |f_4(\lambda^{(1)}, -\lambda^{(1)}, \lambda^{(2)})| d\lambda^{(1)} d\lambda^{(2)} + 2 \int_{R^n} (\cos\langle \lambda, h^{(2)} \rangle - \cos\langle \lambda, h^{(1)} \rangle)^2 f^2(\lambda) d\lambda \Big\}. \tag{4.3.3}$$

For $\lambda \in R^n$ we set $|\lambda|_1 = \sum_{i=1}^n |\lambda_i|$, $|\lambda|_0 = \max_{1 \leq i \leq n} |\lambda_i|$. Let $|\lambda|_1 \geq 2$, $|h^{(1)} - h^{(2)}|_0 < 1$. Then for $\gamma = \frac{1+\beta}{2}$, $\beta \in (0,1]$ it follows from inequality (4.3.2) that

$$|e^{i\langle \lambda, h^{(2)} \rangle} - e^{i\langle \lambda, h^{(1)} \rangle}| = 2\Big|\sin\Big(\frac{|\lambda|_1}{2}\Big\langle \frac{\lambda}{|\lambda|_1}, h^{(2)} - h^{(1)} \Big\rangle\Big)\Big| \leq$$

$$\leq 2(\ln \tfrac{1}{2}e|\lambda|_1)^{(1+\beta)/2} / |\ln |h^{(2)} - h^{(1)}|_0|^{(1+\beta)/2};$$

$$(\cos\langle \lambda, h^{(2)} \rangle - \cos\langle \lambda, h^{(1)} \rangle)^2 \leq 4(\ln \tfrac{1}{2}e|\lambda|_1)^{1+\beta} / |\ln |h^{(2)} - h^{(1)}|_0|^{1+\beta}.$$

For $|\lambda|_1 < 2$, $|h^{(1)} - h^{(2)}|_0 < 1$, the following simple estimates hold:

$$|e^{i\langle\lambda,h^{(2)}\rangle} - e^{i\langle\lambda,h^{(1)}\rangle}| \le (|\lambda|_1|h^{(1)} - h^{(2)}|_0)^{(1+\beta)/2} \le$$

$$\le \left(\frac{2}{e}\right)^{(1+\beta)/2} /|\ln|h^{(1)} - h^{(2)}|_0|^{(1+\beta)/2};$$

$$(\cos\langle\lambda, h^{(2)}\rangle - \cos\langle\lambda, h^{(1)}\rangle)^2 \le \left(\frac{2}{e}\right)^{1+\beta} /|\ln|h^{(1)} - h^{(2)}|_0|^{1+\beta}.$$

Hence

$$E|X(h^{(1)}) - X(h^{(2)})|^2 \le k^*|\ln|h^{(1)} - h^{(2)}|_0|^{-1-\beta}, \qquad (4.3.4)$$

where the number k^* depends on logarithmic moments of the functions $f^2(\lambda)$, $|f_4(\lambda^{(1)}, -\lambda^{(1)}, \lambda^{(2)})|$. ∎

Remark 4.3.1. Inequality (4.3.3) for $\beta \in (0,2]$ provides the obvious bound

$$E|X(h^{(1)}) - |X(h^{(2)})|^2 \le k(\beta)|h^{(1)} - h^{(2)}|_0^\beta, \qquad (4.3.5)$$

where

$$k(\beta) = (2\pi)^n \left\{ \int_{R^{2n}} (|\lambda^{(1)}|_1|\lambda^{(2)}|_1)^{\beta/2}|f_4(\lambda^{(1)}, -\lambda^{(1)}, \lambda^{(2)})|d\lambda^{(1)}d\lambda^{(2)} + \right.$$

$$\left. + 2\int_{R^n} |\lambda|_1^\beta f^2(\lambda)d\lambda \right\}.$$

If for some $\beta > 0$, $k(\beta) < \infty$, then (4.3.5) also implies that the field $X(h)$, $h \in K$ is a.s. continuous. In any event, assumptions VII and VIII ensure the a.s. continuity of $X(h)$, $h \in K$.

Thus, the field $X(h)$, $h \in \Pi(H)$ induces a Gaussian measure in $C(\Pi(H))$. Denote this measure by μ.

We shall introduce an assumption analogous to assumption IX in §1.7.

IX. For some $\delta > 0$, $E|\xi(0)|^{2(2+\delta)} < \infty$ and the mixing rate $\alpha(r) = O(r^{-n-\epsilon})$ as $r \to \infty$, $\epsilon\delta > 2n$.

In the theorems of this chapter, assumption IX can be replaced by the other assumptions of §1.7 for which the corresponding version of the central limit theorem is valid.

Theorem 4.3.1. If assumptions VI–IX are fulfilled, then $\mu_\Delta^{(1)} \Rightarrow \mu$ in $C(\Pi(H))$ as $\Delta \xrightarrow{V.H.} \infty$.

We subdivide the proof of the theorem into several steps.

Lemma 4.3.2. Let assumptions VII, VIII hold and $f_0 = \sup\limits_{\lambda \in R^n} f(\lambda) < \infty$. Then the family of measures $\{\mu_\Delta^{(1)}\}$ is weakly compact in $C(\Pi(H))$.

Proof. Consider the cor. f.

$$\rho_\Delta^{(1)}(h^{(1)}, h^{(2)}) = EX_\Delta^{(1)}(h^{(1)})X_\Delta^{(1)}(h^{(2)}) =$$

$$|\Delta|^{-1} \int_\Delta \int_\Delta [s_4(t^{(1)} - t^{(2)}, \, t^{(1)} - t^{(2)} + h^{(1)}, h^{(2)}, 0) + B(t^{(1)} - t^{(2)}) \times$$

$$\times B(t^{(1)} - t^{(2)} + h^{(1)} - h^{(2)}) + B(t^{(1)} - t^{(2)} + h^{(1)})B(t^{(1)} - t^{(2)} - h^{(2)})]dt^{(1)}dt^{(2)} =$$

$$= \int_{R^{3n}} \exp\{i(\langle\lambda^{(2)}, h^{(1)}\rangle + \langle\lambda^{(3)}, h^{(2)}\rangle)\}\Psi_\Delta(\lambda^{(1)} + \lambda^{(2)}) \times$$

$$\times f_4(\lambda^{(1)}, \lambda^{(2)}, \lambda^{(3)})d\lambda^{(1)}d\lambda^{(2)}d\lambda^{(3)} + \int_{R^{2n}} [\exp\{i\langle\lambda^{(2)}, h^{(1)} - h^{(2)}\rangle\}+$$

$$+ \exp\{i(\langle\lambda^{(1)}, h^{(1)}\rangle - \langle\lambda^{(2)}, h^{(2)}\rangle)\}]\Psi_\Delta(\lambda^{(1)} + \lambda^{(2)}) \times$$

$$\times f(\lambda^{(1)})f(\lambda^{(2)})d\lambda^{(1)}d\lambda^{(2)} = {}_1\rho_\Delta^{(1)}(h^{(1)}, h^{(2)}) + {}_2\rho_\Delta^{(1)}(h^{(1)}, h^{(2)}). \qquad (4.3.6)$$

We shall verify the validity of the conditions of N.N. Chentsov's theorem 1.1.4 for the fields $X_\Delta^{(1)}$. For any pair of vectors $h^{(1)}, h^{(2)} \in \Pi(H)$, using (4.3.6) we obtain

$$E(\delta_{X_\Delta^{(1)}}^{(n)}[h^{(2)}, h^{(1)}])^2 = \sum_{\nu,\nu' \in \mathcal{D}} (-1)^{|\nu|+|\nu'|}\rho_\Delta^{(1)}(h^{(1)} + \nu \otimes (h^{(2)} - h^{(1)}),$$

$$h^{(1)} + \nu' \otimes (h^{(2)} - h^{(1)})) = \int_{R^{3n}} T_1^{(n)}(\lambda^{(2)}, \lambda^{(3)}, h^{(1)}, h^{(2)})\Psi_\Delta(\lambda^{(1)} + \lambda^{(2)}) \times$$

$$\times f_4(\lambda^{(1)}, \lambda^{(2)}, \lambda^{(3)})d\lambda^{(1)}d\lambda^{(2)}d\lambda^{(3)} + \int_{R^{2n}} T_2^{(n)}(\lambda^{(1)}, \lambda^{(2)}, h^{(1)}, h^{(2)}) \times$$

$$\times \Psi_\Delta(\lambda^{(1)} + \lambda^{(2)})f(\lambda^{(1)})f(\lambda^{(2)})d\lambda^{(1)}d\lambda^{(2)},$$

where

$$T_1^{(n)} = \delta_{e^{i\langle\lambda^{(2)},\cdot\rangle}}^{(n)}[h^{(2)}, h^{(1)}]\delta_{e^{i\langle\lambda^{(3)},\cdot\rangle}}^{(n)}[h^{(2)}, h^{(1)}],$$

$$T_2^{(n)} = \sum_{\nu,\nu' \in \mathcal{D}} (-1)^{|\nu|+|\nu'|}[\exp\{i\langle\lambda^{(2)}, (\nu-\nu')\otimes(h^{(2)}-h^{(1)})\rangle\}+$$

$$+ \exp\{i((\langle\lambda^{(1)}, h^{(1)}+\nu\otimes(h^{(2)}-h^{(1)})\rangle) - \langle\lambda^{(2)}, h^{(1)}+\nu'\otimes(h^{(2)}-h^{(1)})\rangle)\}] =$$

$$= |\delta^{(n)}_{e^{i\langle\lambda^{(2)},\cdot\rangle}}[h^{(2)}, h^{(1)}]|^2 + \delta^{(n)}_{e^{i\langle\lambda^{(1)},\cdot\rangle}}[h^{(2)}, h^{(1)}]\overline{\delta^{(n)}_{e^{i\langle\lambda^{(2)},\cdot\rangle}}[h^{(2)}, h^{(1)}]}.$$

Since $\delta^{(n)}_{e^{i\langle\lambda,\cdot\rangle}}[h^{(2)}, h^{(1)}] = \prod_{j=1}^{n}(e^{i\lambda_j h_j^{(2)}} - e^{i\lambda_j h_j^{(1)}})$, the trigonometric polynomials $T_1^{(n)}$ and $T_2^{(n)}$ admit the bounds

$$|T_1^{(n)}| \le 2^{(1-\beta)n} \prod_{j=1}^{n} |\lambda_j^{(2)}\lambda_j^{(3)}|^{(1+\beta)/2}|h_j^{(2)} - h_j^{(1)}|^{1+\beta};$$

$$|T_2^{(n)}| \le 2^{(1-\beta)n} \left(\prod_{j=1}^{n}|\lambda_j^{(2)}|^{1+\beta} + \prod_{j=1}^{n}|\lambda_j^{(1)}\lambda_j^{(2)}|^{(1+\beta)/2}\right)\prod_{j=1}^{n}|h_j^{(2)} - h_j^{(1)}|^{1+\beta}$$

and thus:

$$E(\delta^{(n)}_{X_\Delta^{(1)}}[h^{(2)}, h^{(1)}])^2 \le 2^{(1-\beta)n}\kappa_\Delta \prod_{j=1}^{n}|h_j^{(2)} - h_j^{(1)}|^{1+\beta}, \qquad (4.3.7)$$

where

$$\kappa_\Delta = \int_{R^{3n}} \prod_{j=1}^{n}|\lambda_j^{(2)}\lambda_j^{(3)}|^{(1+\beta)/2}|f_4(\lambda^{(1)}, \lambda^{(2)}, \lambda^{(3)})|\times$$

$$\times\Psi_\Delta(\lambda^{(1)} + \lambda^{(2)})d\lambda^{(1)}d\lambda^{(2)}d\lambda^{(3)} + \int_{R^{2n}}\left(\prod_{j=1}^{n}|\lambda_j^{(2)}|^{1+\beta} + \prod_{j=1}^{n}|\lambda_j^{(1)}\lambda_j^{(2)}|^{(1+\beta)/2}\right)\times$$

$$\times f(\lambda^{(1)})f(\lambda^{(2)})\Psi_\Delta(\lambda^{(1)} + \lambda^{(2)})d\lambda^{(1)}d\lambda^{(2)} =$$

$$= {}_1\kappa_\Delta + {}_2\kappa_\Delta + {}_3\kappa_\Delta.$$

In view of assumption VIII,

$${}_1\kappa_\Delta \le (2\pi)^n \sup_{\lambda\in R^n} \int_{R^{2n}} \prod_{j=1}^{n}|\lambda_j^{(2)}\lambda_j^{(3)}|^{(1+\beta/2}|f_4(\lambda-\lambda^{(2)}, \lambda^{(2)}, \lambda^{(3)})|d\lambda^{(2)}d\lambda^{(3)} < \infty.$$

Denote $I_2(\beta) = \int_{\mathbf{R}^n} \prod_{j=1}^n |\lambda_j|^{1+\beta} f(\lambda) d\lambda$. Then by assumption VII, $_2\kappa_\Delta \leq (2\pi)^n f_0 I_2(\beta) < \infty$. To estimate $_3\kappa_\Delta$ note that

$$_3\kappa_\Delta = |\Delta|^{-1} \int_\Delta \int_\Delta \left| \int_{\mathbf{R}^n} f(\lambda) \prod_{j=1}^n |\lambda_j|^{(1+\beta)/2} e^{i\langle \lambda, x^{(1)} - x^{(2)} \rangle} d\lambda \right|^2 dx^{(1)} dx^{(2)} \leq$$

$$\leq \int_{\mathbf{R}^n} \left| \int_{\mathbf{R}^n} f(\lambda) \prod_{j=1}^n |\lambda_j|^{(1+\beta)/2} e^{i\langle \lambda, x \rangle} d\lambda \right|^2 dx =$$

$$= (2\pi)^n \int_{\mathbf{R}^n} f^2(\lambda) \prod_{j=1}^n |\lambda_j|^{1+\beta} d\lambda \leq (2\pi)^n f_0 I_2(\beta) < \infty.$$

Hence, the quantity κ_Δ in (4.3.7) is bounded and condition 1) of Theorem 1.1.4 is valid. The validity of condition 2) of Theorem 1.1.4 is verified in similar fashion.

Lemma 4.3.3. Under assumptions VI and IX, $\rho_\Delta^{(1)}(h^{(1)}, h^{(2)}) \to \rho(h^{(1)}, h^{(2)})$ as $\Delta \xrightarrow{V.H.} \infty$; $h^{(1)}, h^{(2)} \in K$.

Proof. Using the representation (4.3.6) for $\rho_\Delta^{(1)}(h^{(1)}, h^{(2)})$, we write

$$_1\rho_\Delta^{(1)}(h^{(1)}, h^{(2)}) = \int_{\mathbf{R}^n} f_4^*(\lambda; h^{(1)}, h^{(2)}) \Psi_\Delta(\lambda) d\lambda; \qquad (4.3.8)$$

$$f_4^*(\lambda; h^{(1)}, h^{(2)}) = \int_{\mathbf{R}^{2n}} f_4(\lambda - \lambda^{(2)}, \lambda^{(2)}, \lambda^{(3)}) \exp\{i(\langle \lambda^{(2)}, h^{(1)} \rangle +$$

$$+ \langle \lambda^{(3)}, h^{(2)} \rangle)\} d\lambda^{(2)} d\lambda^{(3)}. \qquad (4.3.9)$$

We shall show that the function of $x \in \mathbf{R}^n$

$$s_4^*(x; h^{(1)}, h^{(2)}) = s_4(x, x + h^{(1)}, h^{(2)}, 0) = \int_{\mathbf{R}^n} e^{i\langle \lambda, x \rangle} f_4^*(\lambda; h^{(1)}, h^{(2)}) d\lambda$$

is integrable over $\mathbf{R}^n$. Indeed,

$$s_4^*(x; h^{(1)}, h^{(2)}) = E\xi(x)\xi(x + h^{(1)})\xi(0)\xi(h^{(2)}) - B(h^{(1)})B(h^{(2)} -$$

$$- B(x)B(x + h^{(1)} - h^{(2)}) - B(x - h^{(1)})B(x + h^{(2)})$$

and assumption IX implies that for some $\epsilon' > 0$ and $|x|$ sufficiently large, the inequality

$$|E\xi(x)\xi(x + h^{(1)})\xi(0)\xi(h^{(2)}) - B(h^{(1)})B(h^{(2)})| \leq k_1 |x|^{-n-\epsilon'}$$

holds with constant k_1 independent of $h^{(1)}$ and $h^{(2)}$. Consequently, in view of the equality

$$f_4^*(\lambda; h^{(1)}, h^{(2)}) = (2\pi)^{-n} \int_{\mathbf{R}^n} e^{-i(\lambda, x)} s_4^*(x; h^{(1)}, h^{(2)}) dx,$$

the function f_4^* is bounded uniformly in $h^{(1)}, h^{(2)} \in K$ and is uniformly continuous with respect to λ on $\mathbf{R}^n$. Now (4.3.8) implies that $_1\rho_\Delta^{(1)}(h^{(1)}, h^{(2)}) \to (2\pi)^n f_4^*(0; h^{(1)}, h^{(2)}) = \rho_1(h^{(1)}, h^{(2)}), \Delta \xrightarrow{V.H.} \infty$. The convergence of $_2\rho_\Delta(h^{(1)}, h^{(2)})$ to $\rho_2((h^{(1)}, h^{(2)})$ is obvious. ∎

Lemma 4.3.4. If assumptions VI and IX are fulfilled, the finite-dimensional distributions of the fields $X_\Delta^{(1)}(h)$, $h \in K$ converge as $\Delta \xrightarrow{V.H.} \infty$ to the finite-dimensional distributions of the Gaussian field $X(h)$, $h \in K$.

Proof. The random field $w_h(x) = \xi(x)\xi(x+h) - B(h)$, $x \in \mathbf{R}^n$, $h \in K$, for fixed h is homogeneous in x and for $h^{(1)}, h^{(2)} \in K$, the fields $w_{h^{(1)}}$ and $w_{h^{(2)}}$ are jointly homogeneous. Therefore for any $m \geq 1$ and a collection of distinct vectors $h^{(1)}, \ldots, h^{(m)} \in K$, the vector $\eta(x) = (w_{h^{(1)}}(x), \ldots, w_{h^{(m)}}(x))$, $x \in \mathbf{R}^n$, is an m-dimensional strictly homogeneous random field. Clearly, for $\eta(x)$ the condition of Theorem 1.7.6 is valid. If $b_m(x) = (b_{ij}(x))_{i,j=1,\ldots,m}$ is the correlation matrix of $\eta(x)$ then, as was shown in the proof of the preceding lemma, $b_{ij}(x) \in L_1(\mathbf{R}^n)$, $i, j = 1, \ldots, m$. Therefore the field $\eta(x)$ has continuous and bounded spectral density function $\phi_m(\lambda) = (\phi_{ij}(\lambda))_{i,j=1,\ldots,m}$ with $\phi_m(0) = (2\pi)^{-n}(\rho(h^{(i)}, h^{(j)}))_{i,j=1,\ldots,m}$. We clarify the last equality. It was shown above that $\rho_1(h^{(i)}, h^{(j)}) = \int_{\mathbf{R}^n} s_4^*(x; h^{(i)}, h^{(j)}) dx$. On the other hand, $\rho_2(h^{(i)}, h^{(j)}) = \int_{\mathbf{R}^n}(B(x)B(x + h^{(i)} - h^{(j)}) + B(x - h^{(i)})B(x + h^{(j)})) dx$. Thus $\rho(h^{(i)}, h^{(j)}) = \int_{\mathbf{R}^n} b_{ij}(x) dx$. The proof of the lemma is completed by referring to Theorem 1.7.6. ∎

Theorem 4.3.1 follows from Lemmas 4.3.1–4.3.4 and Theorem 1.1.4.

Suppose that the field $\xi(x)$ is observed in a rectangular domain $\Delta = \Pi(T)$, $T = (T_1, \ldots, T_n)$. In this case the estimator $\hat{B}_\Delta^{(2)}(h), h \in \Pi(H)$, is defined in an especially simple manner:

$$\hat{B}_\Delta^{(2)}(h) = |\Pi(T)|^{-1} \int_0^{T_1-h_1} \cdots \int_0^{T_n-h_n} u_h(x) dx, \quad h \in \Pi(H). \qquad (4.3.10)$$

The small difference between $\hat{B}_\Delta^{(2)}$ and $\hat{B}_\Delta^{(1)}$ permits us to expect that an assertion analogous to Theorem 4.3.1 will be valid for the measures $\mu_\Delta^{(2)}$ on $C(\Pi(H))$ corresponding to the fields $X_\Delta^{(2)}(h)$. To verify this, we introduce an assumption concerning the function

$$\phi_4^*(\lambda) = \int_{R^{2n}} \prod_{i=1}^n (1 + |\lambda_i^{(1)}|^{(1+\beta)/2})(1 + |\lambda_i^{(2)}|^{(1+\beta)/2}) \times$$

$$\times |f_4(\lambda^{(1)}, \lambda - \lambda^{(1)}, \lambda^{(2)})| d\lambda^{(1)} d\lambda^{(2)}.$$

X. The integral $\int_{R^{n_1}} \phi_4^*(\lambda) \prod_{j \in i(n_1)} d\lambda_j$ over any subset of variables $i^{(n_1)} \subseteq I_n = \{1, \ldots, n\}$, is a bounded function which is continuous at the origin in the space R^{n_2}, $n_1 + n_2 = n$. Note that assumption VIII is a condition of boundedness of $\phi_4^*(\lambda)$ on $\mathbf{R}^n$ for some $\beta \in (0, 1]$.

Theorem 4.3.2. If assumptions IX, VII and X are fulfilled, then $\mu_\Delta^{(2)} \Longrightarrow \mu$ with $\Delta = \Pi(T) \to \infty$ in $C(\Pi(H))$.

The proof of this theorem is analogous to that of the preceding one. Denote $Y_D(h) = X_D^{(1)}(h) - X_D^{(2)}(h)$, $h \in \Pi(H)$, $D \in \mathcal{B}^n$.

Lemma 4.3.5. Suppose that: 1) $f_0 = \sup\limits_{\lambda \in R^n} f(\lambda) < \infty$; 2) the function $f_4^*(\lambda; h, h)$ defined by equality (4.3.9) is bounded in $\lambda \in \mathbf{R}^n$ for each fixed $h \in \Pi(H)$; 3) the family of sets $\{D\} \subset \mathcal{B}^n$ approaches infinity so that for $h \in \Pi(H)$, $|D|^{-1}|D\backslash(D-h)| \xrightarrow[D\to\infty]{} 0$. Then $EY_D^2(h) \xrightarrow[D\to\infty]{} 0$, $h \in \Pi(H)$.

Proof. As in (4.3.6) we obtain

$$EY_D^2(h) = |D|^{-1}E\left[\int_{D\backslash(D-h)} w_h(x)dx\right]^2 = |D|^{-1}|D\backslash(D-h)|\times$$

$$\times\left[\int_{R^n} f_4^*(\lambda; h, h)\Psi_{D\backslash(D-h)}(\lambda)d\lambda + \int_{R^{2n}} (1 + \exp\{i\langle\lambda - 2\lambda^{(1)}, h\rangle\})\times\right.$$

$$\left.\times f(\lambda - \lambda^{(1)})f(\lambda^{(1)})\Psi_{D\backslash(D-h)}(\lambda)d\lambda d\lambda^{(1)}\right] \leq$$

$$\leq (2\pi)^n\left[\sup\limits_{\lambda \in R^n} |f_4^*(\lambda; h, h)| + 2f_0 B(0)\right]|D|^{-1}|D\backslash(D-h)|. \quad \blacksquare$$

Clearly, Lemma 4.3.5 includes the case $D = \Pi(T)$:

$$|D|^{-1}|D\backslash(D-h)| = 1 - \prod_{j=1}^n (1 - h_j T_j^{-1}) \to 0, \quad \min\limits_{1 \leq j \leq n} T_j \to \infty.$$

Lemma 4.3.6. If assumptions VII and X hold, the family of measures $\{\mu_\Delta^{(2)}\}$ is weakly compact in $C(\Pi(H))$.

Proof. The proof consists in verifying the validity of the conditions of Theorem 1.1.4. Consider the estimator (4.3.10) of the cor.f. It is easy to show that

$$\rho_\Delta^{(2)}(h^{(1)}, h^{(2)}) = EX_\Delta^{(2)}(h^{(1)})X_\Delta^{(2)}(h^{(2)}) = |\Delta|^{-1}\int_{R^{3n}} f_4(\lambda^{(1)} - \lambda^{(2)}, \lambda^{(2)}, \lambda^{(3)})\times$$

$$\times\left(e^{i(\lambda^{(2)},h^{(1)})}\int_{\Pi(T-h^{(1)})}e^{i(\lambda^{(1)},x)}dx\right)\left(e^{i(\lambda^{(3)},h^{(2)})}\int_{\Pi(T-h^{(2)})}e^{-i(\lambda^{(1)},x)}dx\right)d\lambda^{(1)}d\lambda^{(2)}d\lambda^{(3)}+$$

$$+|\Delta|^{-1}\int_{R^{2n}}f(\lambda^{(1)})f(\lambda^{(2)})\Big\{(e^{i(\lambda^{(2)},h^{(1)})}+e^{i(\lambda^{(1)},h^{(1)})})\times$$

$$\times\int_{\Pi(T-h^{(1)})}e^{i(\lambda^{(1)}+\lambda^{(2)},x)}dx\left(e^{-i(\lambda^{(2)},h^{(2)})}\int_{\Pi(T-h^{(2)})}e^{-i(\lambda^{(1)}+\lambda^{(2)},x)}dx\right)\Big\}d\lambda^{(1)}d\lambda^{(2)}=$$

$$={}_1\rho_\Delta^{(2)}(h^{(1)},h^{(2)})+{}_2\rho_\Delta^{(2)}(h^{(1)},h^{(2)}). \tag{4.3.11}$$

Consequently,

$$E(\delta_{X_\Delta^{(2)}}^{(n)}[h^{(2)},h^{(1)}])^2=|\Delta|^{-1}\int_{R^{3n}}f_4(\lambda^{(1)}-\lambda^{(2)},\lambda^{(2)},\lambda^{(3)})\times$$

$$\times T_3^{(n)}(\lambda^{(2)},\lambda^{(1)},h^{(1)},h^{(2)})\times$$

$$\times T_3^{(n)}(\lambda^{(2)},-\lambda^{(1)},h^{(1)},h^{(2)})d\lambda^{(1)}d\lambda^{(2)}d\lambda^{(3)}+$$

$$+|\Delta|^{-1}\int_{R^{2n}}f(\lambda^{(1)})f(\lambda^{(2)})\{|T_3^{(n)}(\lambda^{(2)},\lambda^{(1)}+\lambda^{(2)},h^{(1)},h^{(2)})|^2+$$

$$+T_3^{(n)}(\lambda^{(1)},\lambda^{(1)}+\lambda^{(2)},h^{(1)},h^{(2)})\times$$

$$\times T_3^{(n)}(-\lambda^{(2)},-\lambda^{(1)}-\lambda^{(2)},h^{(1)},h^{(2)})\}d\lambda^{(1)}d\lambda^{(2)}=E_1+E_2, \tag{4.3.12}$$

where $T_3^{(n)}(\lambda,\mu,h^{(1)},h^{(2)})=\delta^{(n)}e^{i(\lambda,\cdot)}g(\mu,\cdot)[h^{(2)},h^{(1)}]$,

$$g(\mu,h)=\int_{\Pi(T-h)}e^{i(\mu,x)}dx=\prod_{j=1}^{n}\int_0^{T_j-h_j}e^{i\mu_j x_j}dx_j=\prod_{j=1}^{n}g_j(\mu_j,h_j). \tag{4.3.13}$$

We estimate the function $T_3^{(n)}$. Evidently,

$$T_3^{(n)}(\lambda,\mu,h^{(1)},h^{(2)})=\prod_{j=1}^{n}(e^{i\lambda_j h_j^{(2)}}g_j(\mu_j,h_j^{(2)})-e^{i\lambda_j h_j^{(1)}}g_j(\mu_j,h_j^{(1)}))=\prod_{j=1}^{n}T_{j3};$$

$$|T_{j3}|=\left|(e^{i\lambda_j h_j^{(2)}}\int_{T_j-h_j^{(1)}}^{T_j-h_j^{(2)}}e^{i\mu_j x_j}dx_j+(e^{i\lambda_j h_j^{(2)}}-e^{i\lambda_j h_j^{(1)}})g_j(\mu_j,h_j^{(1)}))\right|\leq$$

$$\leq |h_j^{(2)} - h_j^{(1)}| + 2^{(1-\beta)/2} |\lambda_j|^{(1+\beta)/2} |\sin((T_j - h_j^{(1)})\mu_j/2)/(\mu_j/2)| \, |h_j^{(2)} - h_j^{(1)}|^{(1+\beta)/2}.$$

Therefore the term E_1 in equality (4.3.12) is bounded by the quantity

$$E_1 \leq k_2 |\Delta|^{-1} \int_{R^n} \phi_4^*(\mu) \prod_{j=1}^{n} (1 + \sin^2((T_j - h_j^{(1)})\mu_j/2)/(\mu_j/2)^2) d\mu \times$$

$$\times \prod_{j=1}^{n} |h_j^{(2)} - h_j^{(1)}|^{1+\beta}, \quad k_2 = 2^n(|H|_0^{n(1-\beta)} \vee 2^{n(1-\beta)}). \qquad (4.3.14)$$

In view of assumption X,

$$|\Delta|^{-1} \int_{R^n} \phi_4^*(\mu) \prod_{j=1}^{n} (1 + \sin^2((T_j - h_j^{(1)})\mu_j/2)/(\mu_j/2)^2) d\mu =$$

$$= |\Delta|^{-1} \sum{}^{*} (2\pi)^{n_1} \prod_{j \in i^{(n_1)}} (T_j - h_j^{(1)}) \int_{R^{n_1}} \prod_{j \in i^{(n_1)}} \Phi_{T_j - h_j^{(1)}}(\mu_j) \times$$

$$\times \left(\int_{R^{n_2}} \phi_4^*(\mu) \prod_{j \in i^{(n_2)}} d\mu_j \right) \prod_{j \in i^{(n_1)}} d\mu_j \xrightarrow[\Delta \to \infty]{} (2\pi)^n \phi_4^*(0), \qquad (4.3.15)$$

where $\sum^{*}$ denotes summation over all partitions of the set $I_n = i^{(n_1)} \cup i^{(n_2)}$, $n_1 + n_2 = n$.

To estimate the term E_2 in (4.3.12), we note that

$$|T_3^{(n)}(\lambda, \mu, h^{(1)}, h^{(2)})|^2 \leq k_2 \prod_{j=1}^{n} (1 + \sin^2((T_j - h_j^{(1)})\mu_j/2)/(\mu_j/2)^2) \times$$

$$\times (1 + |\lambda_j|^{(1+\beta)}) |h_j^{(2)} - h_j^{(1)}|^{1+\beta}$$

and hence analogously to (4.3.14), we obtain

$$E_2 \leq 2k_2 |\Delta|^{-1} \int_{R^n} \phi_2^*(\mu) \prod_{j=1}^{n} (1 + \sin^2((T_j - h_j^{(1)})\mu_j/2)/(\mu_j/2)^2) d\mu \times$$

$$\prod_{j=1}^{n} |h_j^{(2)} - h_j^{(1)}|^{1+\beta}, \quad \phi_2^*(\mu) = \int_{R^n} f(\mu - \lambda) f(\lambda) \prod_{j=1}^{n} (1 + |\lambda_j|^{1+\beta}) d\lambda.$$

To derive a relation similar to (4.3.15) it suffices to impose on the function $\phi_2^*(\mu)$ the properties: a) the integral of $\phi_2^*(\mu)$ over any set of variables μ_j, $j \in i^{(n_1)} \subseteq I_n$, is a bounded function on R^{n_2}; b) $\phi_2^*(\mu)$ is continuous at the origin of the space R^{n_2}, $n_1 + n_2$. In turn, in order for a) and b) to hold it suffices that: 1) assumption VII be valid; 2) for any set of variables μ_j, $j \in i^{(n_1)}$, the function $\int_{R^{n_1}} f(\mu) \prod_{j \in i(n_1)} d\mu_j$ be uniformly continuous and bounded on R^{n_2}. However, property 2) follows from the mixing assumption IX. Indeed, suppose, for example, that $i^{(n_2)} = I_{n_2} = \{1, \ldots, n_2\}$. Then by virtue of the integrability of the function $B(x_1, \ldots, x_{n_2}, 0, \ldots, 0)$, which follows from assumption IX, for $x = (x_1, \ldots, x_{n_2}, 0, \ldots, 0)$ we have

$$\int_{R^{n_1}} f(\lambda) d\lambda_{n_2+1} \ldots d\lambda_n = (2\pi)^{-n_2} \int_{R^{n_2}} e^{-i\langle \lambda, x \rangle} B(x) dx.$$

Thus

$$|\Delta|^{-1} \int_{R^n} \phi_2^*(\mu) \prod_{j=1}^{n} (1 + \sin^2((T_j - h_j^{(1)})\mu_j/2)/(\mu_j/2)^2) d\mu \xrightarrow[\Delta \to \infty]{}$$

$$\xrightarrow[\Delta \to \infty]{} (2\pi)^n \phi_2^*(0) = (2\pi)^n \int_{R^n} f^2(\lambda) \prod_{j=1}^{n} (1 + |\lambda_j|^{1+\beta}) d\lambda \le$$

$$\le (2\pi)^n f_0 I_1(\beta). \tag{4.3.16}$$

Since the convergences in (4.3.15) and (4.3.16) are actually uniform in $h \in \Pi(H)$, the first of the conditions of Theorem 1.1.4 is fulfilled. Verification of the validity of the second condition is similar. ∎

Proof of Theorem 4.3.2. The assertion of Theorem 4.3.2 follows from Lemmas 4.3.5 and 4.3.6. Indeed, Lemma 4.3.5 ensures the convergence of the cor. f. and finite-dimensional distributions of the fields $X_\Delta^{(2)}(h)$ to the cor. f. and finite-dimensional distributions of the field $X(h)$, since such a convergence was proved for $X_\Delta^{(1)}(h)$. ∎

Corollary 4.3.2. Let $\xi(x)$ be a Gaussian field and suppose that assumptions VI, IX and VII hold. Then $\mu_\Delta^{(1)} \Longrightarrow \mu$ as $\Delta \xrightarrow{V.H.} \infty$, where μ is the measure in $C(\Pi(H))$ corresponding to a Gaussian field with cor. f.

$$\rho(h^{(1)}, h^{(2)}) = \rho_2(h^{(1)}, h^{(2)}) = 2(2\pi)^n \int_{R^n} \cos\langle \lambda, h^{(1)} \rangle \cos\langle \lambda, h^{(2)} \rangle f^2(\lambda) d\lambda.$$

If $\Delta = \Pi(T) \to \infty$, then under assumptions VII and IX for a Gaussian field $\xi(x)$, $\mu_\Delta^{(2)} \Longrightarrow \mu$.

Remark 4.3.2. In Theorem 4.3.1 a random field $\xi(x)$, $x \in \mathbf{R}^n$, with a known mean $E\xi(x) = 0$ is considered. If, however, the mean value of a homogeneous field $m = E\xi(x)$ is unknown, the estimator

$$\hat{B}_\Delta^{(3)}(h) = |\Delta|^{-1} \int_\Delta [\xi(x) - m_\Delta][\xi(x+h) - m_\Delta]dx$$

may be used in place of the estimator $\hat{B}_\Delta^{(1)}(h)$, where

$$m_\Delta = |\Delta|^{-1} \int_\Delta \xi(y)dy$$

is the least-squares estimator of the unknown mean m (cf. §3.1). Under the conditions of Theorem 4.3.1, the weak convergence of the measures $\mu_\Delta^{(3)} \Rightarrow \mu$ in $C(\Pi(H))$ is then valid as $\Delta \xrightarrow{V.H.} \infty$, where $\mu_\Delta^{(3)}$ are the measures in $C(\Pi(H))$ induced by the random fields

$$X_\Delta^{(3)}(h) = |\Delta|^{1/2}(\hat{B}_\Delta^{(3)}(h) - B(h)), \quad h \in \Pi(H).$$

For a proof, see [71].

Remark 4.3.3. Analogues of assertions of §4.3 for random sequences are presented in [162].

4.4. Asymptotic Normality.
The Case of a Homogeneous Isotropic Field

Suppose that $\xi(x)$, $x \in \mathbf{R}^n$ is a homogeneous isotropic random field. To estimate the cor.f. $B(h)$, $h \in [0, H]$ of the field $\xi(x)$, we shall use the statistics

$$\hat{B}_\Delta(h, l) = |\Delta|^{-1} \int_\Delta \xi(x)\xi(x+hl)dx, \qquad (4.4.1)$$

where $h \in [0, H]$, $\Delta \in \mathcal{B}^n, l \in s(1)$. The estimators $\hat{B}_\Delta(h, l)$ are similar to the estimators of type $\hat{B}_\Delta^{(1)}$ presented in previous sections. To determine $\hat{B}_\Delta(h, l)$ one needs to have observations of the field $\xi(x)$ on the set Δ_H. Define a random process $X_\Delta(h, l) = |\Delta|^{1/2}(\hat{B}_\Delta(h, l) - B(h))$, $h \in [0, H]$. We shall show that if assumption VI of §4.3 holds and a number of requirements are imposed on the spectral characteristics of the field $\xi(x)$ satisfying the mixing condition, then the measures $\mu_\Delta(l)$ corresponding to the a.s. continuous processes $X_\Delta(h, l)$, $h \in [0, H]$, converge weakly to a Gaussian measure on $C[0, H]$ to be described below.

Suppose that the field $\xi(x) \in \Xi^{(4)}$ (cf. §1.3), that is, the 4th order cumulant of $\xi(x)$ admits the spectral representation

$$s_4(\rho_1, \rho_2, \rho_3, 0) = |s(1)|^3 \int_{R_+^3} \prod_{j=1}^{3} Y_n(\nu_j \rho_j) \nu_j^{n-1} g_4(\nu_1, \nu_2, \nu_3) d\nu_1 d\nu_2 d\nu_3,$$

where $g_4(\nu_1, \nu_2, \nu_3)$ is a homogeneous isotropic spectral density function of the 4th order. If $\xi(x) \in \Xi^{(4)}$, then evidently the field $\xi(x)$ has an "ordinary" homogeneous isotropic density function $g(\nu) : dG/d\nu = |s(1)|\nu^{n-1}g(\nu)$; the measure G arises in the representation (1.2.4) of the cor.f. $B(|x|)$. Let $\xi(x), x \in \mathbf{R}^n$, satisfy assumption IX. The field $\xi(x)$ has a bounded and continuous isotropic spectral density function $g(\nu), \nu \in R_+^1$ (cf. Remark 1.6.1). We introduce assumptions similar to assumptions VII and VIII. For some $\beta \in (0, 1]$:

XI. $\qquad I_3(\beta) = \int_0^\infty \nu^{n+\beta} g(\nu) d\nu < \infty;$

XII. $\qquad \sup_{\lambda \in \mathbf{R}^n} \int_{R^2} \nu^{n-1}(1 + \nu_1\nu_2)^{(1+\beta)/2} \times$

$$\times \int_{s(\nu_1)} |g_4(\nu_1, |\lambda - \lambda^{(1)}|, \nu_2)| dm d\nu_1 d\nu_2 < \infty.$$

Consider a separable Gaussian random process $X(h), h \in [0, H]$ with zero mean and cor. f.

$$\rho(h_1, h_2) = (2\pi)^n \{|s(1)|^2 \int_0^\infty \int_0^\infty Y_n(\nu_1 h_1) Y_n(\nu_2 h_2)(\nu_1\nu_2)^{n-1} \times$$

$$\times g_4(\nu_1, \nu_1, \nu_2) d\nu_1 d\nu_2 + |s(1)| \int_0^\infty (Y_n(\nu|h_1 - h_2|) + Y_n(\nu|h_1 + h_2|))\nu^{n-1} g^2(\nu) d\nu =$$

$$= {}_1\rho(h_1, h_2) + {}_2\rho(h_1, h_2). \qquad (4.4.2)$$

Cor.f. (4.4.2) is an isotropic analogue to cor.f. (4.3.1).

Lemma 4.4.1. If XI and the condition

$$I_4(\beta) = \int_{R_+^2} (\nu_1\nu_2)^{n+(\beta-1)/2} |g_4(\nu_1, \nu_1, \nu_2)| d\nu_1 d\nu_2 < \infty,$$

which follows from XII, are valid, then the process $X(h), h \in [0, H]$, is a.s. continuous.

Proof. Utilizing the Poisson integral formula for the Bessel functions (1.2.9) we obtain:

$$E(X(h_1) - X(h_2))^2 = k_1(n) \int_0^\infty \int_0^\infty \int_{-1}^1 \int_{-1}^1 T_4(1 - t_1^2)^{(n-3)/2} \times$$

$$\times (1 - t_2^2)^{(n-3)/2}(\nu_1\nu_2)^{n-1} g_4(\nu_1, \nu_1, \nu_2) dt_1 dt_2 d\nu_1 d\nu_2 +$$

$$+ k_2(n) \int_0^\infty \int_{-1}^1 T_5(1 - t^2)^{(n-3)/2}\nu^{n-1} g^2(\nu) dt d\nu,$$

where

$$k_1(n) = 2^{n+2}\pi^{2n-1}/\Gamma^2(\tfrac{n-1}{2}), \quad k_2(n) = 2^{n+1}\pi^{(3n-1)/2}/\Gamma(\tfrac{n-1}{2});$$

$$|T_4| = |(\cos(h_1\nu_1 t_1) - \cos(h_2\nu_1 l_1))(\cos(h_1\nu_2 t_2) - \cos(h_2\nu_2 t_2))| \le$$

$$\le 2^{1-\beta}|t_1 t_2|^{(1+\beta)/2}(\nu_1\nu_2)^{(1+\beta)/2}|h_1 - h_2|^{1+\beta};$$

$$T_5 = 2(\cos(h_1\nu t) - \cos(h_2\nu t))^2 \le 2^{2-\beta}|t|^{1+\beta}|\nu|^{1+\beta}|h_1 - h_2|^{1+\beta}.$$

Consequently,

$$E(X(h_1) - X(h_2))^2 \le (k_3(\beta, n)I_3(\beta) + k_4(\beta, n)I_4(\beta))|h_1 - h_2|^{1+\beta}. \quad \blacksquare$$

Remark 4.4.1.   The assertion of Lemma 4.4.1 remains valid also under weaker assumptions such as the existence of logarithmic moments of the functions $\nu^{n-1}g(\nu)$ and $(\nu_1\nu_2)^{n-1}|g_4(\nu_1, \nu_1, \nu_2)|$. This fact is analogous to Lemma 4.3.1.

Lemma 4.4.2.   Let assumptions XI and XII hold and suppose that $g_0 = \sup_{\rho \le 0} g(\rho) < \infty$. Then the family of measures $\{\mu_\Delta(l)\}$ is compact in $C[0, H]$ for each $l \in s(1)$.

Proof.   Note that

$$\rho_\Delta(h_1, h_2, l) = E X_\Delta(h_1, l)X_\Delta(h_2, l) = \int_{R^{3n}} \Psi_\Delta(\lambda^{(1)} + \lambda^{(2)}) \times$$

$$\times \exp\{i(\langle\lambda^{(2)}, h_1 l\rangle + \langle\lambda^{(3)}, h_2 l\rangle)\} g_4(|\lambda^{(1)}|, |\lambda^{(2)}|, |\lambda^{(3)}|)d\lambda^{(1)}d\lambda^{(2)}d\lambda^{(3)} +$$

$$+ \int_{R^{2n}} \Psi_\Delta(\lambda^{(1)} + \lambda^{(2)})[\exp\{i\langle\lambda^{(2)}, (h_1 - h_2)l\rangle\} +$$

$$+ \exp\{i((\lambda^{(1)}, h_1 l) - (\lambda^{(2)}, h_2 l))\}]g(|\lambda^{(1)}|)g(|\lambda^{(2)}|)d\lambda^{(1)}d\lambda^{(2)}.$$

Therefore

$$E(X_\Delta(h_1, l) - X_\Delta(h_2, l))^2 = \int_{R^{3n}} \Psi_\Delta(\lambda^{(1)} + \lambda^{(2)})g_4(|\lambda^{(1)}|, |\lambda^{(2)}|\lambda^{(3)}|) \times$$

$$\times T_6 d\lambda^{(1)}d\lambda^{(2)}d\lambda^{(3)} + \int_{R^{2n}} \Psi_\Delta(\lambda^{(1)} + \lambda^{(2)})g(|\lambda^{(1)}|)g(|\lambda^{(2)}|)T_7 d\lambda^{(1)}d\lambda^{(2)},$$

where

$$|T_6| = |(e^{i(\lambda^{(2)}, h_2 l)} - e^{i(\lambda^{(2)}, h_1 l)})(e^{i(\lambda^{(3)}, h_2 l)} - e^{i(\lambda^{(3)}, h_1 l)})| \le$$

$$\le 2^{1-\beta}|\lambda^{(2)}|^{(1+\beta)/2}|\lambda^{(3)}|^{(1+\beta)/2}|h_1 - h_2|^{1+\beta};$$

$$|T_7| = |2(1 - \cos(\lambda^{(2)}, (h_1 - h_2)l)) + (\exp\{i(\lambda^{(1)}, h_2 l)\} -$$

$$- \exp\{i(\lambda^{(1)}, h_1 l)\})(\exp\{-i(\lambda^{(2)}, h_2 l)\} - \exp\{(\lambda^{(2)}, h_1 l)\})| \le$$

$$\le 2^{1-\beta}(|\lambda^{(2)}|^{1+\beta} + |\lambda^{(1)}|^{(1+\beta)/2}|\lambda^{(2)}|^{(1+\beta)/2})|h_1 - h_2|^{1+\beta}.$$

Using the above bounds and the conditions of the lemma in the same manner as in Lemma 4.3.2, one obtains the inequality

$$E|X_\Delta(h_1, l) - X_\Delta(h_2, l)|^2 \le k_5(\beta, n)|h_1 - h_2|^{1+\beta},$$

where

$$k_5(\beta, n) = (2\pi)^n 2^{1-\beta}(2g_0 I_3(\beta) + \sup_{\lambda \in R^n} I_5(\beta, \lambda))$$

and

$$I_5(\beta, \lambda) = \int_{R_+^2} \nu_1^{(1+\beta)/2}\nu_2^{n+(\beta-1)/2} \int_{s(\nu_1)} |g_4(\nu_1, |\lambda - \lambda^{(1)}|, \nu_2)|dmd\nu_1 d\nu_2. \quad \blacksquare$$

Denote by μ the measure in $C[0, H]$ corresponding to the Gaussian process $X(h)$, $h \in [0, H]$.

Theorem 4.4.1. If assumptions VI, IX, XI and XII hold, then for any $l \in s(1)$, $\mu_\Delta(l) \Longrightarrow \mu$ as $\Delta \xrightarrow{V.H.} \infty$ in $C[0, H]$.

Proof. To prove the theorem (cf. §4.3) it remains to show that: 1) $\rho_\Delta(h_1, h_2) \to \rho(h_1, h_2)$, $\Delta \xrightarrow{V.H.} \infty$, $h_1, h_2 \in [0, H]$; 2) finite-dimensional distributions of the random processes $X_\Delta(h, l)$ converge to finite-dimensional distributions of $X(h)$ as $\Delta \xrightarrow{V.H.} \infty$. But these facts are established analogously to Lemmas 4.3.3 and 4.3.4. ∎

Corollary 4.4.1. Let $\xi(x)$ be a Gaussian field and suppose that assumptions VI, IX and XI hold. Then $\mu_\Delta(l) \Longrightarrow \mu$, $\Delta \xrightarrow{V.H.} \infty$, where μ is the measure in $C[0, H]$ corresponding to a Gaussian process with zero mean and cor.f.

$$\rho(h_1, h_2) = {}_2\rho(h_1, h_2) = (2\pi)^n |s(1)| \int_0^\infty (Y_n(\nu|h_1 - h_2|) +$$

$$+ Y_n(\nu|h_1 + h_2|)\nu^{n-1} g^2(\nu) d\nu.$$

Remark 4.4.2. Let a homogeneous isotropic Gaussian random field $\xi(x)$, $x \in \mathbf{R}^n$, $n \geq 2$, with $E\xi(x) = 0$ and unknown cor.f. $B(h)$ be observed on the ball $v(r + h)$, $h \in [0, H]$, $r \to \infty$. Based on M.I. Yadrenko's ideas, reference [69] and papers [45, 46], we consider an estimator for the cor.f. $B(h)$ of the form

$$\hat{B}_r^*(h) = |v(r)|^{-1} \int_{v(r)} \xi(x)\eta_h(x) dx,$$

$$\eta_h(x) = |s_{n-1}(x, h)|^{-1} \int_{s_{n-1}(x,h)} \xi(y) dm_h(y), \qquad (4.4.3)$$

where $s_{n-1}(x, h) = \{y \in \mathbf{R}^n : |y - x| = h\}$ is a sphere of radius h centred at the point $x \in \mathbf{R}^n$, $|s_{n-1}(x, h)| = |s(1)| h^{n-1}$ and $m_h(\cdot)$ is the Lebesgue measure on the sphere $s_{n-1}(x, h)$. If assumption IX is valid, then the field $\xi(x)$, $x \in \mathbf{R}^n$, has a continuous bounded isotropic spectral density function $g(\nu)$, $\nu \in R_+^1$ (cf. Remark 1.6.1). Suppose that for some $\beta > 0$

$$\int_0^\infty \nu^{n+\beta} g^2(\nu) d\nu < \infty.$$

The measures $\tilde{\mu}_r$ induced by the processes $\tilde{X}_r(h) = |v(r)|^{1/2}(\hat{B}_r^*(h) - B(h))$, $h \in [0, H]$, in the space $C[0, H]$ converge weakly to the measure $\tilde{\mu}$ induced in $C[0, H]$ by the Gaussian process $\tilde{X}(h)$, $h \in [0, H]$, with $E\tilde{X}(h) = 0$ and cor.f.

$$\tilde{\rho}(h_1, h_2) = (2\pi)^n |s(1)|^2 2 \int_0^\infty Y_n(\nu h_1) Y_n(\nu h_2) \nu^{n-1} g^2(\nu) d\nu, \qquad (4.4.4)$$

$h_1, h_2 \in [0, H]$ (cf. Corollary 4.4.1)[45, 46]. If the field $\xi(x)$, $x \in \mathbf{R}^n$, is observed on the ball $v(r)$, $r \to \infty$, then the statistic

$$\hat{B}_r^{**}(h) = |v(r)|^{-1} \int_{v(r-h)} \xi(x)\eta_h(x)dx$$

may be taken as an estimator of the cor.f. $B(h)$, where $\eta_h(x)$ is defined in accordance with (4.4.3). This estimator is asymptotically unbiased and

$$\lim_{r \to \infty} r(E\hat{B}_r^{**}(h) - B(h)) = -nhB(h). \tag{4.4.5}$$

Thus, the bias decreases not faster than $1/r$. If

$$\int_0^\infty \nu^{n-1}g^2(\nu)d\nu < \infty,$$

then for $h_1, h_2 \in [0, H]$

$$\lim_{r \to \infty} |v(r)| \operatorname{cov}(\hat{B}_r^{**}(h_1), \hat{B}_r^{**}(h_2)) =$$

$$= \lim_{r \to \infty} |v(r)| \operatorname{cov}(\hat{B}_r^{*}(h_1), \hat{B}_r^{*}(h_2)) = \tilde{\rho}(h_1, h_2). \tag{4.4.6}$$

Formulas (4.4.5) and (4.4.6) imply that the mean-square error of the estimator $\hat{B}_r^{*}(h)$ decreases as $O(r^{-n})$ whereas that of the estimator $\hat{B}_r^{**}(h)$ decreases as $O(r^{-2})$; however, most of the asymptotic properties of $\hat{B}_r^{*}(h)$ and $\hat{B}_r^{**}(h)$ are the same. The use of the spherical average $\eta_h(x)$ (cf. (4.4.3)) for estimators of the cor .f.'s $\hat{B}_r^{*}(h)$ and $\hat{B}_r^{**}(h)$ sometimes reduces the asymptotic variance in comparison to the asymptotic variance of the estimator $\hat{B}_\Delta(x, l)$ as the parameter h increases. Indeed, (4.4.4) yields as $h \to \infty$:

$$\tilde{\rho}(h, h) = (2\pi)^n |s(1)|^2 \, 2 \int_0^\infty Y_n^2(\nu h)\nu^{n-1}g^2(\nu)d\nu = O\left(h^{-n+1} \int_0^\infty g^2(\nu)d\nu\right),$$

whereas in the Gaussian case ($g_4(\nu_1, \nu_2, \nu_3) = 0$), the asymptotic variance of the estimator $\hat{B}_\Delta(h, l)$ equals

$$_2\rho(h, h) = (2\pi)^n |s(1)| \int_0^\infty [1 + Y_n(2\nu h)]\nu^{n-1}g^2(\nu)d\nu \to$$

$$\to (2\pi)^n |s(1)|2 \int_0^\infty \nu^{n-1}g^2(\nu)d\nu, \quad h \to \infty.$$

For example, for $n = 2$ and $g(\nu) = \exp\{-\frac{\nu^2}{4}\}$ we have

$$\int_0^\infty [1 + Y_2(2\nu h)]\nu g^2(\nu)d\nu = 1 + 2e^{-2h^2} > 1, \quad h \geq 0,$$

however,

$$2 \int_0^\infty Y_2^2(\nu h)\nu g^2(\nu)d\nu = 2e^{-h^2} I_0(h^2) < 1, \ h > \sqrt{0.9},$$

where $I_0(h) = \sum_{k=0}^\infty (h/2)^{2k}/(k!)^2$ is the modified Bessel function of the first kind of zero order. Note that for $h = 2$, we have $1 + 2e^{-2h^2} \cong 1.03$ while $2e^{-h^2} I_0(h^2) \cong 0.4$.

Remark 4.4.3. If a homogeneous isotropic random field $\xi(x)$, $x \in \mathbf{R}^n$ is observed on the sphere $s_{n-1}(0, r + h) = s_{n-1}(r + h)$, $h \in [0, H]$, $r \to \infty$, then the statistic

$$_s\hat{B}_r^{(1)}(h) = |s_{n-1}(r)|^{-1} \int_{s_{n-1}(r)} \xi(x)\xi(x + n^{-1/2}h)dm_r(x) \qquad (4.4.7)$$

is an analogue of the statistic $\hat{B}_\Delta(h, l)$ and the statistic

$$_s\hat{B}_r^*(h) = |s_{n-1}(r)|^{-1} \int_{s_{n-1}(r)} \xi(x)\tilde{\eta}_h(x)dm_r(x) \qquad (4.4.8)$$

is one of the analogues of the statistic $\hat{B}_r^*(h)$ with $n \geq 3$, where

$$\tilde{\eta}_h(x) = |s_{n-2}(x, h, r)|^{-1} \int_{s_{n-2}(x,h,r)} \xi(y)dm_{h'}^{(n-2)}(y)$$

is the spherical average of codimension $n - 2$, that is, $s_{n-2}(x, h, r)$ is the sphere of codimension $n - 2$ obtained by intersecting the sphere $s_{n-1}(r)$ with the plane perpendicular to the line in $\mathbf{R}^n$ drawn through the points $x \in s_{n-1}(r)$ and $0 \in \mathbf{R}^n$ so that $\rho_{xy}(x) = h$ for $y \in s_{n-2}(x, h, r)$; $m_{h'}^{(n-2)}(\cdot)$ is the Lebesgue measure on the sphere $s_{n-2}(x, h, r)$ (h' is the radius of the sphere $s_{n-2}(x, h, r)$), $|s_{n-2}(x, h, r)| = (h')^{n-2}2\pi^{(n-1)/2}/\Gamma(\frac{n-1}{2})$.

The estimators (4.4.7) and (4.4.8) are unbiased. If the field $\xi(x)$, $x \in \mathbf{R}^n$ is Gaussian and

$$I = \int_0^\infty z^{n-2}B^2(z)dz < \infty,$$

then it can be shown that these estimates are also consistent. For example, for $n \geq 3$ the relation

$$\lim_{r \to \infty} r^{n-1} \operatorname{var} {}_s\hat{B}_r^*(h) = I$$

holds.

4.5. Estimation by Means of Several Independent Sample Functions

In the preceding sections, properties of estimators of the cor.f. $B(h)$ of a homogeneous random field $\xi(x)$, $x \in \mathbf{R}^n$ from observation of a single sample function under infinite expansion of the observation set Δ were presented. Of no lesser interest is a variation of the problem of estimating $B(h)$ to be studied in this section which is based on a different statistical assumption: the set Δ is fixed but there are many independent sample functions of the field $\xi(x)$ available.

Let Δ, $K \subset \mathbf{R}^n$ be a compact set containing the origin and $\{\xi_j(x),\ x \in \Delta+K\}_{j\geq 1}$, a sequence of independent copies of a m.s. continuous homogeneous separable measurable random field $\xi(x)$, $x \in \Delta + K$. Set

$$u_h^{(j)}(x) = \xi_j(x)\xi_j(x+h),\ {}_j\hat{B}_\Delta^{(1)}(h) = |\Delta|^{-1}\int_\Delta u_h^{(j)}(x)dx;$$

$$_j\hat{B}_\Delta^{(2)}(h) = |\Delta|^{-1}\int_{\Delta\cap(\Delta-h)} u_h^{(j)}(x)dx;$$

$$\hat{B}_m^{(i)}(h) = m^{-1}\sum_{j=1}^m {}_j\hat{B}_\Delta^{(i)}(h),\ h \in K,\ i=1,2.$$

Below, the consistency and asymptotic normality of the estimators $\hat{B}_m^{(i)}(h)$ of the cor.f. $B(h)$ of the field $\xi(x)$ will be considered as $m \to \infty$. Evidently, to determine $\hat{B}_m^{(2)}(h)$ it suffices to have available the sequence $\{\xi_j(x),\ x \in \Delta\}_{j\geq 1}$. Since

$$\int_\Delta |u_h(x)|dx \leq \tfrac{1}{2}\int_\Delta \xi^2(x)dx + \tfrac{1}{2}\int_{\Delta+K} \xi^2(x)dx,\ \int_{\Delta\cap(\Delta-h)} |u_h(x)|dx \leq$$

$$\leq \int_\Delta \xi^2(x)dx,$$

utilizing the Fubini-Tonelli theorem and simple properties of Lebesgue integrals we may choose a.s. continuous versions of the fields $\hat{B}_\Delta^{(i)}(h)$, $h \in K$, $i=1,2$. Therefore in the statements of the theorems below, the fields ${}_j\hat{B}_\Delta^{(i)}$ together with the fields $\hat{B}_m^{(i)}$ can be assumed to be a.s. continuous.

Theorem 4.5.1. The estimators $\hat{B}_m^{(i)}(h)$ have the following properties:

1) $\sup\limits_{h\in K} |\hat{B}_m^{(1)}(h) - B(h)| \xrightarrow[m\to\infty]{} 0$ a.s.;

2) $\sup\limits_{h\in K} |\hat{B}_m^{(2)}(h) - |\Delta|^{-1}|\Delta\cap(\Delta-h)|B(h)| \xrightarrow[m\to\infty]{} 0$ a.s.

Proof. Since

1) $\sup_{h \in K} |\hat{B}_m^{(2)}(h)| \leq \frac{1}{2}(E(|\Delta|^{-1} \int_\Delta \xi^2(x)dx) +$

   $+ E(|\Delta|^{-1} \int_{\Delta+K} \xi^2(x)dx)) = \frac{1}{2}|\Delta|^{-1}(|\Delta| + |\Delta + K|)B(0);$

2) $E \sup_{h \in K} |\hat{B}_\Delta^{(2)}(h)| \leq E(|\Delta|^{-1} \int_\Delta \xi^2(x)dx) = B(0),$

the assertion of the theorem follows from the law of large numbers for random elements in Banach spaces [173]. $\blacksquare$

Thus, the unbiased estimator $\hat{B}_m^{(1)}(h)$ is consistent while the biased estimator $\hat{B}_m^{(2)}(h)$ is not. Nevertheless if the set Δ is convex, the arguments in §4.2 demonstrate that

$$|B(h)(1 - |\Delta|^{-1}|\Delta \cap (\Delta - h)|) \leq k_*|h||\Delta|^{-1/n}|B(h)|. \qquad (4.5.1)$$

Before proceeding to a discussion of conditions for the asymptotic normality of the estimators $\hat{B}_m^{(i)}(h), h \in K$ we state several assertions dealing with a.s. continuity of random fields.

Let $u_0 > 0$ be a number. A function $\phi : [0, u_0] \to R_+^1$ is said to be a modulus on the interval $[0, u_0]$ if: (a) ϕ is continuous, $\phi(0) = 0$;   (b) $\phi(x) \leq \phi(x + y) \leq \phi(x) + \phi(y), x, y \geq 0, x + y \leq u_0.$

Note that if $\phi(u_0) \neq 0$, then $\sup_{u \leq u_0} u/\phi(u) < \infty$. Indeed, for an integer $m > 0$ such that $mu \leq u_0$, we have $\phi(mu) \leq m\phi(u)$. Let $\lambda \geq 1$ be a real number and m an integer satisfying the inequality $m - 1 < \lambda \leq m$. Then $\phi(\lambda u) \leq \phi(mu) \leq (\lambda + 1)\phi(u) \leq 2\lambda\phi(u)$ and thus, $\phi(w) \leq 2\lambda\phi(w/\lambda), w \leq u_0$. The last inequality yields $\phi(u_0)/u_0 \leq 2\phi(v)/v$ for $v \leq u_0$.

Let an a.s. continuous random field $x(h)$ be defined on a set $T \subset \mathbf{R}^n$. A modulus ϕ on $[0, u_0]$ is said to be a sample modulus of the field $x(h)$ if there exists a r.v. η such that for $|h^{(1)} - h^{(2)}| \leq u_0$, $|x(h^{(1)}) - x(h^{(2)})| \leq \eta\phi(|h^{(1)} - h^{(2)}|)$.

Below we shall use a central limit theorem for a.s. continuous random fields defined on a compact set.

Theorem 4.5.2 [184]. Let $\{x_j(h), h \in T\}_{j \geq 1}$ be a sequence of independent a.s. continuous random fields defined on a compact set $T \subset \mathbf{R}^n$ which are copies of an a.s. continuous field $x(h), h \in T$, and suppose that $Ex(h) = 0, Ex^2(h) < \infty, h \in T$. Let the modulus $\phi(u)$ on $[0, u_0]$ be a sample modulus of the field $x(h)$ and let the conditions

(a) $\int_0^{u_0} \frac{\phi(u)}{u|\ln u|^{1/2}} du < \infty,$ (b) $E\eta^2 < \infty$

hold. Then the measures μ_m on $C(T)$ corresponding to the fields $m^{-1/2} \times \sum_{j=1}^{m} x_j(h)$, $h \in T$, converge weakly as $m \to \infty$ to a Gaussian measure μ on $C(T)$ corresponding to a Gaussian field with zero mean and cor.f. $E\{x(h^{(1)}) \times x(h^{(2)})\}$, $h^{(1)}, h^{(2)} \in T : \mu_m \Longrightarrow \mu$.

The following theorem implies integrability of any positive power of the supremum of a bounded Gaussian field.

Theorem 4.5.3 [188, 170]. Let $x(h)$, $h \in T \subset \mathbf{R}^n$, be a separable Gaussian field and $P\{\sup_{h \in T} |x(h)| < \infty\} = 1$. Then for some $\alpha > 0$, $E \exp\{\alpha \sup_{h \in T} x^2(h)\}$ is finite.

Corollary 4.5.1. Let $x(h), h \in T$, (T is a compact set in $\mathbf{R}^n$) be an a.s. continuous Gaussian field and the modulus ϕ on $[0, u_0]$ its sample modulus. Then, for some $\alpha > 0$, $E \exp\{\alpha \tilde{\eta}^2\} < \infty$, where

$$\tilde{\eta} = \sup_{0 < |h^{(1)} - h^{(2)}| \leq u_0} (|x(h^{(1)}) - x(h^{(2)})| / \phi(|h^{(1)} - h^{(2)}|)).$$

Proof. Indeed, $\tilde{\eta} \leq \eta$ and thus, the separable Gaussian field $\tilde{x}(\tilde{h}) = x(h^{(1)}) - x(h^{(2)}))/\phi(|h^{(1)} - h^{(2)}|)$, $\tilde{h} \in \tilde{T} = \{(h^{(1)}, h^{(2)}) \in T \times T : 0 < |h^{(1)} - h^{(2)}| \leq u_0\}$, is a.s. bounded. ∎

The following statement is a particular case of a general theorem due to R.M. Dudley.

Theorem 4.5.4 [166]. Let $x(h), h \in T$, (T is a compact set in $\mathbf{R}^n$) be a separable Gaussian field and suppose that for $h^{(1)}, h^{(2)} \in T$ and some $s \geq 1$, $\epsilon > 0$ and $k_1 < \infty$

$$E|x(h^{(1)}) - x(h^{(2)})|^2 \leq k_1 |\log |h^{(1)} - h^{(2)}||^{-s-\epsilon}. \tag{4.5.2}$$

Then the function $\phi(u) = |\log u|^{-(s-1+\epsilon)/2}$ defined on $[0, u_0]$ ($u_0 \in (0, 1)$) is a sample modulus of the field $x(h)$, $h \in T$.

Below we shall assume that $\xi(x)$, $x \in \Delta + K$ is an a.s. continuous field and that the following assumptions are valid.

XIII. The modulus ϕ on $[0, u_0]$ is a sample modulus of the field $\xi(x)$, $x \in \Delta + K$, that is, for $\eta_0 = \eta_0(\omega) < \infty$ we have a.s.

$$|\xi(h + \tau) - \xi(h)| \leq \eta_0 \phi(|\tau|), h, h + \tau \in \Delta + K, |\tau| \leq u_0$$

and (a) $\int_0^{u_0}(\phi(u)/(u|\ln u|^{1/2}))du < \infty$; (b) η_0 is a r.v. and $E\eta_0^{2+\delta} < \infty$, $E|\xi(0)|^{2+4/\delta} < \infty$ for some $\delta > 0$.

Let

$$w_h^{(j)} = u_h^{(j)}(x) - B(h), \quad \Delta^{(1)} = \Delta, \quad \Delta^{(2)} = \Delta \cap (\Delta - h), \quad {}_jz_\Delta^{(i)}(h) =$$

$$= |\Delta|^{-1}\int_{\Delta^{(i)}} w_h^{(j)}(x)dx, \quad i = 1,2; \quad \rho^{(i)}(h^{(1)}, h^{(2)}) = E z_\Delta^{(i)}(h^{(1)})z_\Delta^{(i)}(h^{(2)}) =$$

$$= |\Delta|^{-2}\int_{\tilde\Delta^{(i)}} E w_{h^{(1)}}(x^{(1)})w_{h^{(2)}}(x^{(2)})dx^{(1)}dx^{(2)}, \quad i = 1,2,$$

$$\tilde\Delta^{(1)} = \Delta \times \Delta, \quad \tilde\Delta^{(2)} = (\Delta \cap (\Delta - h^{(1)})) \times (\Delta \cap (\Delta - h^{(2)})).$$

We introduce the random fields

$$X_m^{(1)}(h) = m^{1/2}(\hat B_m^{(1)}(h) - B(h)) = m^{-1/2}\sum_{j=1}^m {}_jz_\Delta^{(1)}(h), \quad h \in K,$$

$$X_m^{(2)}(h) = m^{1/2}(\hat B_m^{(2)}(h) - |\Delta|^{-1}|\Delta \cap (\Delta - h)|B(h)) = m^{-1/2}\sum_{j=1}^m {}_jz_\Delta^{(2)}(h), \quad h \in K.$$

Denote by $\mu_m^{(i)}$ the probability measures on $C(K)$ corresponding to the fields $X_m^{(i)}$ and by $\mu^{(i)}$ those corresponding to Gaussian fields $X^{(i)}$ with zero mean and cor.f. $\rho^{(i)}(h^{(1)}, h^{(2)})$.

Theorem 4.5.5. If assumption XIII is valid, then $\mu_m^{(1)} \Longrightarrow \mu^{(1)}$. If the set Δ is convex, then XIII holds and $\delta = 2$ in condition (b), then also $\mu_m^{(2)} \Longrightarrow \mu^{(2)}$ as $m \to \infty$.

Proof. Let $|\tau| \le u_0$. Then in view of assumption XIII,

$$|z_\Delta^{(1)}(h + \tau) - z_\Delta^{(1)}(h)| \le |\Delta|^{-1}\int_\Delta |\xi(x)| |\xi(x + h + \tau) - \xi(x + h)|dx +$$

$$+|B(h + \tau) - B(h)| \le$$

$$\le (\eta_0|\Delta|^{-1}\int_\Delta |\xi(x)|dx + E\eta_0|\xi(0)|)\phi(|\tau|).$$

By Hölder's inequality,

$$E(|\eta_0|\Delta|^{-1}\int_\Delta |\xi(x)|dx)^2 \le (E\eta_0^{2+\delta})^{2/(2+\delta)}\times$$

$$\times (E(|\Delta|^{-1} \int_\Delta |\xi(x)| dx)^{2+4/\delta})^{\delta/(2+\delta)} \le (E\eta_0^{2+\delta})^{2/(2+\delta)} \times$$

$$\times (E|\xi(0)|^{2+4/\delta})^{\delta/(2+\delta)} < \infty.$$

Hence Theorem 4.5.2 is applicable to the fields $x_j(h) = {}_j z_\Delta^{(1)}(h)$, $h \in K$, since the r.v. $\eta^{(1)} = \eta_0 |\Delta|^{-1} \int_\Delta |\xi(x)| dx + E\eta_0 |\xi(0)|$ satisfies its condition. Therefore $\mu_m^{(1)} \Rightarrow \mu^{(1)}$, $m \to \infty$.

2.   By analogy with (4.1.1) and (4.1.2) we obtain

$$|z_\Delta^{(2)}(h+\tau) - z_\Delta^{(2)}(h)| \le |\Delta|^{-1} \int_\Delta |\xi(x)| |\xi(x+h+\tau)-$$

$$-\xi(x+h)| dx + |B(h+\tau) - B(h)|+$$

$$+|\Delta|^{-1} \int_{\Delta \cap ((\Delta-h-\tau)\ominus(\Delta-h))} |u_h(x)| dx + |B(h)| |\Delta|^{-1} |\Delta \ominus (\Delta - \tau)|.$$

Let $d = \max_{x \in \Delta} |x|$, $N = [\frac{d}{u_0}] + 1$. Then by assumption XIII and (4.5.1),

$$|\Delta|^{-1} \int_{\Delta \cap ((\Delta-h-\tau)\ominus(\Delta-h))} |u_h(x)| dx \le (N\phi(u_0)\eta_0 + |\xi(0)|)^2 \times$$

$$\times |\Delta \ominus (\Delta - t)| \le 2k_* |\Delta|^{-1/n} (N\phi(u_0)\eta_0 + |\xi(0)|)^2 |\tau| \text{ a.s.};$$

$$|B(h)| |\Delta|^{-1} |\Delta \ominus (\Delta - \tau)| \le 2k_* B(0) |\Delta|^{-1/n} |\tau|.$$

Therefore

$$|z_\Delta^{(2)}(h+\tau) - z_\Delta^{(2)}(h)| \le (\eta^{(1)} + \eta^{(2)})\phi(|\tau|) \text{ a.s.},$$

where $\eta^{(2)} = 2k_* |\Delta|^{-1/n} \sup_{u \le u_0} (u/\phi(u))((N\phi(u_0)\eta_0 + |\xi(0)|)^2 + B(0))$. Hence the field $z_\Delta^{(2)}$ satisfies the conditions of Theorem 4.5.2 provided $E\eta_0^4 < \infty$ and $E|\xi(0)|^4 < \infty$. ∎

Remark 4.5.1.   It is evident from the proof of Theorem 4.5.5 that the condition of convexity of Δ may be replaced by the condition

$$\sup_{|\tau| \le u_0} \phi^{-1}(|\tau|) |\Delta \ominus (\Delta - \tau)| < \infty. \tag{4.5.3}$$

In assumption XIII the integrability condition for the r.v. η_0, which is difficult to verify, is present. If $\xi(x)$ is a Gaussian field, this difficulty can be overcome and one obtains a more complete assertion of weak convergence of the measures $\mu_m^{(i)}$ as compared to Theorem 4.5.5.

Theorem 4.5.6. Let $\xi(x)$, $x \in \Delta + K$ be a Gaussian a.s. continuous field having a sample modulus $\phi(u)$, $u \in [0, u_0]$ which satisfies part (a) of assumption XIII. Then $\mu_m^{(1)} \Rightarrow \mu^{(1)}$. If, moreover, (4.5.3) holds, then $\mu_m^{(2)} \Rightarrow \mu^{(2)}$ as $m \to \infty$.

Proof. Corollary 4.5.1 implies that for any $\beta > 0$,

$$E\left(\sup_{0<|h^{(1)}-h^{(2)}|\leq u_0} (|\xi(h^{(1)}) - \xi(h^{(2)})|/\phi(|h^{(1)} - h^{(2)}|))\right)^\beta < \infty.$$

Therefore one can repeat the proof of Theorem 4.5.5, setting

$$\eta_0 = \sup_{0<|h^{(1)}-h^{(2)}|\leq u_0} (|\xi(h^{(1)}) - \xi(h^{(2)})|/\phi(|h^{(1)} - h^{(2)}|)). \quad \blacksquare$$

Corollary 4.5.2. Let $\xi(x)$, $x \in \Delta+K$, be a separable homogeneous Gaussian field and for some $\epsilon > 0$, $k_2 < \infty$ and $\tau \in v(u_0)$,

$$B(0) - B(\tau) \leq k_2 |\log |\tau||^{-2-\epsilon}. \tag{4.5.4}$$

Then $\mu_m^{(1)} \Rightarrow \mu^{(1)}$ as $m \to \infty$.

Proof. Indeed, in view of Theorem 4.5.4, the modulus $\phi(u) = |\log u|^{-(1+\epsilon)/2}$ is a sample modulus of the field $\xi(x)$. For the function $\phi(u)$, the integral in part (a) of assumption XIII is finite if $u_0 < 1$. $\quad \blacksquare$

Corollary 4.5.3. If (4.5.4) holds and

$$\sup_{|\tau|\leq u_0} |\log |\tau||^{(1+\epsilon)/2} |\Delta \ominus (\Delta - \tau)| < \infty,$$

then $\mu_m^{(2)} \Rightarrow \mu^{(2)}$.

We note in conclusion that if $\xi(x)$, $x \in \Delta + K$, is a homogeneous isotropic field, then the condition $\int_0^\infty \ln^{2+\epsilon}(1+\lambda)G(d\lambda) < \infty$ is sufficient for (4.5.4) to be valid. Here G is the measure appearing in representation (1.2.4) for the cor.f. $B(h)$.

4.6.   Confidence Intervals

Using the notation of the preceding sections, we set

$$F(t) = P\{\sup_{h\in\Pi(H)} |X_\Delta^{(i)}(h)| > t\}, \quad F_m^{(i)}(t) = P\{\sup_{h\in K} |X_m^{(i)}(h)| > t\}$$

and for the Gaussian fields $X(h)$, $X^{(i)}(h)$ we set

$$F(t) = P\Big\{ \sup_{h \in \Pi(H)} |X(h)| > t \Big\}, \quad F^{(i)}(t) = P\Big\{ \sup_{h \in K} |X^{(i)}(h)| > t \Big\}, \quad i = 1, 2.$$

The theorems in §§4.3 and 4.5 imply that uniformly in $t > 0$,

$$|F_\Delta^{(i)}(t) - F(t)| \xrightarrow[\Delta \to \infty]{} 0; \quad |F_m^{(i)}(t) - F^{(i)}(t)| \xrightarrow[m \to \infty]{} 0, \quad i = 1, 2. \tag{4.6.1}$$

Relations (4.6.1) may serve as a basis for constructing asymptotic functional confidence intervals for the cor.f. $B(h)$. To estimate the functions $F(t)$ and $F^{(i)}(t)$, various modifications of the well-known Fernique's inequality can be used.

Let $x(h)$, $h \in \Pi = [0, H]^n$, be a separable Gaussian field with cor.f. $\rho(h^{(1)}, h^{(2)})$; let $\phi(h) = \sup_{h^{(1)}, h^{(2)} \in \Pi, |h^{(1)} - h^{(2)}|_0 \leq h} (E|x(h^{(1)}) - x(h^{(2)})|^2)^{1/2}$ be the m.s. modulus of the field $x(h)$.

Theorem 4.6.1 [170]. If the integral $\int_1^\infty \phi(e^{-x^2})dx$ is finite, the field $x(h)$ is a.s. continuous. Moreover, for any integer $p \geq 2$ and any $t > (1 + 4n \log p)^{1/2}$, the estimate:

$$P\Big\{ \sup_{h \in \Pi} |x(h)| \geq t \Big(\sup_{h^{(1)}, h^{(2)} \in \Pi} \rho^{1/2}(h^{(1)}, h^{(2)}) +$$

$$+ (2 + \sqrt{2}) \int_1^\infty \phi\Big(\frac{H}{2}p^{-u^2}\Big) du \Big) \Big\} \leq 2.5 p^{2n} \int_t^\infty e^{-u^2/2} du \tag{4.6.2}$$

is valid.

Note that by virtue of Theorem 4.6.1, a.s. continuity of the field $x(h)$ follows from (4.5.2).

This section provides several observations dealing with the application of Fernique's inequality to the Gaussian fields $x(h)$ and $x^{(i)}(h)$ with particular cor.f. $\rho(h^{(1)}, h^{(2)})$ and $\rho^{(i)}(h^{(1)}, h^{(2)})$.

The case of a Gaussian homogeneous isotropic random field $\xi(x)$ is considered separately. Here the tail of the distribution of the supremum of the limiting process is estimated on the basis of a comparison theorem for Gaussian random functions which is an extension of the comparison theorem due to D. Slepian [218].

Let two centred Gaussian functions $x(h)$ and $z(h)$ be defined on the at most countable set T, $h \in T$; let $a(h) \geq 0$, $h \in T$, be a non-random function.

Theorem 4.6.2 [16]. If $E|x(h^{(1)}) - x(h^{(2)})|^2 \leq E|z(h^{(1)}) - z(h^{(2)})|^2$, $h^{(1)}, h^{(2)} \in T$, then

$$P\{\sup_{h \in T}(|x(h)| - a(h)) > 0\} \leq 2P\{\sup_{h \in T}(z(h) + \gamma g(h) - a(h)) > 0\}, \quad (4.6.3)$$

where $g^2(h) = \alpha^2 - Ez^2(h) + Ex^2(h)$, $h \in T$, $\alpha^2 = \sup_{h \in T}(0 \vee (Ez^2(h) - Ex^2(h)))$ and γ is a standard Gaussian r.v. independent of $z(h)$.

Consider first a Gaussian field $X(h)$ with cor.f. $\rho(h^{(1)}, h^{(2)})$ defined by formula (4.3.1). Denote

$$\nu_{st}(\lambda) = \int_{R^{2n}} |\lambda^{(1)}|^s |\lambda^{(2)}|^t |f_4(\lambda^{(1)}, \lambda - \lambda^{(1)}, \lambda^{(2)})| d\lambda^{(1)} d\lambda^{(2)}.$$

Lemma 4.6.1. Suppose that $f_0 = \sup_{\lambda \in R^n} f(\lambda) < \infty$, $\nu_{00}(0), \nu_{11}(0) < \infty$, $\mu_2 = \int_{R^n} |\lambda|^2 f(\lambda) d\lambda < \infty$. Then for any $t > (1 + 4n)^{1/2} a$ and $a = (2\pi)^{n/2}([\nu_{00}(0) + 2B(0)f_0]^{1/2} + (Hn^{1/2}/4e)(2 + \sqrt{2})[\nu_{11}(0) + 2f_0\mu_2]^{1/2}$,

$$F(t) \leq G(t) = 2.5\frac{a}{t} \exp\left\{-\frac{t^2}{2a^2} + 2n\right\}. \quad (4.6.4)$$

Proof. The m.s. modulus $\phi(h)$ of the field $X(h)$ admits the bound

$$\phi(h) \leq [(2\pi)^n n(\nu_{11}(0) + 2f_0\mu_2)]^{1/2} h \quad (4.6.5)$$

in view of (4.3.5) and the condition of the lemma. It is also evident that

$$\sup_{h^{(1)}, h^{(2)} \in \Pi(H)} \rho^{1/2}(h^{(1)}, h^{(2)}) \leq (2\pi)^{n/2}[\nu_{00}(0) + 2B(0)f_0]^{1/2}.$$

Inequality (4.6.4) now follows readily from (4.6.2) and (3.3.32). ∎

We set $\nu_{ss}^* = \sup_{\lambda \in R^n} \nu_{ss}(\lambda)$.

Lemma 4.6.2. Suppose that the quantities ν_{11}^*, ν_{00}^*, f_0 and μ_2 are finite. Then for any $t > (1 + 4n)^{1/2} a_1$, $a_1 = (2\pi)^{n/2}([\nu_{00}^* + 2f_0 B(0)]^{1/2} + (Hn^{1/2}/4e) \times (2 + \sqrt{2})[\nu_{11}^* + 2f_0\mu_2]^{1/2})|\Delta|^{-1/2}$,

$$F^{(1)}(t) \leq G^{(1)}(t) = 2.5\frac{a_1}{t} \exp\left\{-\frac{t^2}{2a_1^2} + 2n\right\}.$$

Proof. Clearly, $\rho^{(1)}(h^{(1)}, h^{(2)}) = |\Delta|^{-1}\rho_\Delta^{(1)}(h^{(1)}, h^{(2)})$, where $\rho_\Delta^{(1)}(h^{(1)}, h^{(2)})$ is defined by the equality (4.3.6). Therefore in view of the conditions of the lemma, we obtain a bound on the m.s. modulus of the field $X^{(1)}(h)$ analogous to (4.6.5):

$$\phi(h) \le [(2\pi)^n n(\nu_{11}^* + 2f_0\mu_2)]^{1/2}|\Delta|^{-1/2}h. \tag{4.6.6}$$

In the same manner,

$$\sup_{h^{(1)},h^{(2)}\in\Pi(H)} (\rho^{(1)}(h^{(1)}, h^{(2)})^{1/2} \le (2\pi)^{n/2}|\Delta|^{-1/2}[\nu_{00}^* + 2f_0 B(0)]^{1/2}. \tag{4.6.7}$$

The remainder of the proof is as in Lemma 4.6.1. ■

Denote

$$\nu_0 = \int_{R^{3n}} |f_4(\lambda^{(1)}, \lambda^{(2)}, \lambda^{(3)})|d\lambda^{(1)}d\lambda^{(2)}d\lambda^{(3)} < \infty,$$

$$\nu_1 = \int_{R^{3n}} |\lambda^{(1)}||f_4(\lambda^{(1)}, \lambda^{(2)}, \lambda^{(3)})|d\lambda^{(1)}d\lambda^{(2)}d\lambda^{(3)}.$$

The estimate (4.5.1) implies the following inequality to be utilized below:
$|\Delta \ominus (\Delta - \tau)| \le k|\Delta|^{1-1/n}|\tau|$, where one may choose $k = 2k_*$.

Lemma 4.6.3. Suppose the quantities f_0, p_2, ν_{00}^*, $\nu_{10}^* = \nu_{01}^*$, ν_{11}^*, ν_1 are finite and that the set Δ is convex. Then for any $t > (1 + 4n)^{1/2}$,

$$a_2 = (2\pi)^{n/2}(\nu_{00}^* + 2f_0 B(0))^{1/2}|\Delta|^{-1/2} + (Hn^{1/2}/4e)(2 + \sqrt{2})\times$$

$$\times[k^2(\nu_0 + 2B^2(0)) + 2k(2\pi)^{n/2}((\nu_1\nu_{01}^*)^{1/2} + 2B(0)(f_0\mu_2)^{1/2})|\Delta|^{-1/2+1/n}+$$

$$+(2\pi)^n(\nu_{11}^* + 2f_0\mu_2)|\Delta|^{-1+2/n}]^{1/2}$$

$$F^{(2)}(t) \le G^{(2)}(t) = 2.5\exp\left\{-\frac{t^2}{2a_2^2} + 2n\right\}.$$

Proof. Utilizing equality (4.3.11) and the notation of (4.3.13), where $\Pi(T-h)$ is replaced by $\Delta \cap (\Delta - h)$, we obtain

$$E|X^{(2)}(h^{(1)}) - X^{(2)}(h^{(2)})|^2 = e_1 + e_2; \quad e_1 = |\Delta|^{-2}\int_{R^{3n}} f_4(\lambda^{(1)} - \lambda^{(2)}, \lambda^{(2)}, \lambda^{(3)})\times$$

$$\times(e^{i\langle\lambda^{(2)},h^{(2)}\rangle}g(\lambda^{(1)}, h^{(2)}) - e^{i\langle\lambda^{(2)},h^{(1)}\rangle}g(\lambda^{(1)}, h^{(1)}))(e^{i\langle\lambda^{(3)},h^{(2)}\rangle}g(-\lambda^{(1)}, h^{(2)})-$$

$$-e^{i\langle\lambda^{(3)},h^{(1)}\rangle}g(-\lambda^{(1)}, h^{(1)}))d\lambda^{(1)}d\lambda^{(2)}d\lambda^{(3)};$$

$$e_2 = |\Delta|^{-2} \int_{R^{2n}} f(\lambda^{(1)}) f(\lambda^{(2)}) \{ [e^{i\langle \lambda^{(2)}, h^{(2)} \rangle} g(\lambda^{(1)} + \lambda^{(2)}, h^{(2)}) -$$

$$-e^{i\langle \lambda^{(2)}, h^{(1)} \rangle} g(\lambda^{(1)} + \lambda^{(2)}, h^{(1)})]^2 + [e^{i\langle \lambda^{(1)}, h^{(2)} \rangle} g(\lambda^{(1)} + \lambda^{(2)}, h^{(2)}) -$$

$$-e^{i\langle \lambda^{(1)}, h^{(1)} \rangle} g(\lambda^{(1)} + \lambda^{(2)}, h^{(1)})][e^{-i\langle \lambda^{(2)}, h^{(2)} \rangle} g(-\lambda^{(1)} - \lambda^{(2)}, h^{(2)}) -$$

$$-e^{-i\langle \lambda^{(2)}, h^{(1)} \rangle} g(-\lambda^{(1)} - \lambda^{(2)}, h^{(1)})]\} d\lambda^{(1)} d\lambda^{(2)}.$$

Note that

$$|e^{i\langle \lambda, h^{(2)} \rangle} g(\mu, h^{(2)}) - e^{i\langle \lambda, h^{(1)} \rangle} g(\mu, h^{(1)})| \leq$$

$$\leq |\Delta \ominus (\Delta - (h^{(1)} - h^{(2)}))| + |\lambda| |(h^{(1)} - h^{(2)}| |g(\mu, h^{(1)})|.$$

Therefore it follows from the assumptions of the lemma that

$$e_1 \leq \{ |\Delta|^{-2} \int_{R^{3n}} |f_4(\lambda^{(1)}, \lambda^{(2)}, \lambda^{(3)})| (k|\Delta|^{1-1/n} + |\lambda^{(2)}| |g(\lambda^{(1)} + \lambda^{(2)}, h^{(1)})|) \times$$

$$\times (k|\Delta|^{1-1/n} + |\lambda^{(3)}| |g(\lambda^{(1)} + \lambda^{(2)}, h^{(1)})|) d\lambda^{(1)} d\lambda^{(2)} d\lambda^{(3)} \} |h^{(1)} - h^{(2)}|^2. \quad (4.6.8)$$

It is easy to verify that

$$\int_{R^{3n}} |f_4| (|\lambda^{(2)}| + |\lambda^{(3)}|) |g| d\lambda^{(1)} d\lambda^{(2)} d\lambda^{(3)} \leq 2(\nu_1 \nu_{01}^*)^{1/2} (2\pi)^{n/2} |\Delta|^{1/2}. \quad (4.6.9)$$

The bound

$$\int_{R^{3n}} |f_4| |\lambda^{(2)}| + |\lambda^{(3)}| |g|^2 d\lambda^{(1)} d\lambda^{(2)} d\lambda^{(3)} \leq \nu_{11}^* (2\pi)^n |\Delta| \qquad (4.6.10)$$

has been utilized to derive (4.6.6). It follows from (4.6.8)–(4.6.10) that

$$e_1 \leq (k^2 \nu_0 |\Delta|^{-2/n} + 2k(\nu_1 \nu_{01}^*)^{1/2} (2\pi)^{n/2} |\Delta|^{-1/2-1/n} +$$

$$+ \nu_{11}^* (2\pi)^n |\Delta|^{-1}) n |h^{(2)} - h^{(1)}|_0^2.$$

In the same manner we obtain that

$$e_2 \leq 2|\Delta|^{-2} \int_{R^{2n}} f(\lambda^{(1)}) f(\lambda^{(2)}) [k|\Delta|^{1-1/n} + |\lambda^{(2)}| \times$$

$$\times |g(\lambda^{(1)} + \lambda^{(2)}, h^{(1)})|]^2 d\lambda^{(1)} d\lambda^{(2)} \le$$

$$\le [2k^2 B^2(0)|\Delta|^{-2/n} + 4k(2\pi)^{n/2} B(0)(f_0\mu_2)^{1/2}|\Delta|^{-1/2-1/n} +$$

$$+2(2\pi)^n f_0\mu_2|\Delta|^{-1}]n|h^{(1)} - h^{(2)}|_0^2.$$

Since the bound (4.6.7) is valid for the quantity $\quad \sup\limits_{h^{(1)},h^{(2)} \in K} (\rho^{(2)}(h^{(1)}, h^{(2)}))^{1/2},$

we arrive at the assertion of the lemma. ∎

Theorem 4.6.3. Suppose (4.6.1) holds. Then for any $t > 0$ stipulated by the conditions of Lemmas 4.6.1–4.6.3,

$$\varlimsup_{\Delta \to \infty} \frac{F_\Delta^{(i)}(t)}{G(t)} \le 1, \quad \varlimsup_{m \to \infty} \frac{F_m^{(i)}(t)}{G^{(i)}(t)} \le 1, \quad i = 1, 2, \qquad (4.6.11)$$

The proof is obvious.

Note that in the case of a Gaussian field $\xi(x)$, the density function $f_4 \equiv 0$ and the constants a, a_1 and a_2 are considerably simplified and depend on the quantities $k, |\Delta|, B(0), f_0$ and μ_2. The last three are estimated statistically and the constant k may be directly computed for a given set Δ. However, if $\xi(x)$ is a Gaussian field, to obtain relations analogous to (4.6.11), one can use different arguments which result in more accurate bounds.

Assume that $\xi(x)$, $x \in \mathbf{R}^n$, is a homogeneous and isotropic a.s. continuous Gaussian random field. Consider the estimator $\hat{B}_\Delta(h, l)$, $h \in [0, H]$, of the cor.f. $B(h)$, $h \in [0, H]$, of the field $\xi(x)$ defined by equality (4.4.1). In this case, under the conditions stipulated in Corollary 4.4.1, the measures $\mu_\Delta(l)$ corresponding to the random processes $X_\Delta(h, l) = |\Delta|^{1/2}(\hat{B}_\Delta(h, l) - B(h))$, $h \in [0, H]$, converge weakly to a measure μ on $C[0, H]$ corresponding to an a.s. continuous Gaussian process $X(h)$, $h \in [0, H]$. The cor.f. of the process $X(h)$ is of the form

$$\rho(h_1, h_2) = 2^{n+1}\pi^n \int_{\mathbf{R}^n} \cos(h_1\langle\lambda, l\rangle) \cos(h_2\langle\lambda, l\rangle) f^2(\lambda) d\lambda =$$

$$= (2\pi)^n |s(1)| \int_0^\infty (Y_n(\nu|h_1 - h_2|) + Y_n(\nu|h_1 + h_2|))\nu^{n-1} g^2(\nu) d\nu.$$

Note that

$$E|X(h_1) - X(h_2)|^2 = 2^{n+1}\pi^n \int_{\mathbf{R}^n} |\cos(h_1\langle\lambda, l\rangle) - \cos(h_2\langle\lambda, l\rangle)|^2 f^2(\lambda) d\lambda \le$$

$$\leq 2^{n+3}\pi^n f_0 \int_{\mathbf{R}^n} \sin^2((h_1 - h_2)\langle \lambda, l \rangle/2) f(\lambda) d\lambda =$$

$$= 2^{n+1}\pi^n f_0 E|\xi(h_1 l) - \xi(h_2 l)|^2; \tag{4.6.12}$$

$$E|X(h)|^2 \leq 2^{n+1}\pi^n f_0 B(0) = b^2. \tag{4.6.13}$$

Inequality (4.6.12) permits us to apply Theorem 4.6.2 to prove the following assertion.

Theorem 4.6.4. Let $f_0 < \infty$ and let $B(h)$ be a twice differentiable function. Then for each $t > 0$

$$\overline{\lim_{\Delta \to \infty}} P\left\{ \sup_{h \in [0,H]} |X_\Delta(h,l)| > t \right\}/\tilde{G}(t) \leq 1,$$

$$\tilde{G}(t) = \left(2 + \frac{H}{\pi}\left(\frac{B''(0)}{B(0)}\right)^{1/2}\right)e^{-t^2/2b^2} +$$

$$+ \left(\frac{1}{\sqrt{2}} + \frac{1}{\sqrt{2\pi}}\left(\frac{B''(0)}{B(0)}\right)^{1/2}\right)e^{-t^2/4b^2}.$$

Proof. Suppose that in Theorem 4.6.2, $x(h) = X(h)$, $z(h) = 2^{(n+1)/2}\pi^{n/2} \times f_0^{1/2}\xi(hl)$, $h \in [0,H]$, $a(h) \equiv t$. Since (4.6.13) implies $E|z(h)|^2 = b^2 \geq E|X(h)|^2$, we have in (4.6.3),

$$\alpha^2 = b^2 - \inf_{h \in [0,H]} E|X(h)|^2, \quad g^2(h) = E|X(h)|^2 - \inf_{h \in [0,H]} E|X(h)|^2.$$

Set $g_+^2 = \sup_{h \in [0,H]} g^2(h) \leq b^2$.

Utilizing simple estimates, we obtain

$$P\left\{ \sup_{h \in [0,H]} |X(h)| > t \right\} \leq 2P\left\{ \sup_{h \in [0,H]} (z(h) - \gamma g(h)) > t \right\} \leq$$

$$\leq 2P\{ \sup_{h \in [0,H]} z(h) > t\} + \left(\frac{2}{\pi}\right)^{1/2} \int_0^\infty P\left\{ \sup_{h \in [0,H]} z(h) > \right.$$

$$\left. > t - ug_+ \right\}e^{-u^2/2}du \leq 2P\left\{ \sup_{h \in [0,H]} z(h) > t \right\} +$$

$$+ \left(\frac{2}{\pi}\right)^{1/2} \int_0^{t/b} P\left\{ \sup_{h \in [0,A]} z(h) > t - ub \right\}e^{-u^2/2}du + \left(\frac{2}{\pi}\right)^{1/2} \int_{t/b}^\infty e^{-u^2/2}du. \tag{4.6.14}$$

Let $U_w[0, H]$ be the number of upcrossings of a level $w \geq 0$ for the stationary Gaussian process $b^{-1}z(h) = \xi(hl)/B(0))^{1/2}$ on the interval $0 \leq h \leq H$. The cor.f. of the process $\xi(hl)/(B(0))^{1/2}$ equals $B(h)/B(0)$. Therefore by the Rice formula [161],

$$P\left\{ \sup_{h\in[0,H]} z(h) > v\right\} \leq P\{z(0) > v\} + P\{U_{v/b}[0, H] \geq 1\} \leq$$

$$\leq (2\pi)^{-1/2} \int_{v/b}^{\infty} e^{-u^2/2}du + EU_{v/b}[0, H] = (2\pi)^{-1/2} \int_{v/b}^{\infty} e^{-u^2/2}du +$$

$$+ \frac{H}{2\pi}\left(\frac{B''(0)}{B(0)}\right)^{1/2} e^{-v^2/2b^2}. \tag{4.6.15}$$

Note that it is easy to verify the inequality

$$(2\pi)^{-1/2} \int_{\tau}^{\infty} e^{-u^2/2}du \leq \tfrac{1}{2}e^{-\tau^2/2}, \tag{4.6.16}$$

valid for all $\tau \geq 0$, which is clearly less accurate than (3.3.34) for large τ. From (4.6.15) and (4.6.16) we obtain the bound

$$2P\left\{ \sup_{h\in[0,H]} z(h) > t\right\} + \left(\frac{2}{\pi}\right)^{1/2} \int_{t/b}^{\infty} e^{-u^2/2}du \leq$$

$$\leq \left(2 + \frac{H}{\pi}\left(\frac{B''(0)}{B(0)}\right)^{1/2}\right)e^{-t^2/2b^2}. \tag{4.6.17}$$

Proceeding analogously we obtain

$$\left(\frac{2}{\pi}\right)^{1/2} \int_0^{t/b} P\left\{ \sup_{h\in[0,H]} z(h) > t - ub\right\}e^{-u^2/2}du \leq$$

$$\leq \left(\frac{2}{\pi}\right)^{1/2}\left[\tfrac{1}{2} + \frac{H}{\pi}\left(\frac{B''(0)}{B(0)}\right)^{1/2}\right]\int_0^{t/b} e^{-(t-ub)^2/2b^2 - u^2/2}du \leq$$

$$\leq \left[\frac{1}{\sqrt{2}} + \frac{H}{\pi\sqrt{2}}\left(\frac{B''(0)}{B(0)}\right)^{1/2}\right]e^{-t^2/4b^2}. \tag{4.6.18}$$

The estimates (4.6.14), (4.6.17) and (4.6.18) prove the theorem. ∎

Note that for $n \geq 5$, the differentiability of $B(h)$ in the statement of Theorem 4.6.4 may be dropped since in this case, the 2nd derivative of $B(h)$ does exist (cf. item 2 of Lemma 1.2.1).

References

1. Z.S. Antoshevsky, 'On the Asymptotics of an Estimate for the Correlation Function of a Stationary Gaussian Sequence', *Litovsk. Mat. Sb.* **16**:4 (1976),21–26 (Russian).

2. O.P. Banina and E.I. Ostrovsky, 'Invariant Statistics and Confidence Intervals for the Correlation Function of a Gaussian Process Relative to Different Metrics', In: *Proc. All-Union Symp. on Statist. Rand. Proc., Math Inst. of the Acad. Sci. of the Ukrain. SSR and Kiev State University,*(1975),148–149 (Russian).

3. T.A. Bardadym and A.V. Ivanov, 'An Asymptotic Expansion Related to an Empirical Regression Function', *Teor. Veroyatnost. i Mat. Statist.* **30** (1984),8–15 (Russian).

4. T.A. Bardadym and A.V. Ivanov, 'Asymptotic Expansions Related to the Estimation of the Error Variance of Observations in the "Signal plus Noise" Model', *Teor. Veroyatnost. i Mat. Statist.* **33**(1985),11–20 (Russian).

5. Yu.K. Belayev and V.I. Piterbarg, 'Asymptotics of the Mean Number of A-points of Excursions above a High Level for a Gaussian Field', In: *Topics on Random Fields,*Vol.29,(1972), Moscow State University, 62–89 (Russian).

6. V. Bentkus and D. Surgailis, 'On Certain Classes of Automodel Random Fields', *Litovsk. Mat. Sb.* **21**:2(1981),53–66 (Russian).

7. R.Yu. Bentkus, 'On the Error of an Estimate of the Correlation Function of a Stationary Process', *Litovsk. Mat. Sb.* **12**:1(1972),55–71 (Russian).

8. R. Yu. Bentkus, 'On Cumulants of the Spectrum of a Stationary Sequence', *Litovsk. Mat. Sb.* **16**:4(1976),37–61 (Russian).

9. R.Yu.Bentkus and I.G. Zhurbenko, 'Asymptotic Normality of Spectral Estimates', *Dokl. Akad. Nauk SSSR* **229**:1(1976),11–16 (Russian).

10. R. Bentkus and R. Rutkauskas, 'On the Asymptotics of the First Two Moments of the Second Order Spectral Estimates', *Litovsk. Mat. Sb.* **13**:1 (1973),29–45 (Russian).

11. R. Benktus, R. Rudskis and Yu. Sušinskas, 'On the Mean Value of Spectral Estimates of a Homogeneous Random Field", *Litovsk. Mat. Sb.* **14**:3 (1974),67–72 (Russian).

12. S.N. Bernstein, 'An Extension of a Limit Theorem in Probability Theory to Sums of Dependent Variables', *Uspekhi Mat. Nauk* **10**(1944),65–114 (Russian).

13. A.A. Borovkov, 'Convergence of Measures and Random Processes', *Uspekhi Mat. Nauk* **31**:2(1976),3–68 (Russian).

14. A.V. Borshchevsky and A.V. Ivanov, 'On the Normal Approximation of the Distribution of the Optimum Point in a Problem of Data Processing by the Least Moduli Method', *Kibernetika* **6**(1985),86–92 (Russian).

15. V.V. Buldygin, 'Limit Theorems in Functional Spaces and a Problem in Random Processes Statistics', In: *Probability Methods of Infinite-Dimensional Analysis*, Math. Inst. of the Acad. Sci. of the Ukrain. SSR, Kiev, (1980), 24–36 (Russian).

16. V.V. Buldygin, 'On a Comparison Inequality for the Distribution of the Gaussian Process Maximum', *Teor. Veroyatnost. i Mat. Statist.* **28**(1983), 9–14 (Russian).

17. V.V. Buldygin and E.V. Ilarionov, 'On a Problem of Random Fields Statistics', In: *Probabilistic Infinite-Dimensional Analysis*, Math. Inst. of the Acad. Sci. of the Ukrain. SSR, Kiev, (1981),6–14 (Russian).

18. A.V. Bulinsky, *Limit Theorems for Random Processes and Fields*, Izdat. MGU, Moscow, 1981 (Russian).

19. A.V. Bulinsky, 'On Dependence Measures Close to the Maximal Correlation Coefficient', *Dokl. Akad. Nauk SSSR* **275**:4(1984),789–792 (Russian).

20. A.V. Bulinsky and I.G. Zhurbenko, 'A Central Limit Theorem for Additive Random Functions', *Teor. Veroyatnost. i Primenen.* **21**:4(1976),707–717 (Russian). English transl. in: *Theory Probab. Appl.* **21**,687–697.

21. A.V Bulinsky and I.G. Zhurbenko, 'A Central Limit Theorem for Random Fields', *Dokl. Akad. Nauk SSSR* **226**:1(1976),23–25 (Russian).

22. M.V. Burnashev, 'Asymptotic Expansions of Estimates of a Parameter of a Signal in the White Gaussian Noise', *Mat Sb.* **104**:2(1977),179–206 (Russian).

23. M.V. Burnashev, 'The Investigation of the Second Order Properties of Statistical Estimates in the Scheme of Independent Observations', *Izv. Akad. Nauk SSSR, Ser. Mat.* **45**:3(1981),509–539 (Russian).

24. N.Ya. Vilenkin, *Special Functions and the Representation Theory of Groups*,

Nauka,Moscow,1965 (Russian).

25. V.A. Volkonsky and Yu.A. Rozanov, 'Some Limit Theorems for Random Functions' 1, *Teor. Veroyatnost. i Primenen.* 4:2(1959),186–207 (Russian). English transl. in: *Theory Probab. Appl.* 4,178–197.

26. V.F. Gaposhkin, 'Criteria for the Strong Law of Large Numbers for Classes of Second Order Stationary Processes and Homogeneous Random Fields', *Teor. Veroyatnost. i Primenen.* 22:2(1977),295–319 (Russian). English transl. in: *Theory Probab. Appl.* 22,286–310.

27. I.M. Gelfand and N.Ya. Vilenkin, *Generalized Functions. Some Applications of Harmonic Analysis. Equipped Hilbert Spaces*, Vol.4,Fizmatgiz, Moscow,1961.

28. L.O. Giraitis, 'On Convergence of Non-Linear Transformations of a Gaussian Sequence to Automodel Processes', *Litovsk. Mat. Sb.* 23:1(1983),58–68 (Russian).

29. L.O. Giraitis, 'Limit Theorems for Local Functionals', In: *Proc. XXIV Litovsk. Math. Soc.*, Vilnius,1983,52–53 (Russian).

30. I.I. Gikhman and A.V. Skorokhod, *Theory of Random Processes* Vol 1, Nauka, Moscow, 1971 (Russian). (English transl.: Springer-Verlag, 1974).

31. B.V. Gnedenko and A.N. Kolmogorov, Limit Distributions for Sums of Independent Random Variables, Gostekh Izdat, Leningrad–Moscow, 1949 (Russian). (English transl.: Addison Wesley,1954).

32. V.V. Gorodetskii, 'On the Strong Mixing Property for Linear Sequences', *Teor. Veroyatnost. i Primenen.* 22:2(1977),421–423 (Russian). English transl. in: *Theory Probab. Appl.* 22,411–413.

33. V.V. Gorodetskii, 'On Convergence to Semistable Gaussian Processes', *Teor. Veroyatnost. i Primenen.* 22:3(1977),513–522; corrigenda, *ibid.* 24:2 (1979),444 (Russian). English transl. in: *Theory Probab. Appl.* 22,498–508.

34. V.V. Gorodetskii, 'Invariance Principle for Functions of Jointly Stationary Gaussian Variables', *Zap. Nauchn. Sem. LOMI* 50(1980),32–34 (Russian).

35. V.V. Gorodetskii, 'The Invariance Principle for Stationary Random Fields with a Strong Mixing Condition', *Teor. Veroyatnost. i Primenen.* 27:2 (1982),358–364 (Russian). English transl. in: *Theory Probab. Appl.* 27,380–385.

36. V.V. Gorodetskii, 'Moment Inequalities and Central Limit Theorem for Integrals of Mixing Random Fields', *Zap. Nauchn. Sem. LOMI* 142(1985), 39–47 (Russian).

37. I.S. Gradstein and I.M. Ryzhik, *Tables of Integrals, Sums, Series and Prod-*

ucts, Fizmatgiz, Moscow, 1962 (Russian).

38. V.A. Gurevich, *The Least Moduli Method for the Non-Linear Regression Model*. In:*Applied Statistics*, Nauka, Moscow, 1983 (Russian).

39. S.I. Gusev, 'Asymptotic Expansions associated with some Statistical Estimators in the Smooth Case', *Teor. Veroyatnost. i Primenen.* I **20**:3 (1975),488–514; II **21**:1(1976),16–33 (Russian). English transl. in: *Theory Probab. Appl.* I **20**,470–498; II ibid: **21**,14–33.

40. E. Gečauskas, 'Distribution Function of the Distance between Two Points inside an Ovaloid', *Litovsk. Mat. Sb.* **6**:2(1966),245–248 (Russian).

41. Yu.A. Davydov, 'Convergence of Distributions Generated by Stationary Stochastic Processes', *Teor. Veroyatnost. i Primenen.* **13**:4(1968),730–737 (Russian). English transl. in: *Theory Probab. Appl.* **13**,691–696.

42. Yu.A. Davydov, 'The Invariance Principle for Stationary Processes', *Teor. Veroyatnost. i Primenen.* **15**:3(1970),498–509 Russian. English transl. in: *Theory Probab. Appl.* **15**,487–498.

43. R.L. Dobrushin, 'The Description of a Random Field by Means of Conditional Probabilities and Conditions of its Regularity', *Teor. Veroyatnost. i Primenen.* **13**:2(1968),201–229 (Russian). English transl. in: *Theory Probab. Appl.* **13**,197–224.

44. A.Ya. Dorogovtsev, *Theory of Parameter Estimation of Random Processes*, Vyshcha Shkola,Kiev,1982 (Russian).

45. A.A. Dykhovichny, 'On the Estimation of the Correlation Function of a Homogeneous Isotropic Gaussian Field', *Teor. Veroyatnost. i Mat. Statist.* **29**(1983),37–40 (Russian).

46. A.A. Dykhovichny, 'On Estimates of the Correlation and Spectral Functions of a Homogeneous Isotropic Random Field', *Teor. Veroyatnost. i Mat. Statist.* **32**(1985),17–27 (Russian).

47. I.G. Zhurbenko, *Spectral Analysis of Time Series*, Izdat. MGU,Moscow, 1982 (Russian).

48. I.G. Zhurbenko, *Analysis of Stationary and Homogeneous Systems*, Izdat. MGU,Moscow,1987 (Russian).

49. I.G. Zhurbenko and N.N. Trush, 'Spectral Analysis of Random Stationary Processes', *Vest. BGU*, 1981 No.1, 58–61 (Russian).

50. V.M. Zolotaryov, *One-dimensional Stable Distributions*, Nauka,Moscow, 1983 (Russian).

51. I.A. Ibragimov and Yu.V. Linnik, *Independent and Stationary Sequences of Random Variables*, Nauka, Moscow,1965 (Russian). Transl. Wolters-

Noordhoff, Groningen, 1971)

52. I.A. Ibragimov and Yu.A.Rozanov, *Gaussian Random Processes*, Nauka, Moscow,1970 (Russian).

53. I.A. Ibragimov and R.Z. Hasminsky, *Asymptotic Theory of Estimation*, Nauka,Moscow,1979 (Russian). English transl. Springer-Verlag,1984.

54. A.V. Ivanov, *Statistical Properties of Parameter Estimates in Non-linear Regression Models*, Candidate of Sciences Thesis, Kiev, 1973 (Russian).

55. A.V. Ivanov, 'Berry-Esseen Inequality for the Distribution of the Least Squares Estimates of Parameters of a Non-linear Regression Function', *Mat. Zam.* **20**:2(1976),293–303 (Russian).

56. A.V. Ivanov, 'An Asymptotic Expansion for the Distribution of the Least Squares Estimator of a Parameter of the Non-linear Regression Function', *Teor. Veroyatnost. i Primenen.* **21**:3(1976),571–583 (Russian). English transl. in: *Theory Probab. Appl.* **21**,557–570.

57. A.V. Ivanov, 'A Solution of the Problem of Detection of Hidden Periodicity', *Teor. Veroyatnost. i Mat. Statist.* **20**(1979),44–59 (Russian).

58. A.V. Ivanov, 'Asymptotic Expansion of Moments of the Least Squares Estimate for a Vector Parameter of Non-linear Regression', *Ukrain. Mat. Zh.***34**:2(1982)164–170 (Russian).

59. A.V. Ivanov, 'Two Theorems on the Consistency of the Least Squares Estimate', *Teor. Veroyatnost. i Mat. Statist.* **28**(1983),21–31 (Russian).

60. A.V. Ivanov, 'On the Consistency and Asymptotic Normality of the Least Moduli Estimate', *Ukrain. Mat. Zh.***36**:3(1984),297–303 (Russian).

61. A.V. Ivanov and N.N. Leonenko, 'On the Convergence of the Distributions of Functionals of Correlation Function Estimates', *Litovsk. Mat. Sb.***18** (1978),35–44 (Russian).

62. A.V. Ivanov and N.N. Leonenko, 'On the Invariance Principle for an Estimate of the Correlation Function of a Homogeneous Random Field', *Ukrain. Mat. Zh.* **32**:3(1980),323–331 (Russian).

63. A.V. Ivanov and N.N. Leonenko, 'On the Invariance Principle for an Estimate of the Correlation Function of a Homogeneous Isotropic Random Field', *Ukrain. Mat. Zh.* **33**:3(1981),313–323 (Russian).

64. A.V. Ivanov and N.N. Leonenko, 'Invariance Principle for an Estimate of the Correlation Function of a Homogeneous Random Field', *Dokl. Akad. Nauk Ukrain. SSR Ser. A*, 1(1981),17–20 (Russian).

65. A.V. Ivanov and N.N. Leonenko, 'On the Invariance Principle for Quadric Functionals of Random Fields', In: *Mathematical Models of Statistical*

Physics, Tyumen,1982,49–54 (Russian).

66. A.V. Ivanov and S. Zwanzig, 'Asymptotic Expansion for the Distribution of the Least Squares Estimate of a Vector Parameter of Non-linear Regression', *Dokl. Akad. Nauk SSSR* **256**:4(1981),784–787 (Russian).

67. A.V. Ivanov and S. Zwanzig, 'Asymptotic Expansion of the Least Squares Estimate of a Vector Parameter of Non-linear Regreesion', *Teor. Veroyatnost. i Mat. Statist.* **26**(1982),41–48 (Russian).

68. S.G. Kalandarashvili, 'On the Consistency of Estimates for Homogeneous Isotropic Random Fields', *Soobshcheniya Akad. Nauk GSSR* **82**:2(1976), 273–276 (Russian).

69. P.S. Knopov, 'On Optimal Estimates of Certain Characteristics of Isotropic Gaussian Random Fields', In: *Theory of Optimal Decisions*, Izdat. Ukrain. Akad. Nauk USSR, Kiev,1969,134–140 (Russian).

70. P.S. Knopov, *Optimal Parameters Estimates for Stochastic Systems*, Naukova Dumka, Kiev,1981 (Russian).

71. T.V. Kovalchuk, 'On the Distributions Convergence for the Functional of a Correlogram of a Homogeneous Random Field with an Unknown Mean', *Teor. Veroyatnost. i Mat. Statist.* **35**(1986),44–51 (Russian).

72. Yu.V. Kozachenko, 'Local Properties of Sample Paths of Certain Random Functions', *Ukrain. Mat. Zh.***19**:2(1967),109–116 (Russian).

73. A.N. Kolmogorov, *Basic Concepts of Probability Theory*, ONTI, Moscow-Leningrad, 1936 (Russian).

74. A.N. Kolmogorov, 'The Wiener Helix and Other Interesting Curves in Hilbert Space', *Dokl. Akad. Nauk SSSR* **26**:2(1940),115–118 (Russian).

75. V.S. Korolyuk, N.I. Portenko, A.V. Skorokhod and A.F. Turbin, *Handbook on Probability Theory and Mathematical Statistics*, Naukova Dumka, Kiev, 1978 (Russian).

76. Yu.A. Kutoyants, 'Estimation of Parameters of Random Processes', *Akad Armyan.Nauk SSR Dokl.*, 1980 (Russian).

77. N.N. Leonenko, 'Central Limit Theorem for Homogeneous Random Fields and Asymptotic Normality of Estimates of Regression Coefficients', *Dopovidi Akad. Nauk Ukrain. SSR*, Ser. A,8(1974),699–702 (Ukrainian).

78. N.N. Leonenko, 'Central Limit Theorem for Random Fields with a Weighting Function', *Kibernetika*, 5(1975),153–155 (Russian).

79. N.N. Leonenko, 'On the Convergence Rate in the Central Limit Theorem for m-Dependent Random Fields', *Mat. Zametki* **17**:1(1975),129–132 (Russian).

80. N.N. Leonenko, 'Limit Theorems for Additive Random Functions', In: *Studies in the Theory of Random Processes*, Math. Inst. of the Acad. Sci. of the Ukrain. SSR, Kiev, 1976,94–105 (Russian).

81. N.N. Leonenko, 'On the Convergence of Probability Measures Generated by Homogeneous Random Fields', *Dokl. Akad. Nauk Ukrain. SSR Ser. A*,11(1978),970–971 (Russian).

82. N.N. Leonenko, 'On Estimation of Linear Regression Coefficients for a Homogeneous Random Field', *Ukrain. Mat. Zh.* 30:6(1978), 749–756 (Russian).

83. N.N. Leonenko, 'On Measures of the Excess over a Level for a Gaussian Isotropic Random Field', *Teor. Veroyatnost. i Mat. Statist.* 31(1984),64–82 (Russian).

84. N.N. Leonenko, 'Reduction Conditions for Measures of the Excess above a Moving Level for Strongly Dependent Homogeneous Isotropic Gaussian Fields', *Teor. Veroyatnost. i Primenen.* 29:3(1984),611–612 (Russian). English transl. in: *Theory Probab. Appl.* 29.

85. N.N. Leonenko, 'Limit Distributions for Characteristics of the Excess above a Level for a Gaussian Random Field', *Mat. Zametki* 41:4(1987),608–618 (Russian).

86. N.N. Leonenko and K.V. Rybasov, 'Conditions for the Convergence of Spherical Averages of Local Functionals of Gaussian Fields to a Wiener Process', *Teor. Veroyatnost. i Mat. Statist.* 34(1986),85–93 (Russian).

87. N.N. Leonenko and A.H. El Bassioni, 'Limit Theorems for Certain Characteristics of the Excess above a Level for a Strongly Dependent Gaussian Field' *Teor. Veroyatnost. i Mat. Statist.* 35(1986),54–60 (Russian).

88. N.N. Leonenko and M.I. Yadrenko, 'On Asymptotic Normality of the Least Squares Estimates of Regression Coefficients for Homogeneous Isotropic Random Fields', *Kibernetika* 2(1977), 108–112 (Russian).

89. N.N. Leonenko and M.I. Yadrenko, 'On the Invariance Principle for Certain Classes of Random Fields', *Ukrain. Mat. Zh.* 31:5(1979), 559–565 (Russian).

90. N.N. Leonenko and M.I. Yadrenko, 'Limit Theorems for Homogeneous and Isotropic Random Fields', *Teor. Veroyatnost. i Mat. Statist.* 21(1979),97–109 (Russian).

91. N.N. Leonenko and M.I. Yadrenko, 'On the Invariance Principle for Homogeneous and Isotropic Random Fields', *Teor. Veroyatnost. i Primenen.* 24:1(1979),175–181 (Russian). English transl. in: *Theory Probab. Appl.* 24,

175-181.

92. V.P. Leonov, *Some Applications of Higher Cumulants to the Theory of Stationary Random Processes*, Nauka,Moscow,1964 (Russian).

93. Yu.V. Linnik and N.M. Mitrofanova, 'On the Asymptotic Distribution of the Least Squares Estimates', *Dokl. Akad. Nauk SSSR* **149**:3(1963), 518–520 (Russian).

94. V.A. Malyshev, 'Central Limit Theorem for Hilbert Random Fields', *Dokl. Akad. Nauk SSSR* **224**:1(1975), 35–38 (Russian).

95. V.A. Malyshev, 'Cluster Decompositions in Models of Statistical Physics and Quantum Field Theory', *Uspekhi Mat. Nauk* **35**:2(1980), 3–53 (Russian).

96. N.M. Mitrofanova, 'An Asymptotic Expansion for the Maximum Likelihood Estimate of a Vector Parameter', *Teor. Veroyatnost. i Primenen.* **12**:3(1967), 418–425 (Russian). English transl. in: *Theory Probab. Appl.* **12**, 364–372.

97. S.A. Molchanov and A.K. Stepanov, 'V.A. Malyshev's Bound for Moments of Wick's Polynomials and Some of its Applications', In: *Random Processes and Fields*, MGU, Moscow,1979, 39–48 (Russian).

98. S.A. Molchanov and A.K. Stepanov, 'Percolation of a Gaussian Field through a High Level', *Dokl. Akad. Nauk SSSR* **20**:6(1979), 1177–1281 (Russian).

99. A.S. Monin and A.M.Yaglom, *Statistical Fluid Mechanics* I, Nauka, Moscow,1965; II, Nauka, Moscow,1967 (Russian).

100. B.S. Nakhapetyan, 'Central Limit Theorem for Random Fields Meeting the Strong Mixing Condition', In: *Multicomponent Random Systems*, Nauka, Moscow,1978, 276–288 (Russian).

101. V.P. Nosko, *On Bursts of Gaussian Random Fields*, Vestnik MGU, Ser. Mat., Mech., 1970, No.4, 18–22 (Russian).

102. V.P. Nosko, 'The Horizon of a Random Field of Cones on a Plane. The Mean Number of the Horizon Corners', *Teor. Veroyatnost. i Primenen.* **24**:2(1982), 259–265 (Russian). English transl. in: *Theory Probab. Appl.* **24**, 269–279.

103. V.P. Nosko, 'Weak Convergence of the Horizon of a Random Field of Cones in an Expanding Strip', *Teor. Veroyatnost. i Primenen.* **24**:4(1982), 693–706 (Russian). English transl. in: *Theory Probab. Appl.* **24**, 688–702.

104. I.N. Pak, 'On Sums of Trigonometric Series', *Uspekhi Mat. Nauk* **35**:2 (1980), 91–144 (Russian).

105. S. Panchev, *Random Functions and Turbulence*, Gidrometizdat, Leningrad, 1967 (Russian).

106. V.V. Petrov, *Sums of Independent Random Variables*, Nauka, Moscow, 1971 (Russian). English transl. Springer, New York, 1975.

107. V.I. Piterbarg, 'Mixing of Sampled Random Processes', In: *Studies in Random Fields*, MGU, Moscow Vol. 50, 1974, 59–77 (Russian).

108. V.I. Piterbarg, 'Gaussian Random Processes. Advances in Science and Technology. Probability Theory. Math. Statistics. Theoret. Cybernetics', **19**(1982), 155–199 (Russian).

109. A. Plikusas, 'Cumulant Estimates and Large Deviations for certain Nonlinear Transformations of a Stationary Gaussian Process', *Litovsk. Mat. Sb.* 2(1980), 119–128 (Russian).

110. Yu.V. Prokhorov, 'The Convergence of Random Processes and Limit Theorems of Probability Theory', *Teor. Veroyatnost. i Primenen.* 1:2(1956), 177–238 (Russian). English transl. in: *Theory Probab. Appl.* 1, 187–214.

111. G.H. Rakhimov, 'On the Estimation of the Correlation Function of a Homogeneous Isotropic Field on a Sphere' I,II *Teor. Veroyatnost. i Mat. Statist.* **24**(1981), 125–127; **28**(1983), 112–115 (Russian).

112. Yu.A. Rozanov, *Stationary Random Processes*, Fizmatgiz, Moscow, 1963, (Russian).

113. Ya.A. Rudzit, *Microgeometry and Contact Interaction of Surfaces*, Zinatne, Riga, 1975 (Russian).

114. K.V. Rybasov, 'Spherical Measures of the Excess above a level for a Gaussian Field', *Fourth Internat. Conf. on Probab. Theory and Math. Statist.* Proc. Vilnius Acad. Sci. of the Lithuan. SSR, 1985, 83–84 (Russian).

115. K.V. Rybasov, 'On the Asymptotic Normality of a Functional of a Homogeneous Isotropic Random Gaussian Field', *Teor. Veroyatnost. i Mat. Statist.* **37**(1987), 111–117 (Russian).

116. S.M. Rytov, Yu.A. Kravtsov and V.I. Tatarsky, *Introduction to Statistical Radiophysics* Vol.II. Random Fields, Nauka, Moscow, 1978 (Russian).

117. S.M. Sadikova, 'Some Inequalities for Characteristic Functions', *Teor. Veroyatnost. i Primenen.* **11**:3(1966), 500–506 (Russian). English transl. in: *Theory Probab. Appl.* 11, 441–447.

118. Ya.G. Sinay, *Theory of Phase Transitions*, Nauka, Moscow, 1980 (Russian).

119. V.A. Statulevičius, 'Theorems on Large Deviations for sums of Dependent Random Variables' I *Litovsk. Mat. Sb.***19**:2(1979), 199–208 (Russian).

120. D. Surgailis, 'On the Convergence of Sums of Non-linear Functions of Mov-

ing Averages to Automodel Process', *Dokl. Akad. Nauk SSSR* **2**:7(1981), 51–54 (Russian).

121. D. Surgailis, 'On Attraction Zones for Automodel Multiple Integrals', *Litovsk. Mat. Sb.***22**:3(1982), 185–201 (Russian).

122. V.I. Tatarsky, *Wave Propagation in Turbulent Atmosphere*, Nauka, Moscow, 1967 (Russian).

123. A.A. Tempelman, *Ergodic Theorems on Groups*, Mokslas, Vilnius, 1986 (Russian).

124. A.N. Tikhomirov, 'On the Convergence Rate in the Central Limit Theorem for Weakly Dependent Random Variables', *Teor. Veroyatnost. i Primenen.* **25**:4(1980), 800–818 (Russian). English transl. in: *Theory Probab. Appl.* **25**, 790–809.

125. S.A. Utev, 'Inequalities for Sums of Weakly Dependent Random Variables and Estimates for Convergence Rate in the Invariance Principle ', In: *Limit Theorems for Sums of Random Variables*, Trudy.Inst.Mat., Vol.3,50–57 (Russian).

126. A. Khalilov, 'Central Limit Theorem for Random Fields in Continuous Time', *Teor. Veroyatnost. i Mat. Statist.* **26**(1982), 151–155 (Russian).

127. A.S. Kholevo, 'On Estimates of Regression Coefficients", *Teor. Veroyatnost. i Primenen.* **14**:1(1969), 78–101 (Russian). English transl. in: *Theory Probab. Appl.* **11**, 79–104.

128. A.S. Kholevo, 'On the Asymptotic Efficiency of Pseudo-optimal Estimates', *Teor. Veroyatnost. i Primenen.* **16**:3(1971), 524–534 (Russian). English transl. in: *Theory Probab. Appl.* **16**, 516–527.

129. A.S. Kholevo, 'On the Asymptotic Normality of Estimates of Regression Coefficients', *Teor. Veroyatnost. i Primenen.* **16**:4(1971), 724–728 (Russian). English transl. in: *Theory Probab. Appl.* **16**, 707–711.

130. A.P. Husu, Yu.R. Viterberg and V.A. Palmov, *The Roughness of Surfaces*, Nauka, Moscow, 1975 (Russian).

131. N.N. Chentsov, *Validation of Statistical Criteria by Methods of the Theory of Random Processes*, Math. Inst. Acad. Sci. of the USSR, Moscow,1958 (Russian).

132. N.N. Chentsov, 'Limit Theorems for certain Classes of Random Functions', Trans. of the All-Union Workshop on Probab. Theory and Math. Statist., Yerevan, 1960, 280–284 (Russian).

133. D.M. Chibisov, 'Asymptotic Expansion for Maximum Likelihood Estimates', *Teor. Veroyatnost. i Primenen.* **17**:2(1972), 387–388 (Russian).

English transl. in: *Theory Probab. Appl.* **17**, 368–369.

134. D.M. Chibisov, 'An Asymptotic Expansion for a Class of Estimators Containing Maximum Likelihood Estimators', *Teor. Veroyatnost. i Primenen.* **18**:2(1973), 303–311 (Russian). English transl. in: *Theory Probab. Appl.* **18**, 295–303.

135. A.N. Shiryaev, 'Some Problems in the Spectral Theory of Higher Moments', I *Teor. Veroyatnost. i Primenen.* **5**:3(1960), 293–313 (Russian). English transl. in: *Theory Probab. Appl.* **5**, 265–284.

136. V.V. Yurinski, 'Bounds for Characteristic Functions of Certain Degenerate Multidimensional Distributions', *Teor. Veroyatnost. i Primenen.* **17**:1 (1972), 99–110 (Russian). English transl. in: *Theory Probab. Appl.* **17**, 101–113.

137. V.V. Yurinski, 'On the Strong Law of Large Numbers for Homogeneous Random Fields', *Mat. Zametki*, **16**:1(1974), 141–149 (Russian).

138. A.M. Yaglom, 'Some Classes of Random Fields in n-dimensional Space Related to Stationary Random Processes', *Teor. Veroyatnost. i Primenen.* **2**:3(1957), 292–338 (Russian). English transl. in: *Theory Probab. Appl.* **2**, 273–320.

139. A.M. Yaglom, 'An Introduction to the Theory of Stationary Random Functions, *Uspekhi Mat. Nauk* **7**:5(51)(1952), 1–168 (Russian).

140. M.I. Yadrenko, *The Spectral Theory of Random Fields*, Vyshcha Shkola, Kiev, 1980 (Russian).

141. R.J. Adler, *Geometry of Random Fields*, Wiley, 1981.

142. T.W. Anderson, *The Statistical Analysis of Time Series*, Wiley, 1971.

143. M.S. Bartlett, *An Introduction to Stochastic Processes*, Cambridge University Press, 1956.

144. G. Basset and R. Koenker, 'Asymptotic Theory of Least Absolute Error Regression', *J. Amer. Statist. Assoc.* **73**:363(1978), 618–622.

145. H. Bateman and A. Erdelyi, *Higher Transcendental Functions*, Vol.II, McGraw-Hill, New York, 1953.

146. I. Berkes and G.J. Morrow, 'Strong Invariance Principles for Mixing Random Fields', *Zeitschrift für Wahrscheinlichkeitstheorie und verw. Gebiete* **57**:1(1981), 15–37.

147. S.M. Berman, 'High level sojourns for strongly dependent Gaussian processes', *Zeitschrift für Wahrscheinlichkeitstheorie und verw. Gebiete* **50**:2 (1979), 223–236.

148. S.M. Berman, 'Sojourns of vector Gaussian processes inside and outside a

sphere', *Zeitschrift für Wahrscheinlichkeitstheorie und verw. Gebiete* **66**:4 (1984), 529–542.

149. R.N. Bhattacharya and R. Rao Ranga, *Normal Approximation and Asymptotic Expansions*, Wiley, 1976.

150. P. Billingsley, *Convergence of probability measures*, Wiley, 1968.

151. N.H. Bingham, 'A Tauberian theorem for integral transforms of Hankel type', *J. London Math. Soc.* **2**:5(1972), 493–503.

152. P. Bloomfield and W.L. Steiger, *Least absolute deviations*, Birkhauser, Boston, 1983.

153. E. Bolthausen, 'On the central limit theorem for stationary mixing random fields', *Ann. Probab.* **10**:4(1982), 1047–1050.

154. R.C. Bradley, 'On a new measure of dependence of two σ-algebras', *J. Multivariate Anal.* **13**(1983), 167–176.

155. R.C. Bradley, 'Basic properties of strong mixing conditions', In: *Dependence in Probability and Statistics*, eds. E. Eberlein and M.S. Taqqu, Birkhauser, Boston, 1986, 165–192.

156. P. Brener and P. Major, 'Central limit theorems for non-linear functionals of Gaussian fields', *J. Multivariate Anal.* **13**(1983), 425–441.

157. D.R. Brillinger, *Time series. Data analysis and theory*, Holt, Rinehart and Winston, Inc., 1975.

158. D.R. Brillinger and M. Rosenblatt, Asymptotic theory of kth order spectra, *Spectral analysis of time series*, Wiley, New York, 1967, 153–188.

159. K.C. Chanda, 'Strong mixing properties of linear stochastic processes', *J. Appl. Probab.* **11**:2(1974), 401–408.

160. H. Cramér, *Mathematical methods of statistics*, Princeton Univ. Press, Princeton, N.J., 1946.

161. H. Cramér and M.R. Leadbetter, *Stationary and related stochastic processes*, Wiley, New York, 1967.

162. R. Dahlhaus, 'Asymptotic normality of spectral estimates, *J. Multivariate Anal.* **16**(1985), 412–431.

163. C.M. Deo, 'A functional limit theorem for stationary random fields', *Ann. Probab.* **13**:4(1975), 708–715.

164. R.L. Dobrushin, 'Gaussian and their subordinated self-similar random generalized fields', *Ann. Probab.* **7**:1(1979), 1–28.

165. R.L. Dobrushin and P. Major, 'Non-central limit theorems for non-linear functionals of Gaussian fields', *Zeitschrift für Wahrscheinlichkeitstheorie und verw. Gebiete* **50**:1(1979), 27–52.

166. R.M. Dudley, 'Sample functions of the Gaussian process', *Ann. Probab.* **1**:1(1973).

167. D.D. Engel, 'The multiple stochastic integral', *Mem. American Mathematical Society* 265(1982), 1–82.

168. K.G. Esseen, 'Fourier analysis of distribution functions', *Acta Math.* **77**:1 (1944), 1–125.

169. W. Feller, *An introduction to probability theory and its applications*, Vol.II, Wiley, New York, 1966.

170. X. Fernique, 'Régularité des trajectoires des fonctions aléatoires gaussiennes', *Lecture Notes in Mathematics* Vol.408, Springer-Verlag, New York, 1975, 1–96.

171. C.M. Goldie and P.E. Greenwood, 'Characterisations of set-indexed Brownian motion and associated conditions for finite-dimensional convergence', *Ann. Probab.***14**:3(1986), 802–816.

172. C.M. Goldie and P.E. Greenwood, 'Variance of set-indexed sums of mixing random variables and weak convergence of set-indexed processes', *Ann. Probab.* **14**:3(1986), 817–839.

173. V. Grenander, *Probabilities on algebraic structures*, Wiley, New York, 1963.

174. V. Grenander and M. Rosenblatt, *Statistical analysis of stationary time series*, Almqvist and Wiksell, Stockholm, 1956.

175. E.J. Hannan, *Multiple time series*, Wiley, 1970.

176. E.J. Hannan, 'Non-linear time series regresson', *J. Appl. Prob.* **8**(1971) 767–780.

177. C.S. Herz, 'Fourier transform related to convex sets', *Ann. of Math.* **72**:1 (1962), 81–92.

178. F. Harary, *Graph theory*, Addison-Wesley, Reading, Mass., 1969.

179. P.J. Huber, 'The behaviour of maximum likelihood estimates under nonstandard conditions', In: *Proceedings of the Fifth Berkeley Symposium on Mathematical Statistics and Probability*, Berkeley-Los Angeles, Univ. Calif. Press., 1967, 221–233.

180. P.J. Huber, *Robust Statistics*, Wiley, New York, 1981.

181. K. Itô, 'Multiple Wiener integral', *J. Math. Soc. Japan* **3**(1951), 157–169.

182. K. Itô, 'Spectral type of shift transformations for differential process with stationary increments', *Trans. Amer. Math. Soc.* **81**(1956), 253–263.

183. A.V Ivanov and S. Zwanzig, 'An Asymptotic Expansion of the Distribution of Least Squares Estimators in the Non-linear Regression Model', *Math. Operationsforsch. Statist. Ser. Statist.* **14**:1(1983), 7–27.

184. N.C. Jain and M.B. Marcus, 'Central limit theorems for $C(S)$-valued random variables', *J. Funct. Anal.* **19**:3(1975), 216–231.

185. R.I. Jennrich, 'Non-linear least squares estimators', *Ann. Math. Statist.* **40**(1969), 633–643.

186. M.G. Kendall and P.A.P. Moran, *Geometrical Probability*, Statistical Monographs and Courses, No. 5, Griffin, London, 1963.

187. J. Lamperti, 'Semi-stable stochastic processes', *Trans. Amer. Math. Soc.* **104**:1(1962), 62–78.

188. H.J. Landau and L.A. Shepp, 'On the supremum of a Gaussian process', *Sankhyā, Ser. A* **31**(1971), 369–378.

189. N.N. Leonenko, 'The limit distributions of spherical averages of non-linear transformation of Gaussian isotropic random fields', *Fourth USSR-Japan Sympos. on Probab. Theory and Math. Statist.*, Vol. II, 1982, Abstracts and Communications, Tbilisi, 64–65.

190. Yu.V. Linnik and N.M. Mitrofanova, 'Some asymptotic expansions for the distribution of the maximum likelihood estimate' *Sankhyā, Ser. A* **27** (1965), 73–82.

191. M. Loève, *Probability theory*, D. Van Norstrand, Princeton, N.J., 1960.

192. M. Maejima, 'Some sojourn time problems for strongly dependent Gaussian processes', *Zeitschrift für Wahrscheinlichkeitstheorie und verw. Gebiete* **57**(1981), 1–14.

193. M. Maejima, 'A limit theorem for sojourn times for strongly dependent Gaussian processes', *Zeitschrift für Wahrscheinlichkeitstheorie und verw. Gebiete* **60**(1982), 359–380.

194. M. Maejima, 'Sojourns of multi-dimensional Gaussian processes with dependent components' *Yokohama Math. J.* **33**:1–2(1985), 121–130.

195. M. Maejima, 'Some sojourn time problems for 2-dimensional Gaussian processes', *J. Multiv. Analysis* **18**:1(1986), 52–69.

196. M. Maejima, ' Sojourns of multi-dimensional Gaussian processes', In: *Dependence in Probability and Statistics*, eds. E. Eberlein and M.S. Taqqu, Birkhauser, Boston, 165–192.

197. P. Major, *Multiple Wiener-Itô integrals*, Lecture Notes in Mathematics Vol. 849, Springer-Verlag, 1981.

198. E. Malinvaud, 'The consistency of non-linear regression', *Ann. Math. Statist.* **41**(1970), 956–969.

199. R. Michel, 'An asymptotic expansion for the distribution of asymptotic maximum likelihood estimators of vector parameters', *J. Multivariate*

Analysis **15**(1975), 67–82.

200. R. Michel, 'A multi-dimensional Newton-Raphson method and its applications to the existence of asymptotic Fn-estimators and their stochastic expansions', *J. Multivariate Analysis* **7**(1977), 235–248.

201. C.C. Neaderhouser, 'Limit theorems for multiple indexed mixing random variables with application to Gibbs random fields', *Ann. Probab.* **6**:2(1978), 207–215.

202. W. Oberhofer, The consistency of non-linear regression minimizing the L_1-norm', *Ann. Statist.* **10**:1(1982), 316–319.

203. E. Orsingher, 'Some results on geometry of Gaussian random fields', *Rev. Roum. Math. Pures et Appl.* **28**:6(1983), 493–511.

204. M. Peligrad, 'Invariance principle under weak dependence', *J. Multiv. Analysis* **19**(1986), 299–310.

205. J. Pfanzagl, 'Asymptotically optimum estimation and test procedures', *Proc. Prague Symp. Asymptotic Stat.* **1**(1973), 201–272.

206. W. Philipp and F. Staut, 'Almost sure invariance principles for partial sums of weakly dependent random variables', *Mem. Amer. Math. Soc.* 161 (1975), 3–120.

207. B.L.S. Prakasa Rao, 'A non-uniform estimate of the rate of convergence in the central limit theorem for m-dependent random fields', *Zeitschrift für Wahrscheinlichkeitstheorie und verw. Gebiete* **58**:2(1981), 247–256.

208. B. Rosen, 'A note on asymptotic normality of sums of higher dimensionally indexed random variables', *Ark. Math. Stockholm* **8**(1969), 33–43.

209. M. Rosenblatt, 'A central limit theorem and a strong mixing condition', *Proc. Nat. Acad. Sci., USA* **42**(1956), 43–47.

210. M. Rosenblatt, 'Some comments on narrow band pass filters', *Quart. Appl. Math.* **18**(1961), 387–393.

211. M. Rosenblatt, 'Central limit theorem for stationary processes', *Proc. Sixth Berkeley Symp. Math. Statist. Probab.* **2**(1970), 551–561.

212. M. Rosenblatt, 'Limit theorems for Fourier transforms of functional of Gaussian sequences', *Zeitschrift für Wahrscheinlichkeitstheorie und verw. Gebiete* **55**:2(1981), 123–132.

213. M. Rosenblatt, 'Remarks on limit theorems for non-linear functionals of Gaussian sequences', *Prob. Th. Rel. Fields* **75**:1(1987), 1–10.

214. D. Ruelle, *Statistical Mechanics. Rigorous Results*, W.A. Benjamin, New York-Amsterdam, 1969.

215. L.A. Santaló, *Integral geometry and geometric probability*, Addison-Wesley,

Reading, Mass., 1976.

216. R.J. Serfling, 'Contribution to central limit theorem for dependent variables', *Ann. Math. Stat.* **39**(1968), 1158–1175.

217. T.K. Sheng, 'The distance between two random points in plane regions', *Adv. in Appl. Probab.* **17**(1985), 748–773.

218. D. Slepian, 'The one-sided barrier problem for Gaussian noise', *Bell System Tech. J.* **41**(1962), 463–501.

219. T.C. Sun, 'Some further results on central limit theorems for non-linear functions of a normal stationary process', *J. Math. and Mech.* **14**:1(1965), 71–85.

220. N. Takahata, 'On the rate in the central limit theorem for weakly dependent random fields', *Zeitschrift für Wahrscheinlichkeitstheorie und verw. Gebiete* **64**(1983), 445–456.

221. M.S. Taqqu, 'Weak convergence to fractional Brownian motion and to the Rosenblatt process', *Zeitschrift für Wahrscheinlichkeitstheorie und verw. Gebiete* **31**(1975), 287–302.

222. M.S. Taqqu, 'Law of the iterated logarithm for sums of non-linear functions of Gaussian variables that exhibit a long-range dependence', *Zeitschrift für Wahrscheinlichkeitstheorie und verw. Gebiete* **40**(1977), 203–238.

223. M.S. Taqqu, 'A representation for self-similar processes', *Stoch. Proc. Appl.* **17**(1978), 55–64.

224. M.S. Taqqu, Convergence of integrated processes of arbitrary Hermite rank', *Zeitschrift für Wahrscheinlichkeitstheorie und verw. Gebiete* **50**:1 (1979), 53–83.

225. M.S. Taqqu, 'Sojourn in an elliptical domain', *Stoch. Proc. Appl.* **21**(1986), 319–326.

226. M.S. Taqqu, 'A bibliographical guide to self-similar processes and long-range dependence', In: *Dependence in Probability and Statistics*, eds. E. Eberlein and M.S. Taqqu, Birkhauser, Boston, 1986.

227. E. Vanmarcke, *Random Fields. Analysis and Synthesis*, MIT Press, Cambridge, Mass.–London, 1983.

228. A.M. Walker, 'On the estimation of a harmonic component in a time series with stationary independent residuals', *Biometrika* **58**(1971), 21–36.

229. A.M. Walker, 'On the estimation of a harmonic component in a time series with stationary dependent residuals', *Adv. in Appl. Probab.* **15**(1973), 217–241.

230. N. Wiener, 'The Homogeneous Chaos', *Amer. J. Math.* **60**(1938), 897–936.

231. J.H. Wilkinson, *The Algebraic Eigenvalue Problem*, Clarendon Press, Oxford, 1965.

232. C.S. Withers, 'Conditions for linear processes to be strong-mixing', *Zeitschrift für Wahrscheinlichkeitstheorie und verw. Gebiete* **57**(1981), 477–480.

233. C.S. Withers, 'Central limit theorems for dependent variables', II *Prob. Th. Rel. Fields* **76**:1(1987), 1–13.

234. M. Wschebor, 'Surfaces aléatoires', Lecture Notes in Math. Vol. 1147, 1985.

235. Y. Xinjing, 'Discretization of generalized stationary Gaussian random fields', *Chinese Journal of Applied Probability and Statistics* **2**:3(1986), 199–204.

236. A.M. Yaglom, *Correlation Theory of Stationary and Related Random Functions* I, II, Springer Series in Statistics, 1987.

237. R. Yokoyama, 'Moment bounds for stationary mixing sequences', *Zeitschrift für Wahrscheinlichkeitstheorie und verw. Gebiete* **52**:1(1980), 45–57.

238. K. Yoshihara, 'Moment inequalities for mixing sequences', *Kodai Math. J.* **1**:2(1978), 316–328.

Comments

Chapter 1

1.1.   The mode of convergence of sets to infinity discussed herein is adopted from the literature on statistical mechanics [214]. Concepts from probability theory presented in this section are extensively covered in books [73, 30, 191, 150, 53, 161, 169] and also in papers [131, 132, 13].

1.2.   A detailed exposition of results on spectral theory of random fields is contained in books by I.I. Gikhmann and A.V. Skorokhod [30], M.I. Yadrenko [140] and also in the works by A.M. Yaglom [138, 139, 236] and others.

The concept of the stochastic integral with respect to an orthogonal random measure is discussed in the book [30, Ch.IV]. Monograph [140] also deals with other classes of random fields whose distributions are invariant with respect to certain groups of transformations. In this monograph numerous examples of spectral decompositions are presented. The above-mentioned works also outline the history of the problem. The definition and properties of the indicated special functions are presented in the book [145]. The relation between special functions and group representation theory is studied in the monograph by N.Ya. Vilenkin [24]. The above-stated concept of a multi-dimensional random field is not the only one possible. Alternative generalizations to the multi-dimensional case occurring in turbulence theory are proposed in the works [99, 138, 236, 105, 122]. Other generalizations of the correlation theory of random fields are discussed in books [116, 122, 227].

1.3.   The main results on random processes belonging to Fortet-Blanc-Lapierre classes were obtained by V.P. Leonov [92] and A.N. Shiryaev [135]. General properties of cumulants are described in the books [157, 149, 47, 48] and also in [95]. In the statistical analysis of time series, higher order cumulants were used by D. Brillinger, M. Rosenblatt, R. Bentkus, I.G. Zhurbenko and others [158, 7-11, 47-49]. The exposition in the present section is based on articles [63, 65].

233

1.4.   The uniform distribution plays an important role in probability theory [168, 117]. Bounds on the absolute value of the ch.f. (1.4.2) of the uniform
distribution on a convex set are derived in [177, 117, 136] and others. Lemma
1.4.2 is proved in the books [186, 215]. Lemmas 1.4.3, 1.4.4 appear in [89, 90,
40]. For other sets on the plane, the form of the distribution function for the
distance between two uniformly distributed independent vectors is presented
in [217].

1.5.   The exposition is based on works [89–91]. Other versions of Tauberian theorems for the Hankel transforms differing from Lemma 1.5.4 are discussed in [151]. Some results for two-dimensional fields similar to those stated
above are presented in [203].

1.6.   Lemmas 1.6.1, 1.6.2 are basic for the proof of limit theorems for
dependent variables (see [51, 112, 41]). The relation of particular cases of
coefficient (1.6.8) to the maximal correlation coefficient is described in [19].
For a more detailed account on the latest results in this area, refer to the
review by R. Bradley [155] and the paper by Withers [233].

1.7.   The c.l.t. for random processes in the presence of a weight function
was considered in the works by M. Rosenblatt [210] and A.S. Kholevo [129]
and for random fields in [77, 78, 82]. For the proof of the u.c.l.t. we follow
[53]. The sectioning method was proposed by S.N. Bernstein [12] and used by
many authors [51, 112]. The proof of Theorem 1.7.1 is based on the ideas in the
work by A.V. Bulinsky and I.G. Zhurbenko [20]; see also [48], although similar
results may be obtained by performing the sectioning in a slightly different
manner [80, 100, 201]. To prove the c.l.t. for random fields, other methods are
also used [94, 211, 153, 21, 126]. As to Remark 1.7.2, see [18].

Verification of strong mixing conditions for particular random fields is a
difficult task. For linear random sequences, a solution is proposed in [159,
32, 232]. For Gaussian random sequences and processes, similar topics are
discussed in detail in the book [52]. This problem is discussed also in book [48]
and in [155].

A number of papers are devoted to estimation of the convergence rate in
the c.l.t. for random fields [79, 18, 124, 207, 220], etc. The convergence rate
for random processes has been investigated by V.A. Statulevicius [119] and his
students and also by other authors.

1.8.   Here the terminology of the book by F. Harary [178] is used. Classical analogues of results presented in this section are moment bounds for the
sums of independent random variables [106, 53, 149]. Similar inequalities for

various classes of dependent variables have been investigated in [237, 238, 125]. After the book was written, the paper [36] appeared containing a more general result. Assertion 1.8.3 is apparently new.

1.9. The exposition is based on the paper [89, 91]. For random fields similar results were obtained by Yu.A. Davydov [41] (cf. also [150]). To prove Lemma 1.9.1, the Cramér-Wold-Sapogov method is used.

Other forms of the invariance principle for mixing random fields are presented in [163, 81, 146, 35, 171, 172].

Chapter 2

2.1. Variances of local Gaussian functionals for stationary processes were considered in [219, 221, 224, 34] etc. For random fields of a discrete argument, analogous problems were investigated in [165, 156].

The assertion of Lemma 2.1.1 is well known (see, for example, [112]). Lemma 2.1.3 appears in [83], Lemmas 2.1.8, 2.1.9 were obtained by K.V. Rybasov [114]. The remaining assertions in this section have not been presented previously.

2.2. Reduction conditions were first derived in the work by M.S. Taqqu [221]. Reduction conditions for functionals of stationary sequences and fields with a discrete parameter whose correlation function has a regular variation at infinity have been considered in [165, 34, 212, 213] and, for random processes, in [224, 192]. Theorem 2.2.1 is presented in [83, 84] and Theorem 2.2.2 in [114]. The remaining results in this section are new.

2.3. Applications of diagram techniques are discussed in [94, 95, 97, 98] and other works. Theorems 2.3.1, 2.3.2 are derived in [86]. The diagram technique for proving the c.l.t. for random sequences and random fields of a discrete parameter was first used by L. Giraitis [29] and by P. Brener and P. Major [156]. Estimates in terms of cumulants yielding similar results were obtained by A. Plikusas [109].

2.4. The functionals of a geometric nature discussed herein are of major importance in the theory of surface roughness [130] as well as in other applications [113, 227]. Characteristics of excess above a level for stationary processes are covered in §10.8 of monograph [161]. Monographs by R.J. Adler [141] and M. Wschebor [234] are devoted to the geometry of random fields. For $n = 2$ the first two moments for certain geometric functionals are derived in [203]. The remaining results of this section are new.

Interesting results on level crossings for Gaussian fields were obtained by

Yu.K. Belayev and his students (see [5,pp.101–103], etc.). A review of these results was prepared by V.I. Piterbarg [108].

2.5.   The results of this section are new.

2.6.   In such generality, the results are presented here for the first time. The ideas of reduction theorems stated for functionals of the type "measures of level excess" for a Gaussian process are presented in the paper by S. Berman [147] (see also [192, 193]). Lemmas 2.6.5, 2.6.6 extend results by M. Maejima [192, 193] to random fields.

2.7.   The statements in this section extend the results by S.M. Berman [147], M. Maejima [192] to random fields. The exposition follows papers [83, 84]. The exposition of Theorems 2.7.8, 2.7.9 follows [85]. An investigation of measures of sojourn of vector-valued Gaussian processes in certain domains is carried out in [148, 194, 195, 196, 225]. Spherical functionals of a geometric nature have been examined in [114, 115, 87].

2.8.   The results presented here have not been published previously. With $n = 1$ similar results are derived in [193].

2.9.   As to generalized random fields, refer to paper [138] and the monograph [27]. Multiple stochastic integrals were introduced by K. Itô [181, 182] and N. Wiener [230]. A more detailed account on these topics is presented in books [197, 167] and articles [164, 165, 235]. Further references may be found in the review [226]. Bounds on cumulants of multiple stochastic integrals are derived in [109]. Multiple stochastic integrals with respect to a Poisson measure were examined in [120, 121, 6, 167].

2.10.   The first publication on these problems is the work by A.N. Kolmogorov [74]. Semi-stable (Hermitian, automodel) processes were introduced by J. Lamperti [187]. The class of semi-stable distributions generalizes the class of stable distributions [50]; the attraction conditions to the latter were thoroughly examined in [31, 51]. Attraction conditions for the sums of linear random sequences to Gaussian semi-stable processes were investigated by Yu. A. Davydov [42] and V.V. Gorodetsky [33]. Attraction conditions for random sequences and processes to non-Gaussian semi-stable processes were studied by R.L. Dobrushin, P. Major [165], M. Rosenblatt [212, 213], M.S. Taqqu [221–224], V.V. Gorodetsky [34], D. Surgailis [120, 121], L. Giraitis [28] and others. A complete bibliographic survey of works in this field was compiled by M.S. Taqqu [226] containing 256 references. The definition of semi-stable processes by means of multiple integrals is given in [223, 164, 165, 120, 6]. We note that these topics are of major importance in modern mathematical physics [118].

Theorems 2.10.1, 2.10.2, 2.10.4 are presented in such a form for the first time. Similar assertions have been stated in [189]. Theorem 2.10.3 is a continuous analogue of Theorem 1 from [165]. An alternative approach to the investigation of functionals of random fields with a continuous parameter was proposed in [235].

Chapter 3

3.1. The concept of spectral measure of a regression function has been fruitfully utilized in monographs by V. Grenander and M. Rosenblatt [174], I.A. Ibragimov and Yu.A. Rozanov [52], E. Hannan [175] and also in papers by A.S. Kholevo [127-129]. The exposition follows [88, 82]. Paper [68] deals with similar problems. The results on the distribution of regression coefficient estimates in the case when the spectral density of the "noise" is unbounded are apparently new.

3.2. Asymptotical properties of the least squares estimates for a nonlinear parameter in regression functions for sequences of independent and stationary related observations were investigated in [185, 198, 176, 44, 55, 56, 58, 59, 66, 67, 183] and many others. Similar problems under the assumption that observations are either processes or fields with a continuous parameter have been considered by A.Ya. Dorogovtsev [44], A.V. Ivanov [54, 57], P.S. Knopov [70] and others. Asymptotic properties of various statistical estimates for a parameter of a valid signal observed in a random noise are examined in detail in [53, 76].

Theorem 3.2.1 generalizes the result by E. Malinvaud [198]. The statement of Theorem 3.2.3 is based on assertions from the books by M.I. Yadrenko [140] and I.A. Ibragimov and R.Z. Hasminsky [53]. Theorem 3.2.4 extends an assertion from [56]. To prove it, we utilize the approach to the investigation of probabilities of large deviations for statistical estimates given in [53]. Theorem 3.2.5 generalizes a theorem by R. Jennrich [185].

3.3. The theory of asymptotic expansions of statistical estimators and their probability characteristics is being intensively developed at the present time. Asymptotic expansions of maximum likelihood estimators in the classical observations scheme were derived by Yu.V. Linnik and N.M. Mitrofanova [93, 96, 190]. Asymptotic expansions for a number of statistical estimators were considered in references [134, 205, 22, 23, 39, 199, 200] and many others. Theorem 3.3.1 extends a result of paper [67]. The proof is based on an approach due to D.M. Chibisov [134].

3.4. Theorem 3.4.2 extends a result of paper [129] to a non-linear regression model. Efficiency problems for the least-squares estimators and more general pseudo-optimal estimators were considered in [128, 52]. An approach to the construction of estimators analogous to pseudo-optimal ones under a non-linear parametrization of regression functions were proposed by E. Hannan [176].

3.5. Asymptotic properties of the least moduli estimators for a sequence of independent observations and a non-linear regression function were investigated in [38, 202, 5] and others. Similar results for observations with a stationary error sequence were stated in communication [14]. Theorem 3.5.1 is new.

The theory of the least moduli for linear regression models with independent errors is discussed in the book by P. Bloomfield and P. Steiger [152]. In this book the history of the problem is presented; also numerical algorithms are stated to solve optimization problems arising in the course of a study of the least moduli method of estimation.

3.6. Theorem 3.6.1 is new. As to the idea underlying the proof of Lemma 3.6.1, refer to the works by P. Huber [180, 179].

Chapter 4

4.1. A number of relations for characteristics of a correlogram of random processes in discrete and continuous times are presented in the books [142, 143, 175] and others.

4.2. Strong laws of large numbers for random fields were the subject of numerous investigations (see [123], [135] and the references therein). Theorem 4.2.2 expresses a natural property of the estimators of the cor.f. under consideration but the authors are not aware of publications containing such an assertion.

4.3–4.4. Here the authors' results are presented in a revised form [61–64]. Conditions for asymptotic normality of statistics of a form more general than correlograms constructed from observations of a sequence of r.v.'s were examined in papers [8, 9]. An assertion for a stationary Gaussian process similar to the results of §4.3 is stated in communication [2] and for Gaussian sequences in paper [1]. Other estimators for the cor.f. of random fields were considered in papers [45, 111].

4.5. This section provides results of paper [15, 17] in an expanded form.

4.6. A construction of confidence intervals for the estimators of the cor.f.

based on the X. Fernique's inequality was proposed by V.V. Buldygin. Lemmas 4.6.1–4.6.3 based on the results of §4.3 contain new bounds for the cor.f. Theorem 4.6.4 is a modification of an assertion due to V.V. Buldygin [16].

Index